Construction Jobsite Management

William R. Mincks

Hal Johnston

Delmar Publishers

an International Thomson Publishing Company

Albany • Bonn • Boston • Cincinnati • Detroit • London • Madrid
Melbourne • Mexico City • New York • Pacific Grove • Paris • San Francisco
Singapore • Tokyo • Toronto • Washington

NOTICE TO THE READER

Delmar Staff
Publisher: Alar Elken
Senior Administrative Editor: John Anderson
Developmental Editor: Michelle Ruelos Cannistraci
Production Coordinator: Toni Bolognino
Art and Design Coordinator: Cheri Plasse
Editorial Assistant: John Fisher

COPYRIGHT © 1998
By Delmar Publishers
an International Thomson Publishing Company

The ITP logo is a trademark under license.

Printed in the United States of America

For more information, contact:

Delmar Publishers
3 Columbia Circle, Box 15015
Albany, New York 12212-5015

International Thomson Publishing Europe
Berkshire House 168-173
High Holborn
London WC1V 7AA
England

Thomas Nelson Australia
102 Dodds Street
South Melbourne, 3205
Victoria, Australia

Nelson Canada
1120 Birchmont Road
Scarborough, Ontario
Canada M1K 5G4

Online Services

Delmar Online
To access a wide variety of Delmar products and services on the World Wide Web, point your browser to:
   http://www.delmar.com
   or email: info@delmar.com

thomson.com
To access International Thomson Publishing's home site for information on more than 34 publishers and 20,000 products, point your browser to:
   http://www.thomson.com
   or email: findit@kiosk.thomson.com

A service of I(T)P®

International Thomson Editores
Campos Eliseos 385, Piso 7
Col Polanco
11560 Mexico D F Mexico

International Thomson Publishing GmbH
Konigswinterer Strasse 418
53227 Bonn
Germany

International Thomson Publishing Asia
221 Henderson Road
#05–10 Henderson Building
Singapore 0315

International Thomson Publishing—Japan
Hirakawacho Kyowa Building, 3F
2-2-1 Hirakawacho
Chiyoda-ku, Tokyo 102
Japan

All rights reserved. No part of this work covered by the copyright hereon may be reproduced or used in any form or by any means—graphic, electronic, or mechanical, including photocopying, recording, taping, or information storage and retrieval systems—without the written permission of the publisher.

1 2 3 4 5 6 7 8 9 10 XXX 02 01 00 99 98 97

Library of Congress Cataloging-in-Publication Data

Mincks, William R.
      Construction jobsite management / William R. Mincks, Hal Johnston.
         p.   cm.
      Includes index.
      ISBN 0-8273-7152-7 (case)
      1. Building—Superintendence.   I. Johnston, Hal.   II. Title.
TH438.M55   1997   97-21518
690'.068—dc21   CIP

DEDICATION

To our Fathers,
Ralph Mincks and Harold "Bud" Johnston,
who taught us that construction
is a respectable profession.

CONTENTS

PREFACE

Construction Jobsite Management concentrates on the procedures and methods that are used by the construction contractor during the construction and post-construction phases of a project. Construction today involves much more than the physical erection of a project. It is essential for the contractor to systematically plan, organize, manage, control, and document jobsite activities. There is no margin for error on the jobsite in the current construction market; possession of good organizational skills and the ability to anticipate problems is essential. An efficiently managed jobsite should result in a profitable construction project. A good documentation system increases the manager's awareness of problems that develop early on in the construction process, which saves the effort and expense normally expended for claims and litigation. The current legal climate requires a detailed documentation of construction activities and events.

The procedures and methods contained in this book focus on the contractor's operation, however, many of these procedures and methods apply to owner's representatives, architects and engineers, specialty contractors, and construction managers as well. The methods herein are primarily applicable to commercial and industrial building construction, although many can be applied to all types of construction. Each project, depending on its size and specific attributes, will have different jobsite management needs. The constructor should use the procedures that will meet the needs of the project. Small projects normally consolidate several of the functions and activities detailed herein, but they nevertheless need the proper management to maintain profitability.

The five sequential, generally recognized phases of the construction process are pre-design, design, bid/award, construction, and post-construction.

This book will focus primarily on the construction phase of the process, although other phases will be discussed. The construction period begins after the contract for construction is awarded and includes pre-construction meetings and activities and the actual physical construction of the facility. This book also will examine closeout and completion procedures after substantial completion, usually classified in the post-construction phase. Additionally, jobsite management activities associated with project scheduling, project safety, contract documents, and building codes will be addressed, however, for a detailed look at these activities, numerous sources are available for further reference.

The project management system should meet project requirements and blend with company policy. The management system and organization should be designed to optimize efficiency at the jobsite but minimize direct overhead and labor costs. The contractor's primary goal during a construction project is to make a profit while satisfying contractual requirements. Thus, the main objective for a project management system is to facilitate the completion of a project as efficiently as possible.

This book addresses many of the methods involved in the management of construction jobsites. Each project, depending on its size, location, company policy, and contractual requirements, may use varying configurations of project management methods and structure. The contractor should evaluate each particular situation and use the proper tools accordingly. The procedures described in this book are illustrative rather than literal descriptions, however, they do not specifically apply to all situations on the jobsite.

Publisher's Note: The English language does not have a nonspecific gender pronoun. Rather than using the awkward he/she construction to avoid appearing sexist, people throughout this text are referred to as he. Although construction is still a predominantly male profession, the author and Delmar Publishers realize that women make substantial contributions to the profession as well.

Acknowledgments

The authors would like to thank our wives, Rena and Joy, for their patience, inspiration, and help in this undertaking. A special thanks to Joy Johnston for her work in obtaining permissions for material in this book.

The authors would like to acknowledge the encouragement and mentoring of the late Professor Richard Young. He knew we could do it, but didn't get to see the final product.

We also would like to thank our Department Chairs, Rafi Samizay, School of Architecture, Washington State University, and Jim Rodger, Department of Construction Management, California Polytechnic State University, for their support and help. We also appreciate the encouragement of our colleagues Ken Carper, Larry Fisher, Ed Turnquist, and Jim Borland.

Several individuals helped us with information for this book: Colin Matsushima, Ron Warrick, and Bill Davis. Thanks for your help.

We'd also like to thank the following firms and organizations for technical information and documents: American Institute of Architects, Concrete Reinforcing Steel Institute, International Conference of Building Officials, Associated Builders and Contractors, Primavera Systems, Inc., ChemRex, Inc., Meridian Project Systems, Microsoft Corporation, and the Associated General Contractors of America, Inc.

The authors and Delmar Publishers wish to thank the following reviewers for their valuable contributions:

Kenneth Anderson, North Dakota State University, Fargo, ND
David Carns, Central Washington University, Ellensburg, WA
David Goodloe, Clemson University, Clemson, SC
Dr. Nancy Holland, Texas A&M University, College Station, TX
Albert Kulick, Jr., Catonsville Community College, Catonsville, MD
Walter Lehner, Central Connecticut University, Bethel, CT
Michael O'Dea, University of Arkansas, Little Rock, AR
James Rowings, Jr., Iowa State University, Ames, IA
John Schaufelberger, University of Washington, Seattle, WA

The authors would also like to thank the staff at Delmar Publishers, particularly Michelle Ruelos Cannistraci, for their help, in what seemed to be a never-ending project.

CHAPTER 1

THE PROJECT TEAM

CHAPTER OUTLINE Roles, Responsibilities, and Authority of
Project Participants

The Traditional Contract Delivery System
(Owner-Architect-Contractor)

The Construction Management (CM) Delivery
System

The Design-Build Delivery System

Summary

Roles, Responsibilities, and Authority of Project Participants

There are many ways to structure a construction project. The **delivery systems** provide a matrix of organization, with formal and informal contractual relationships between participants. Participants are assigned certain specific responsibilities within their contracts. Standard contractual forms are available for each delivery system, using generally uniform terminology and defining the roles and responsibilities for participants. Professional and trade associations produce and endorse contractual agreements that are widely used throughout the construction industry. The American Institute of Architects (AIA) publishes a complete set of contractual documents, including architect and contractor agreements. The Engineers Joint Contract Documents Committee (EJCDC) has produced a set of contract documents that is primarily used in engineering construction. The Associated General Contractors of America, Inc. (AGC) also produces a set of contractual documents, which includes agreements between contractors and owners and between contractors and subcontractors. Owners who use either of these forms can supplement and modify the standard contract with specific clauses relating to their particular needs. Public agencies such as municipalities, states, and the federal government have custom contract forms that meet the contracting regulations that are legislated specifically for them.

This chapter examines three basic delivery systems:

- traditional system
- construction management
- design-build

There are many variations and hybrid combinations of these three basic systems. Basic roles will be defined, which can be used as a standard, however, these roles will change as the delivery system changes.

The Traditional Contract Delivery System (Owner-Architect-Contractor)

The traditional system has three primary parties:

- the owner
- the architect
- the contractor

The owner and the architect execute a contract for applicable studies, design, production of construction documents, and administration of the construction process at the beginning of the project process. The AIA agreement between the owner and the architect is AIA Form B-141. The owner and the contractor execute a contract for the construction of the project, according to the construction documents prepared by the architect, after the design and construction documents are completed. The architect then administers the contract as an agent of the owner. There is no direct contractual relationship between the architect and the contractor, but an indirect relationship exists because the architect is acting as the agent of the owner during the construction phase of the project.

Since neither the architect or the contractor actually does 100 percent of the work assigned to their own forces, each party makes an agreement with the other business firms involved to accomplish specific areas of the work assigned under their contract. The architect will normally utilize the services of professional consultants, referred to as subconsultants within the contract matrix, such as civil and environmental engineers, structural engineers, mechanical engineers, electrical engineers, and several other specialties contained within the project. These subconsultants have an agreement (AIA Form C-141) with and provide specific services for the architect. There is no contractual relationship between the subconsultant and the owner. The architect is responsible to the owner for the competent completion of his work, including work performed by the subconsultant. For example, if the structural engineer made a serious error in the design of the structural members of the roof, resulting in the roof's collapse, the owner would seek relief from the architect as the responsible party because the owner and the architect have an agreement that requires competent design of the entire project. The architect would then subrogate the claim to the structural engineer, who was contracted by the architect to design the roof system.

In building construction, much of the work is accomplished by specialty contractors who have an agreement with the contractor to complete a specific portion of the work covered by the contractor's contract requirements. This subcontract agreement details specific responsibilities of the work, with requirements added by the contractor. The subcontract agreement is between the contractor and the subcontractor. A common type of this contract is AGC Form 600, however many contractors have their own standard subcontract agreement. Like the architect and subconsultant relationship, there is no direct contractual relationship between the subcontractor and the owner. The contractor is responsible to the owner for complete compliance to the contract documents, despite the fact that portions of the work have been subcontracted. If in the previous example the roof structure was improperly installed by the subcontractor, the owner would seek relief from the contractor for repair and for relief of damages. The contractor would in turn look to the subcontractor for repair and damages, appropriate to the subcontract agreement.

Figure 1–1 illustrates the relationships between parties in the traditional delivery system. Note that the heavy dark lines indicate a **direct**

FIGURE 1–1
The Traditional
Construction
Delivery System

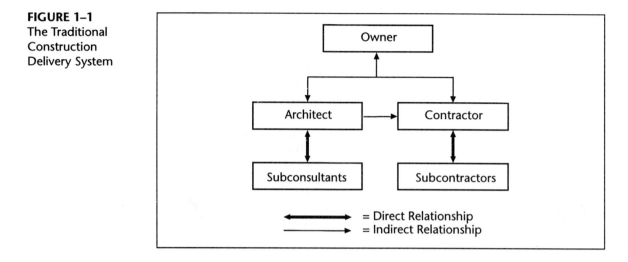

relationship, also known as **privity of contract,** and the light line represents an **indirect relationship,** also known as an **agency relationship.**

The traditional contract delivery system is not always the best delivery system for every situation. It is, however, the most prevalent of the delivery systems in the building construction industry and is used on every
size contract by both private and public owners. Other delivery systems,
such as construction management or design-build, are growing in use, but
the traditional system is still used by the majority of owners for their construction projects.

There are several methods of compensating the contractor for work
completed:

- Lump sum contract: This is the most common method, where the contractor gives the owner a lump sum price to complete the project according to the contract documents, which include the contract
  provisions, drawings, and technical specifications. Changes to the
  scope of work are accomplished through change orders, which adjust
  the lump sum amount during the construction period. The contractor
  is normally paid on a monthly basis during the construction period for
  the work installed and for the materials furnished during the month.
- Cost, plus a fee: In some cases, usually when the scope of the work is
  difficult to define, the contractor is reimbursed for costs on the project,
  plus a fee that includes indirect overhead and profit. "Costs" normally
  refer to labor, material, equipment, subcontracts, and direct (on-site)
  overhead. This method of payment is often referred to as "cost plus" or
  "time and material."
- Cost plus, with a guaranteed maximum price: This method of compensation for the construction contract is a hybrid of the lump sum
  and cost plus contracts. This method is commonly referred to as Guar-

anteed Maximum Price or GMP. In this method, the contractor quotes a maximum price for the scope of work and proceeds on a cost-plus-a-fee for the project, often with an arrangement to split the savings between the contractor and the owner.

- Unit price contract: This method lists quantities for components of the project, which are priced per unit by the contractor. The total of the product of the quantity and the unit price is then added to determine the lump sum price for the bid. Payment is based on the completion of the quantities for each line item. Unit price contracts are not common in the building construction industry, but are quite common in civil engineering projects.

Responsibilities of the Contractual Parties

The responsibilities of the three contractual parties—the owner, the architect, and the contractor—should be well-defined by the contract. A brief summary of the common responsibilities of the participants in the traditional delivery system is discussed next. (Refer to the AIA or AGC agreement forms for a complete definition of the responsibilities of each party.)

The Owner.   The owner is responsible for paying for the work contracted to the architect and to the contractor. He is responsible to the contractor for providing coordination of the project, whether through an architect, an in-house representative, a project manager, or a "clerk of the works." The owner also provides the site for the project, provides the architect with whatever he needs and may determine the scope of the project. Additionally, the owner provides the contractor with documents that adequately describe what the project will entail.

The Architect.   The architect normally provides the owner with the design of the project and the construction documents, based on the owner's needs. He is often engaged in providing construction administration for the project, acting as the owner's agent. The architect also provides an interpretation of the contract documents.

The Contractor.   The contractor is responsible for providing the labor, material, equipment, and expertise to complete the project, as indicated by the documents furnished by the owner, for compensation as stipulated. He is responsible for developing the means and methods of accomplishing the work, including sequencing, labor plan, equipment usage, and schedule. The contractor also is responsible for coordinating the work, including hiring the subcontractors, and paying for all labor, material, and subcontracts contained within the work.

During the actual construction of the project, a number of participants are involved, each having several different roles in the process. The following list describes the major parties that are involved in the construction phase of a project.

Owner
Capital Projects Officer
Financial Officer
Owner's Representative
Owner's Inspector
Testing Agency

Architect
Principal-in-Charge
Project Manager
Project Architect
Contract Administrator
Subconsultants
Specialty Coordinators
—Project Coordinator
—Mechanical Coordinator
—Electrical Coordinator

Contractor
Officer-in-Charge
Project Manager
Superintendent
Project Engineer
Field Engineer
Foremen
Craftspeople
Subcontractors
—Foremen
—Craftspeople

Other
Building Inspector
Plumbing Inspector
Electrical Inspector
Fire Marshall
Elevator Inspector
Safety Inspector

The Owner's Roles During the Construction Phase

The structure of the construction project in the owner's organization will vary greatly, depending upon the size of the project and the level of involvement in managing the project. The portion of an owner's organization chart that deals with capital projects may be similar to the one shown in Figure 1–2, which illustrates some of the roles in the owner's organization that are involved in the construction process.

Capital Projects Officer.   The capital projects officer is the individual responsible to the owners and/or stockholders of the company for the project. This position may be filled by the owner of the company, the chief executive officer (CEO), the president, the vice-president in charge of capital projects, the facility manager, or several other responsible people. This person normally is involved at the inception of the project, but may not intimately participate in day-to-day construction activities. He authorizes major changes and oversees the construction phase on a periodic basis.

Financial Officer.   The financial officer for the owner is primarily concerned with the disbursement of funds for the project. As monies for

FIGURE 1–2
The Owner's
Organization for
the Construction
Project

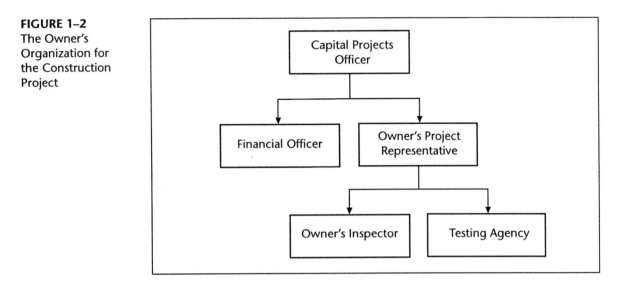

payment of construction activities usually come from sources other than operating funds, the financial officer must anticipate and plan for the financial needs of the project, or what is commonly referred to as project cash flow management. The contractor provides the financial officer with a cash flow projection, which relates the project's schedule of values to the construction schedule.

Owner's Representative.   The owner's representative, sometimes referred to as the owner's project manager, is the owner's daily representative during the construction project. This individual may perform other responsibilities for the owner, but he is the owner's prime contact for the architect and the contractor. The owner's representative conducts business with the project managers for the architect and the contractor. He will be the conduit of information between the architect and the owner's ultimate decision maker, the capital projects officer, and should be knowledgeable about construction practices. For owners who have continuous construction programs, the owner's representative is often a trained construction professional who may be hired solely for the project's duration.

Owner's Inspector.   The owner's inspector, sometimes referred to as the clerk of the works, is an individual who reports to the owner's representative. This person observes the construction process and documents the progress and problems encountered. He is normally concerned with the quality of the construction work as it is installed and with transmitting information on any deviations to the architect and to the owner's representative. The inspector has no authority to direct craftspeople, subcontractors, or the contractor to stop work. Such direction will come from the owner's or architect's representative. Depending upon the conditions of the project, the owner's inspector may be on the jobsite full-time or part-time. Or, the owner may decide not to have an inspector on the project at all, relying instead on an inspection by the architect.

Testing Agency.   An outside testing agency is often contracted by the owner to perform certain quality control tests to verify that the materials are installed to the specified standards. Some of these tests might include soil compaction tests, concrete strength tests, reinforcing steel placement inspections, weld inspections, and bolt-torque inspections. The testing agency is contracted by the owner, and its reports are directed to the owner, with copies normally sent to the architect and contractor. The owner can transfer the responsibility for testing to the contractor in the contract documents, however, a more impartial relationship exists if the testing is contracted by the owner.

The Architect's Roles During the Construction Phase

The complexity of the architect's organization is dependent upon the size of the architectural firm and the size of the construction process. Some architectural firms will combine the roles of project manager, project architect, and contract administrator, relying on one individual to perform all of these roles during the process. Figure 1–3 depicts an organizational strategy for an architectural firm.

Principal-in-Charge.   The principal-in-charge is an upper management level individual who is the ultimate decision maker for the firm on the

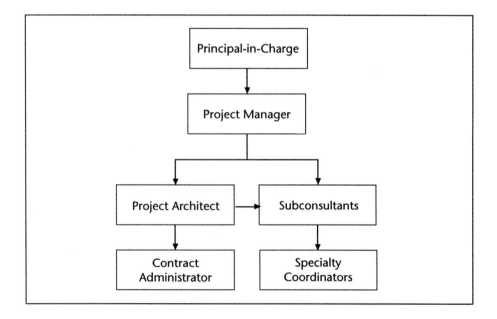

FIGURE 1–3
Architect's
Organization

project. The individual could be the owner of the firm, the CEO, the president, or one of the principals. The principal-in-charge usually has been involved in obtaining the contract for the architectural firm. A business relationship exists between the principal-in-charge and the capital projects officer in the owner's organization. Like the capital projects officer, the principal-in-charge is not involved in the project during the construction phase on a daily basis and limits his participation to major issues only. The principal-in-charge, however, maintains high-level communication between the architect and the owner.

Project Manager. The project manager is in charge of the project from beginning to end. He may be working concurrently on other projects but is still fully responsible for each one. The project manager is the architect's primary contact with the owner's representative. He provides direction to the architect's other employees and to the subconsultants who are working on the project. The project manager is also the direct liaison with the contractor's project manager and is concerned with the project budget and with the costs incurred by the architectural firm. The project manager is responsible for the architect's team composition and its ability to fulfill its contractual obligations. He will be involved in all of the decisions concerning changes in the construction contract and in evaluating the validity of those changes to the project.

Project Architect. The project architect is primarily involved in the design of the project and construction documents. He coordinates the designers, architects, engineers, draftspeople, specification writers, and subconsultants in the process that ultimately produces the construction documents. By being involved in this process from the project's inception, the project architect is considered the expert regarding the intent and interpretation of the contract documents for the architectural firm, although he does not participate on a daily basis in the construction process. The project architect is probably the best source for the review of shop drawings for the project, if time allows. He can also serve as a reference if other parties are to review the shop drawings. Some architectural firms do not have a contract administrator in their organization, but instead utilize the project architect as the liaison during the construction phase. In that case, the contract administrator's duties, explained next, would be the responsibility of the project architect.

Contract Administrator. Many architectural firms employ a contract or construction administrator who is a specialist in projects that are under construction. The contract administrator processes shop drawings, progress payments, requests for information, change orders, and correspondence relating to the project. He conducts meetings with the contractor and issues minutes of the meetings. The contract administrator is the primary day-to-day contact for the contractor's project manager and superintendent and observes construction and relates information about the project to the project manager. The contract administrator may also be responsible for making certain decisions on the project, depending upon the level of involvement of the project manager during the construction phase.

Some architectural firms have an inspector on the jobsite at all times. This inspector is a clerk of the works who fulfills the duties of the owner's inspector, described earlier. On projects that require a full-time inspector, this person either is employed by the architectural firm or the owner, but usually not by both.

Subconsultants. The subconsultants provide design services and portions of the construction documents that are not provided by the architect's in-house staff. Typical subconsultants used on a building project include civil and environmental engineering, structural engineering, mechanical engineering, electrical engineering, and interior design firms. Other specialty subconsultants, such as acoustical, kitchen, detention, and industrial engineering services also may be used, depending on the specifics of the project. During the construction phase, the subconsultant reviews shop drawings and provides input relevant to his portion of the project to the architect and to the contractor (through the architect). Most subconsultants retain some involvement during the project, usually relating to special installations that are needed to execute the work and approve the installation at its completion.

Specialty Coordinators. The specialty coordinators are inspectors and engineers hired by the subconsultants who provide services on the jobsite during the construction phase. As many of the specialty areas, such as mechanical and electrical, are extremely complex and relate to the work of many trades, the specialty coordinator can facilitate the work by becoming frequently involved in the project. These specialists often are intimately involved in equipment start-up and testing.

The Contractor's Roles During the Construction Phase

The organizational chart presented in Figure 1–4 indicates the typical hierarchy of the contractor's organization for the construction of the project. This arrangement may vary, depending upon the project's size, the special characteristics of the project, and the management philosophy of the construction firm. Some firms prefer to maintain a strong management presence on the jobsite, while others prefer to keep management to the minimum.

Officer-in-Charge. The officer-in-charge, like the owner and the architect, is responsible for the firm's performance. This individual may be the construction company's owner, president, CEO, vice-president, or district manager. He has a business level relationship with the owner's capital projects officer and the architect's principal-in-charge. The officer-in-charge normally is not involved with the project on a daily basis, but is involved in matters that affect the success of the project.

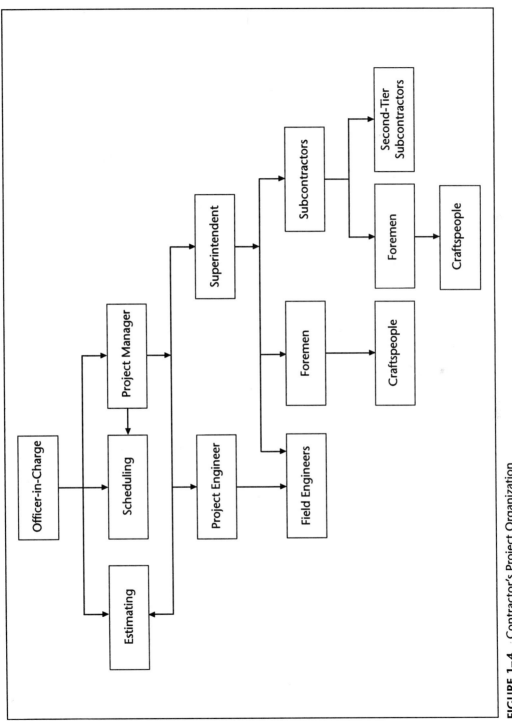

FIGURE 1–4 · Contractor's Project Organization

Project Manager. The contractor's project manager organizes and manages the contractor's project team. His responsibility to upper management is to ensure the project's profitability. The project manager selects and maintains the appropriate team to economically and efficiently complete the project to the owner's standards, as set forth in the plans and specifications, while making a profit for the construction company. The project manager is responsible for implementing time and schedule control, cost control, and quality control during the project. He conducts business with the owner's representative and the architect's project manager. In most contractor organizations, the project manager operates at a higher level than does the superintendent, however the latter has a great deal of autonomy in managing the project's physical construction. In some organizations, the superintendent is considered an equal to the project manager. Normally, the project manager is more involved in the business and formal requirements of the contract, while the superintendent usually is associated with the activities regarding the construction of the facility. While the organizational chart in Figure 1–4 represents the lines of command where the superintendent reports to the project manager, Figure 1–5 illustrates the contractor's project organization where the superintendent and project manager are at equal, or parallel, hierarchy levels. In this case, an operations officer supervises the superintendents.

During the construction phase of the project, specific estimating and scheduling functions need to be accomplished. Estimating functions during the construction phase relates primarily to estimating changes to the contract. The schedule, of course, is a major planning tool of the construction process and provides a standard for measuring success in compliance to the time constraints on the project. Both of these elements fall under the project manager's domain, but for smaller projects, the project manager may actually personally perform each of these functions. Large construction organizations have estimating and scheduling departments, which perform these services for the project manager. Sometimes these duties may be assigned to lower-tier management personnel such as the project or field engineer.

Superintendent. With either organizational structure presented, the superintendent is responsible for the correct, timely, and profitable construction of the project. The superintendent has the necessary skills and understanding of common construction methods and practices. He manages a crew of craftspeople employed by the contractor and subcontractors. It is the superintendent's responsibility to coordinate labor, material, equipment, and subcontractors during project installation. The superintendent determines the labor force, equipment on site, and timing of the delivery of materials and subcontractor work. He is responsible for the jobsite: its safety, efficiency, and compliance to the parameters required by the construction documents and regulatory codes. The superintendent is considered a full-time representative of the contractor at the jobsite, while the project manager may be assigned to several projects. The super-

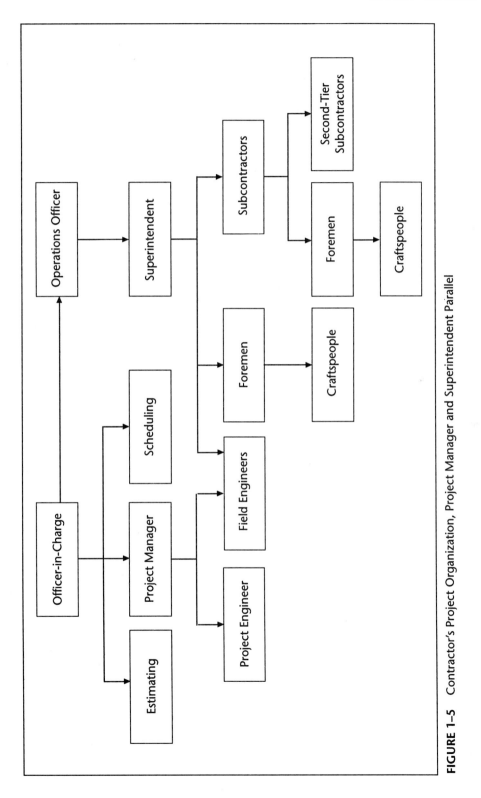

FIGURE 1–5   Contractor's Project Organization, Project Manager and Superintendent Parallel

intendent gives direction to the foremen and craftspeople who are employed by the contractor, coordinates subcontractors, and is the point of contact for subcontractor foremen. The superintendent normally communicates with the architect's contract administrator. If there is an on-site inspector, either for the owner or architect, the inspector's point of contact will be the superintendent. Depending upon the management philosophy of the contractor, field management employees, such as the field engineer, may report to the superintendent or project manager. The superintendent is often the only member of the contractor's management personnel on the jobsite on small- to medium-sized projects. On some small projects, the superintendent actually participates in work activities. In these instances, he is called a "working superintendent."

Project Engineer.   The project engineer usually reports directly to the project manager. He performs paperwork activities for the project manager such as subcontract agreements, material submittals and shop drawings, payment requests, contract change orders, requests for information (RFI's), correspondence with subcontractors and suppliers, and project documentation. The project engineer normally assists the project manager with any activities that are necessary to keep the project flowing and on track. He will informally communicate with the architect's project manager, contract administrator, and inspector, however, formal conversations with these individuals usually are pursued by the project manager or superintendent.

Field Engineer.   The field engineer reports either to the project manager/project engineer or the superintendent. The field engineer is involved in the layout of work and the interpretation of the construction documents. He should be knowledgeable about the contents of the construction documents and is often responsible for documenting jobsite conditions and conversations. The field engineer is responsible for issuing requests for information to the architect about clarifications, differing field conditions, and erroneous information contained in the construction documents. The field engineer may order materials and review and/or process shop drawings and submittals. He is often responsible for quality control and assurance in the project. This position is the lowest tier on the management side of the contractor's employees but the field engineer is responsible for a wide variety of tasks in assisting the superintendent and project engineer. Some firms will divide the responsibilities of the field engineer, such as assignment of the layout and field coordination to them. Office work, such as submittals, payments, and change orders, is assigned to the office engineer. In other firms, the assistant superintendent performs most of the tasks described for the field engineer.

Foremen.   Foremen are supervisory personnel. They usually are paid hourly and receive slightly higher pay than craftspeople. Foremen are responsible for directing the labor crew in their work activities. Foremen, as hourly employees, are not considered management, but labor. Foremen are usually knowledgeable about the installation techniques that are necessary to perform the work of the crew. Foremen create work assignments for craftspeople. They are normally in charge of a crew made up predominantly of their craftspeople, such as carpenters, however a crew can include several different craftspeople such as carpenters, laborers, operating engineers, and cement masons. Foremen are responsible for preparing time and quantity in-place reports for cost and schedule control purposes. They also are responsible for reporting work-ready or complete-for-quality inspections. Foremen are responsible for successfully completing work activities within a specific budget and time frame. Depending upon the size of the crew and the nature of the task, foremen may act in a purely supervisory role or they may actually perform some of the labor for the task, in which case they may be referred to as working foremen.

On large construction projects, a lead craftsperson is used to supervise each crew when the foreman is in charge of several crews. The lead craftsperson normally performs work activities with the crew.

Craftspeople. Craftspeople are hourly employees who are trained to perform specific tasks. Each craft or trade performs its work assignment with a high level of efficiency and quality. The following is a list of some common craftspeople who could be directly employed by the contractor:

Laborer
Carpenter
Operating engineer
Teamster
Ironworker
Cement mason

Construction trades or crafts distinguish between classifications of work performed, resulting in a slight wage differential. The trained craftsperson, receiving the full wage under the work classification, is referred to as a **journeyman. Apprentices** are journeymen-in-training and receive a lower wage than journeymen. Some crafts use a **helper,** who aids the journeyman and is similar to a laborer, but who lacks the training progression and wage increase of the apprentice. Current trends show that the general contractor is employing fewer and fewer craftspeople, relying now on much of the construction labor to be furnished by subcontractors.

Subcontractors. Subcontractors are separate business entities from the contractor. They provide labor, material, equipment, and occasionally second-tier subcontracts to complete a specific portion of the construction. They have agreements with and are responsible to the contractor. All correspondence and requests for clarifications from the subcontractor go to the contractor, who determines which course of action to take. The subcontractor management usually will contact the contractor's project manager or the project engineer for clarifications and contractual discussions. The subcontractor's foreman at the jobsite communicates with the contractor's superintendent concerning work parameters, changes, and directions.

Subcontractor Foreman.   The subcontractor foreman is the subcontractor's site representative. Larger subcontracts will require a superintendent for the subcontractor when there are several crews and a variety of work, for instance, with a large mechanical subcontract. The subcontractor's foreman is responsible for the quality of work accomplished by his crew or crews. The subcontractor's foreman is often a working foreman, in that he also works with the tools. The subcontractor foreman's primary task is to facilitate a profit for the subcontractor on the particular work assigned for the project; he provides direction to his crew or crews to install and complete the work assigned.

Subcontractor Craftspeople.   The subcontractor directly employs craftspeople as hourly laborers on the project. Different training and classifications are assigned to craftspeople who are doing specific tasks. Figure 1–6 lists some of the trades subcontractors utilize on the jobsite. Some subcontractors employ the same craftspeople as the contractor, however, they usually perform different types of tasks.

Other Roles in the Construction Process

Significant roles in the construction process are played by entities other than the owner, architect, and contractor. The majority of these individuals represent regulatory agencies, as required by the municipal, state, or federal governments. There are a number of inspectors from different agencies and levels of government who need access to construction sites. These inspectors examine the installations for compliance to codes that are legislated standards of compliance necessary to protect the public. The codes apply to the design and installation of the particular systems in the project and supersede the contract documents. It is assumed

FIGURE 1–6
List of Trades
Employed By
Subcontractors

Laborer	Glazier	Lather
Carpenter	Plasterer	Taper
Operating Engineer	Ceramic Tile Installer	Floor Covering Installer
Teamster	Terrazzo Mechanic	Painter
Ironworker	Elevator Mechanic	Millwright
Cement Mason	Plumber	Pipe Fitter
Bricklayer	Steamfitter	Sprinkler Fitter
Roofer	Mechanical Insulator	Temperature Control
Sheet Metal Worker	Refrigeration Mechanic	Mechanic
		Electrician

Inspector	From	Items Inspected
Building Inspector	City, County	Concrete Footings Concrete Reinforcing Wood Framing Steel Framing Final Compliance
Plumbing Inspector	City, County	Plumbing Rough-in Sewer Installation Water Line Installation
Electrical Inspector	City, County, State	Electrical Rough-in Electrical Finish
Fire Marshall	City, State	Fire Alarm Systems Fire Protection Systems
Elevator Inspector	City, State	Elevators, Conveyance Systems
Safety Inspector	State, Federal (OSHA)	Safety Compliance of Jobsite

FIGURE 1–7
List of Typical
Inspections

that the contractor has met all of the codes with his installations. State industrial safety agencies and the federal government industrial safety agency, the Occupational Safety and Health Administration (OSHA), also have the right to inspect the jobsite for safety compliance.

Figure 1–7 lists some of the inspections that are common to building construction.

Communications in the Traditional System

The previous discussion indicated that there are distinct hierarchical levels within each organization. Communications, whether they are verbal or written, are normally between individuals at the same level. Figure 1–8 illustrates a communications matrix indicating direct, or contractual lines of communications, and indirect, or nonbinding, communications.

During the construction phase, special care must be taken to avoid communications that transcend the contractual lines of privity. Subcontractors should not communicate directly with the architect, but should approach the contractor's superintendent or project manager. Most problems on the jobsite have a larger impact than just the particular subcontractor's work, and the superintendent or project manager has the responsibility and perspective to discern the full impact of the change or concern. There is often informal communication between parties, however, throughout the construction process and considerable communication between the subconsultants and subcontractors that may be technical in nature and may be misconstrued if transferred through several parties. Informal communication is important for clarifications, but binding clarifications, changes, and directions must go through the

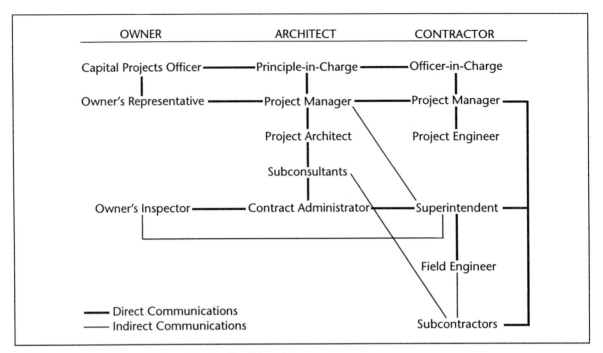

FIGURE 1–8   Communications Matrix, Traditional System

proper contractual channel, preferably, and are often required in writing. The subconsultant and subcontractor should be careful to transmit information that has an impact on the contracts to the architect and contractor, for official and binding communication.

The Construction Management (CM) Delivery System

The traditional delivery system does not always meet all of the owner's needs. On complex projects, where budget, time, and/or quality are exceptional concerns in the project, the construction management (CM) delivery system may be used to accommodate those needs. The CM process applies contractor-based management systems early in the project, providing more tools and controls to contain the project within its scope parameters during the design as well as the construction process.

There are two basic forms of construction management:

- Agency CM
- CM-at-Risk or Guaranteed Maximum Price (GMP)-CM

Agency CM involves the use of a manager, as the agent of the owner, without design or construction responsibilities. The construction manager in the agency arrangement works on a fee basis with the owner. Agency CM enters the process early and acts only as the agent of the owner throughout the process. The Agency construction manager brings management tools to all phases of the work, without having a vested interest in either the design or construction of the building.

Figure 1–9 illustrates the organization of the Agency CM arrangement.

Under the Agency CM system, the contracts for the architects and contractors are written directly with the owner, however the construction manager acts as the owner's agent and manages both the architect and the contractor. Depending upon the level of service offered by the construction manager, the contractors may be trade contractors, execution contractors, or specialty/subcontractors, without a general contractor, and may be managed by the construction manager. Under this system, the architect's position in the construction phase of the project is reduced to an advisory role, with the construction manager providing the construction administration for the owner.

Although the construction management system adds another layer of bureaucracy to the project, it provides a more intense and appropriately focused management of the process than does the traditional process. It is best utilized by an owner who has little or no construction expertise within their organization, relying on the construction management firm to coordinate the project. One prominent CM firm advertises its services as "Extension of Staff." The CM process provides the owner with specialized management to control problems with project duration, project budget and cost, or project quality. Although the fee for the CM firm appears to be an additional cost to the owner, the construction manager is replacing the owner's in-house construction manager, also a significant project cost.

CM-at-Risk is similar to the guaranteed maximum price arrangement in the traditional system, except that the construction manager is involved in the conception and design of the project, rather than entering after the completion of the contract documents. The construction manager in this arrangement provides the owner with a maximum price for

FIGURE 1–9
Organization of the
CM Delivery System
(Agency CM)

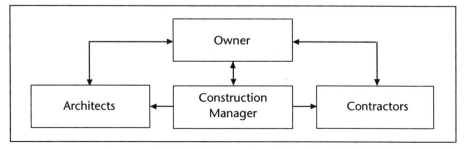

the project, considering the project's initial scope. The GMP construction manager manages the construction phase much like a contractor does under the traditional system, often subcontracting out all or most of the work. The roles during the construction phase under the construction management system are modified from the traditional system.

The roles involved during the construction phase for the construction management system include:

- The owner, who still provides funding for the construction and the site for the project. The owner makes the decisions about the project, but is advised in all matters by the construction manager. As the construction manager handles the entire project, from conception through completion, the owner has little day-to-day involvement.
- The construction manager, who is responsible for the administration of the construction contracts during the construction phase of the project. The construction manager maintains a liaison with the architect for advice on intent of the documents. He schedules and manages the submittal process, while the architect reviews the submittals and shop drawings. The construction manager may provide a detailed schedule for the completion of the construction and acts as the communications conduit for the contractor or contractors on the project. The construction manager also will process progress payments and contract completion.
- The architect, who is engaged to design the project and prepare construction documents. During the construction phase, the architect serves as a reference and an advisor on the intent of the construction documents, but the construction manager is responsible for the administration of the project. The architect is involved in reviewing submittals and shop drawings, but under this system, is not normally involved on a daily basis in the construction of the project.
- The contractors, who, under the CM process, play a role similar to that of the subcontractor in the traditional system. The contractor may be assigned a large element of the work, such as the foundation or building envelope, or may be awarded a small contract, such as caulking and sealants. The smaller subcontracts, which encompass work by a single trade, are usually called **trade contracts**. For larger portions of work, the contractor would employ a superintendent, but would still look to the construction manager's superintendent for coordination of the work.

Roles of the Construction Management Project Team

Figure 1–10 illustrates the roles of personnel within the CM project team. Each project and CM firm will use a variation of this basic arrangement.

The roles involved in the CM team include:

- The project manager, who is the responsible party to the owner for the success of the project. The project manager is in charge of all of the CM

FIGURE 1–10
Construction
Management
Roles

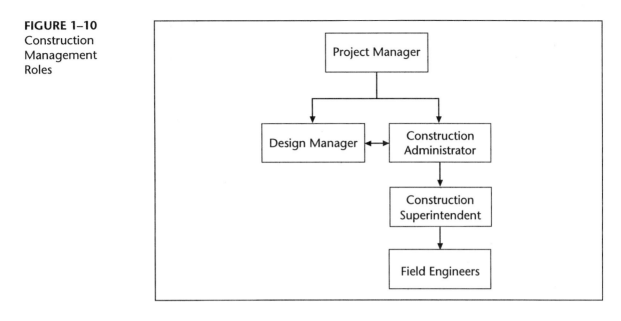

employees on the project. He or she primarily oversees the construction process, however most of the contact with the contractor during the construction phase is handled by the construction administrator.

- The design manager, who is involved with the architect during the design and construction processes. The decisions concerning contract packaging, early purchase of materials, and project strategies are made during the design phase of the project. Some CM firms have the design manager phase into acting as the construction administrator, while other firms will use different individuals who possess specific skills for their phase of the project. Liaison and communication need to continue between the design manager and the construction administrator concerning the intent of the design.

- The construction administrator, or often referred to as the construction manager, who is in charge of the construction process for the CM firm. The construction administrator is the contact person for the contractors' project managers. He is responsible for the entire construction process, delegating part of the field responsibilities to the construction superintendent and the field engineers.

- The construction superintendent, who coordinates the field activities. Unlike the traditional delivery system where the construction superintendent is in charge of direct labor and subcontractors, in the CM delivery system this person manages only the trade or execution contractors, as the construction manager does not use direct labor. In cases where the contractor or contractors do substantial portions of the work, providing coordination of the subcontracts within their scope, the construction manager may utilize a construction administrator,

FIGURE 1–11
Communications in
the Construction
Management
System

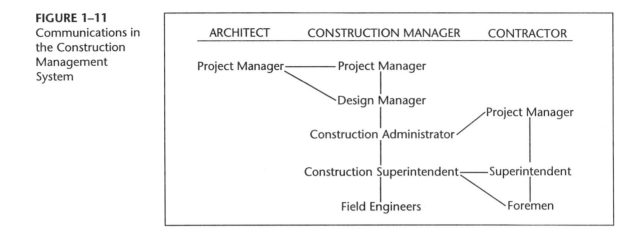

and not a construction superintendent. It is essential however, to have a knowledgeable construction superintendent on the site when using a myriad of trade contractors, in order to provide some order to the process. The construction superintendent is responsible for maintaining the construction schedule, for ensuring compliance of all work to the documents, and for coordinating all of the trade contractors. The construction superintendent may need assistance from the field engineers, depending upon the scope of the project.

• The field engineer, who is responsible for the coordination of shop drawings, submittals, layout, subcontractor organization, payment verification, and whatever duties are assigned by the construction superintendent or construction administrator.

Communications in the Construction Management Delivery System

The matrix of communications lines in a typical Agency CM project for the construction phase is illustrated in Figure 1–11.

Since the construction manager is the controlling entity in the project, all communications flow through him. This system should facilitate prompt responses to communications throughout the construction phase.

The Design-Build Delivery System

Both of the previous systems can be cumbersome for the owner who wants to avoid conflict between the project participants during the project, while still striving for a project that meets their needs in a short period of time. The design-build delivery system is a single-source procurement for the owner. Design and construction, and even occasionally the purchase of the site, are provided to the owner for a "Guaranteed-Maximum Price" (GMP), also known as a "cost not to exceed" price from the design-build firm. There often is a cost savings split clause, which will split the savings between the GMP and the actual cost between the owner and the design-build contractor. Instead of creating an adversarial relationship between the architects and contractors, as in the other two delivery systems, the design-build firm acts as the facilitator for designers and constructors in the design-build team to work together to achieve the owner's objectives.

There are four basic configurations of the design-build firm, with numerous variations. The first is the design-build firm that exclusively does design-build work and has under its direct control both designers and constructors, as shown in Figure 1–12.

This type of organization provides internal control over design and construction, resulting in a high-value package for the owner. Most design-build firms with this type of organization are large.

The second is the design-build firm that contracts out both architectural and construction services. The design-build firm could have its origins in architecture/engineering, construction, or even property development. This is a popular organizational model for developers who are involved in design-build construction. Figure 1–13 illustrates this type of design-build organization.

The third is the contractor-lead venture, with the contractor being the contracting point with the owner and the designer being subcontracted

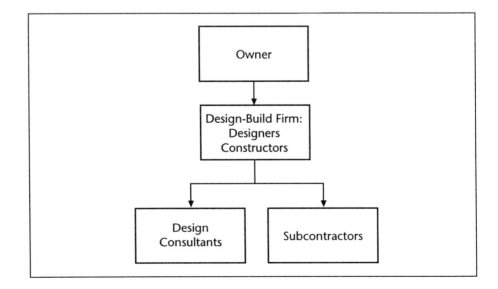

FIGURE 1–12
Design-Build
Organization I

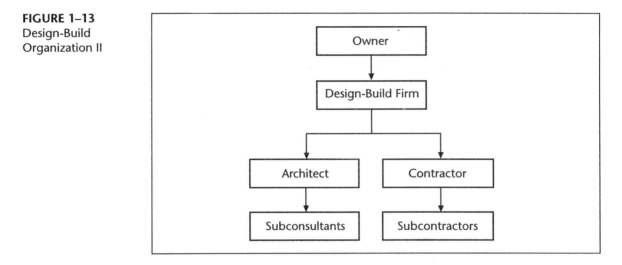

FIGURE 1–13
Design-Build
Organization II

to the contractor: This type of organization is frequently used when determination of the successful design-build team is based on bids. Figure 1–14 illustrates this type of organization.

The fourth is a hybrid of the above-mentioned systems, when an architectural engineering firm and a contracting firm form a joint venture or a partnership for one specific project. This type of organization pools resources and can rely on the reputations of the two firms. Figure 1–15 illustrates this type of design-build organization.

Each of these organizational types provides the design-build firm with funding, leadership, and the resources necessary for the particular type of

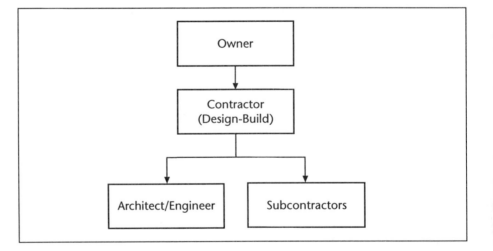

FIGURE 1–14
Design-Build
Organization III

FIGURE 1–15
Design-Build
Organization IV

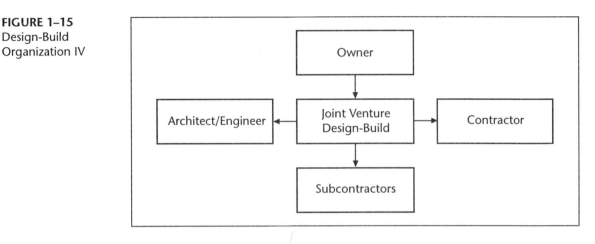

project. These different options are available to firms that are working in the design-build area. The first organization, shown in Figure 1–12, represents a firm that would work exclusively in design-build projects. The other three organizations, shown in Figures 1–13, 1–14, and 1–15, can be composed of firms that would work in both the traditional and construction management delivery systems, as well as in the design-build system.

Communications in the Design-Build Delivery System

Communications between the design-build firm and the owner, regardless of the configuration, are rather simple. The owner's representative communicates with the design-build firm's project manager. The design-build firm, then, needs to determine the best way to achieve its objectives, through its own communications matrix, determined by the exact relationships within the design-build firm. The design-build team acts as an in-house operation, as there should be no conflicts between entities, as in the other processes. The design-build team is focused on producing a solution to the owner's needs in a project within the scope of time, budget, and quality. Figure 1–16 illustrates the typical communications matrix for the design-build delivery system.

Engineering-Procurement-Construction (E-P-C)

Another type of delivery system, actually a form of design-build, is the Engineering-Procurement-Construction system, commonly referred to as E-P-C. E-P-C is a common delivery system for industrial plants, power plants, refineries, and other heavy construction facilities. Under this agreement, the E-P-C firm provides engineering for the facility, procurement of all material and equipment, and construction of the facility, with

FIGURE 1–16
Design-Build
Communications

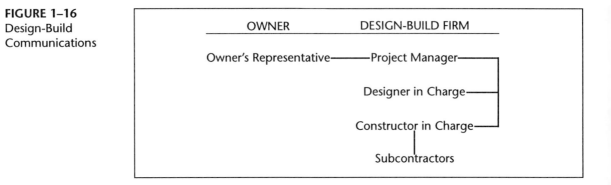

its own forces or those contracted to other firms. As with the design-build system, there is a single point contact with the owner during the entire project.

Summary

Three primary delivery systems—the traditional system, the construction management system, and the design-build system—are used in the construction of building and industrial facilities today. Each of these systems has different roles for the contractor, architect, and owner and each has a different communication network between the parties in the system. Understanding these systems is necessary for effective project management.

Although the traditional system is the most common delivery system currently used, the alternative systems of construction management and design-build offer opportunities to provide more management in the process, controlling cost, time, and quality. Each project situation has unique parameters that need to be examined to determine the most desirable delivery system for the owner.

Each of the three delivery systems is effective in certain situations and project communications vary greatly in each. Comparison of the three communications charts in this chapter (Figures 1–8, 1–11, and 1–16) indicates the relative amount of communication complexity. The challenge then in each delivery system is to recognize where the problems might arise and to use management techniques to eliminate or minimize those problems.

CHAPTER 2

USE OF CONSTRUCTION DOCUMENTS ON THE JOBSITE

CHAPTER OUTLINE   The Construction Documents

Use of the Construction Documents

Summary

In most construction projects, the contractor does not develop the design of the project. The design is envisioned by the architect or engineer and is then transmitted to the contractor through drawings and written specifications. These construction documents describe the nature of the project, the materials desired, the level of the quality of work desired, the connection of materials, and the installation and systems necessary to achieve the project. The construction documents indicate to the contractor what will be built. The documents should be complete enough to construct the intended facility with little or no further clarification.

Construction documents, however, are *not* for the novice. The terminology used in these documents is a special language that accurately describes construction components. The plans, elevations, details, and diagrams used in construction are fairly unique to the industry, not readily understandable to the general public. The construction documents are written with the assumption that the contractor is experienced and knowledgeable and will understand them. It is also assumed that the contractor is experienced and knowledgeable in the means and methods of constructing the particular project. The documents usually do not indicate the way in which the project will be built, that is, the means and methods. The contractor furnishes his expertise for construction of the project, just as the architect and the engineer do in the design of the project. The contractor is expected to know how to construct the different systems composing the entire project—the crafts and subcontractors to use, the quantity and type of material necessary, the type of tools to use, and the type of equipment necessary. The contractor also is expected to possess the expertise to ascertain the quality level required by the documents and implement that quality level into the project, and to construct the project in a safe and legal manner.

Although codes and regulations are not formally included as construction documents, they are the governing regulations for compliance of design and construction. Codes are normally referenced in the construction documents, but the contractor is obligated to comply to codes and regulations even if they are not specified or referenced. The contractor must comply to the drawings and specifications as well. The design is based on the constraints of the codes. Standard codes, such as the Uniform Building Code and the National Building Code are modified by the local municipality or governing body. The contractor is expected to install the project elements in compliance with the locally enforced building, plumbing, electrical, energy, and local fire protection codes. It is assumed, however, that the construction documents comply to code and have been reviewed for such.

Material and equipment attributes also affect the installation of the construction elements. Each material and piece of equipment has installation parameters, instructions, and restrictions that must be followed to obtain a quality, durable installation. Most construction documents state that the installation of the material should be done in compliance with the manufacturer's recommendations and instructions. The latter are complementary documents to the construction documents. The contractor uses the manufacturer's recommendations, the construction documents, and the applicable codes and regulations to provide information for the construction of the project.

The Construction Documents

Traditionally, construction documents consist of two major physical components: the project manual and the drawings. Various formats for both components are used, depending on the construction type and the local customs. The project manual normally includes:

- Invitation or Advertisement To Bid
- Instructions To Bidders
- Bid Forms
- General Conditions of the Contract
- Supplementary Conditions to the Contract
- Additional Information To Bidders (such as soil reports)
- General Requirements (Division 1)
- Technical Specifications (Divisions 2–16)

Construction drawings for a building project normally include:

- Abbreviations and general notes
- Code-related compliance information
- Site surveys
- Site plans
- Utility plans
- Floor plans
- Building sections
- Wall sections
- Exterior elevations
- Interior elevations
- Details
- Equipment plans and details
- Structural engineering drawings
- Mechanical engineering drawings
- Electrical engineering drawings
- Special construction information

The project manual is usually in book form, often bound in plastic bindings. Some projects may include the drawings in the bound volume, particularly when there are few drawings for the project. There may be several volumes of the project manual. Drawings may be in several formats and sizes and can range in size from 8½″ × 11″ sheets to 48″ × 36″

sheets. Some projects reduce the drawings to half their size to conserve paper and to provide a more convenient size drawing for use in the field. Special care should be taken with reduced size prints to determine the scale used on the drawings. The original scale may be $\frac{1}{4}'' = 1'0''$, but in half-size reduction this scale becomes $\frac{1}{8}'' = 1'0''$. The modified scale is rarely noted, so it is important for the user to observe if the drawings have been reduced.

The primary purpose of construction drawings is to communicate the intended elements, systems, and relationships of the materials, equipment, and assemblies to the contractor in the construction of the facility. The documents also provide the necessary information for the contractor to compile a cost estimate and plan the construction process for the project. The construction documents also should provide sufficient information for the preparation of shop drawings and submittals for material suppliers and fabricators, as well as provide information to the owner of the facility after the construction is completed for the maintenance and operation of the facility. Usually the construction documents are revised to reflect changes during the construction phase of the project, referred to as "as-built drawings," for use by the owner during the operation of the facility.

The owner of the project furnishes the construction documents to the contractor for the project. The owner, in most cases, does not prepare these documents, but usually contracts with an architectural or engineering firm to design the project and prepare the documents. Some large corporate or public owners use in-house architects and engineers to design and prepare construction documents. Several different disciplines are involved in the design and document preparation, coordinated by the lead architect or engineer. Consultants employed by architects include: geotechnical, structural, mechanical, and electrical engineers, and specialty designers such as industrial engineers, landscape architects, interior designers, and equipment consultants. The construction documents consist of several different components, some in written narrative and some appearing as graphical documents.

What follows is a discussion describing the information generally found in each section of the construction documents. As previously noted, the information found in these sections varies, due to local customs and the type of construction. Engineering-type construction documents are generally a bit different than building construction documents. The mode of drawing in engineering documents, for engineering components such as bridges or roads, is more mechanical, reflecting critical attributes like location and elevation. Engineering drawings use an engineering scale, graduated in one-tenths of a foot, while architectural drawings primarily use an architectural scale, graduated in multiples of one-eighth of an inch. Many architects and engineers use some standard documents for certain parts of the project manual, such as the Invitation to Bid and General Conditions. The American Institute of Architects (AIA) has a standard set of documents used in building construction documents, just as the Engineers Joint Construction Documents Committee (EJCDC) uses a set of documents for engineering construction. These documents are somewhat similar, but do have some differences, based on custom in the appropriate fields. Many architectural and engineering firms use a master specification guide for technical specifications, resulting in similar specifications for projects.

Advertisement or Invitation To Bid

This section usually appears at the front of the Project Manual. The advertisement or invitation to bid contains information about the project during the bid period. The information normally contained here includes:

- Project name, number, location, and description
- Owner name and address
- Architect/Engineer name, address, and contact individual
- Bid date and time
- Bid security required
- Restrictions on bidders
- Anticipated price range of project (optional)
- Anticipated project duration (optional)

Figure 2–1 illustrates an example of an "Advertisement for Bids" on a public works project.

Instructions To Bidders

This section contains information about the bid procedure. Both the AIA and the EJCDC have standard forms that can be used, with desired modifications, as instructions to bidders. Most of these bid procedures are standard. The bidder should review this section for each project, as specific information concerning the particular project will be contained here. Specific information concerning requests for substitutions, clarifications, bidder prequalification, bid period, and construction period duration can be included.

AIA Form A701, Instructions to Bidders, includes the following types of information:

- Bidder representations, defining the obligations of the bidder
- Bidding documents: where and to whom they are available
- Interpretation: procedures for questions and interpretations of the documents
- Addenda: inclusion of addenda as part of the documents and procedures for addenda
- Bidding procedures, including the form and style of acceptable bids, the bid security required, procedures for submission of bids, and rules concerning modification or withdrawal of bids

---

ADVERTISEMENT FOR BIDS

Sealed proposals will be received for the following project:

PROJECT NO.	98-250
TITLE:	State Office Building Centerville, WA
AGENCY:	Department of Public Works
ESTIMATED BASE BID COST RANGE:	$ 2,500,000 to $ 2,600,000
TIME/DAY/DATE:	BIDS WILL BE RECEIVED UP TO 3:00 PM Thursday, October 15, 1998
LOCATION:	Department of Public Works, Olympia, WA
BID OPENING:	Bid Center, Capital Center Annex 417 West 4th Avenue Olympia, WA 98504-1112

Contractors may obtain plans and specifications from the office of the Consultant, NTG Architects, S. 400 Main Street, Centerville, WA 99345, telephone (509) 439-0478, upon the deposit of $ 100.00, or they may be view at the following locations: Associated Builders & Contractors, Inc., Centerville; Associated General Contractors, Seattle and Centerville; Centerville Plan Center, Centerville; Dodge/Scan, Seattle; Spokane Construction Council, Spokane. The State reserves the right to reject or accept any or all proposals and to waive informalities.

The bidding documents may contain mandatory requirements for Minority and Women Business Enterprises (MWBE) participation. The participating MWBE's must be certified, and the name and dollar amount for each participating MWBE must be included in the "Form of Proposal". MWBE firms qualified for certification but currently not certified are encouraged to contact the Office of Minority and Women's Business Enterprises (OMWBE) immediately regarding the steps necessary for certification. Only certification by OMWBE, State of Washington, will be accepted. For assistance verifying certification contact: OMWBE, 406 South Water, Olympia, WA 98504-4611, telephone (206) 753-9693.

A pre-bid conference and walk-through has been scheduled at the project site, Thursday, October 1, 1998, at 2:00 PM. General Contractors, subcontractors, and interested vendors are encouraged to attend this pre-bid conference.

FIGURE 2–1 Example Advertisement for Bids

- Bid opening, including procedures for bid opening
- Rejection of bids
- Award of project to successful bidder
- Requirements for post-bid information, such as bidder qualification and financial information, and product submittals

- Bonding requirements
- Form of contract to be used for the project

Bid Forms

The necessary bid forms and other certification forms usually are included in the Project Manual. Bid security forms may be included as well. Bid forms include information relevant to qualification as a responsive bidder, such as receipt of addenda, bid price, alternate pricing, proposed completion of construction, subcontractor information, and certifications of compliance to regulations. Many projects, particularly public works projects, require numerous certifications concerning affirmative action, Disadvantaged Business Enterprise utilization, and other compliance issues. Forms that are required to be submitted with the bid normally are included in the Project Manual. Figures 2–2A and B are illustrations of a bid form used in public works construction.

General and Supplementary Conditions of the Contract

The General Conditions of the Contract relate to a construction project as rules would relate to a football game. These conditions establish the relationship between the owner, its agents (architects and engineers) and the contractor and define the relationship and responsibilities of the parties to maintain this relationship. The conditions also establish the terms of the legal contract between the owner and contractor.

The Supplementary Conditions to the Contract are additional provisions or modifications to the General Conditions of the Contract that are specific to the particular owner and project. Together, these form the Conditions of the Contract. All actions of the owner and contractor can be judged for compliance to the Conditions of the Contract. The General Conditions and the Supplementary Conditions of the Contract are the base of reference for determining responsibility, and resulting liability, for actions by either party during the project period.

The use of a standard form of General Conditions of the Contract is common in the construction industry. AIA Form A201, located in the Appendix, is the most commonly used form in building construction. Public works contracts usually use a custom set of conditions that is prepared for the particular branch of government, such as federal, state, or municipal. Corporate owners often will use a custom-prepared contract. The standard form of contract conditions is widely used, as the provisions are tested and well-established. As the General Conditions of the Contract appear in pre-printed form, they are occasionally included only by reference in the Project Manual. A copy of the General Conditions of the Contract and the accompanying contract agreement (AIA Form A101) should be readily available to the project manager and superintendent at all times for reference, whether bound in the Project Manual or not. Because there

STATE OF WASHINGTON
DEPARTMENT OF PUBLIC WORKS
417 WEST 4TH AVENUE
OLYMPIA, WASHINGTON 98504-1112

P R O P O S A L

PROJECT NO. 98-250
State Office Building
Centerville, Washington

I/We, the undersigned, having read all the requirements of this call for bids, together with all the special and supplemental conditions, specifications, do agree thereto in every particular, and will furnish all labor and materials specified herein necessary for and incidental to the completion of the work in a workmanlike manner for the sum of:

_____

_____Dollars ($_____)
(Base Bid shall be shown numerically and in writing)

ALTERNATE BIDS:

Alternate Bid No. 1: Water main and fire hydrants   Add $_____

Alternate Bid No. 2: Brick Pavers at Plaza   Add $_____

Alternate Bid No. 3: Plumbing Fixture Upgrade   Add $_____

TIME FOR COMPLETION:
The undersigned hereby agrees to complete all the work under the Base Bid (and accepted alternates) within 270 calendar days after the date of the Owner's letter of Notice to Proceed.
LIQUIDATED DAMAGES:
The undersigned acknowledges and agrees to abide by all provisions of the "Liquidated Damages" section of the General Conditions as it pertains to the Contractor for all work under this contract. The undersigned further agrees to pay the Owner as liquidated dames the sum of $ 500.00 for each consecutive calendar date that he shall be in default after the time for completion specified herein.

ADDENDUM RECEIPT:
Receipt of the following addenda to the specifications is acknowledged:

Addendum No. _____ Date_____   Addendum No. _____ Date_____

Addendum No. _____ Date_____   Addendum No. _____ Date_____

FIGURE 2–2A   Example Bid (Proposal) Form

Bid Form, Page 2

MINORITY AND WOMEN'S BUSINESS ENTERPRISE UTILIZATION CERTIFICATION:

To be eligible for award of this contract, the bidder must execute and submit with the bid the certification relation to MWBE Participation. This certification shall be deemed a part of the resulting contract.

The undersigned acknowledges that the goals for the new base bid work have been established for this contract in the amount of:

| Minority Business Enterprise (MBE) | $ 225,000.00 |
| Women Business Enterprise (WBE) | $ 125,000.00 |

The undersigned certifies that if they are the successful bidder on this project, the following MWBE firms will be utilized on the project and compensated in the amounts shown:

| | | DOLLAR AMOUNT | |
| FIRM NAME | DESCRIPTION OF UTILIZATION | MBE | WBE |

TOTAL PARTICIPATION $_____ $_____

NOTIFICATION:

If written notice of acceptance of this bid is mailed, telegraphed, or delivered to the undersigned within the time limit noted in the contract documents after the date of bid opening, or any time thereafter before the bid is withdrawn, the undersigned will, within ten (10) days after the date of such ,mailing, telegraphing, or delivering of such notice, execute and deliver a contract on the State of Washington Public Works Contract Form.

This bid may be withdrawn at any time prior to the scheduled time for the opening of bids, or any authorized postponement thereof.

Enclosed is a certified check, cashier's check, or bid bond in the amount of 5% of the Base Bid. Bid deposits in cash will not be accepted.

BIDDER INFORMATION:

NAME OF FIRM: _____

SIGNATURE   _____

OFFICIAL CAPACITY   _____

ADDRESS   _____

CITY AND STATE   _____ZIP_____

DATE   _____TELEPHONE_____

STATE OF WASHINGTON CONTRACTOR'S LICENSE NO. _____

NOTE: If bidder is a corporation, write State of Corporation below; if a partnership, give full names and addresses of all partners below:

FIGURE 2–2B   Example Bid (Proposal) Form continued

are several standard General Conditions of the Contract, the project management team must be aware of the particular form for their project. All project performance must comply with the General Conditions of the Contract and the modifications contained in the Supplementary Conditions.

As the basis for the legal contract between the owner and contractor, the General Conditions of the Contract is a very important document. Each sentence in the General Conditions and the Supplementary Conditions can have an impact on the construction project. Complete understanding of these documents is essential to project management. Because the purpose of this section of the book is to introduce the reader to various documents, a detailed discussion of the Conditions of the Contract is not included herein. The reader is encouraged, however, to carefully read and study the Conditions of the Contract that will be used, such as AIA Form A201.

The following is a list of the type of information contained in the General Conditions of the Contract:

- Definitions of the documents and their use in the construction project
- The rights and responsibilities of the owner, including information to furnish the contractor, fee payment, right to stop work, and the right to carry out the work
- The rights and responsibilities of the contractor, including the scope of work, supervision responsibilities, schedules, shop drawings and submittals, cutting/patching, and project clean-up
- Administration of the contract, including the responsibilities of the architect, engineer, and construction manager
- Procedures for claims and disputes during the contract, including method of dispute resolution
- Provisions regarding the use of subcontracts, including the contractor's responsibility for subcontractors
- Provisions allowing the owner use of separate contracts
- Provisions concerning changes in the work, including procedures for facilitating changes
- Provisions involving the timely completion of the project
- Payment procedures and methods
- Provisions regarding the completion of the contract
- Protection of persons and property related to the execution of the contract
- Provisions requiring insurance, including liability and property coverage
- Provisions requiring performance and payment bonds
- Provisions relating to testing, inspection, and correction of work
- Provisions for the termination of the contract

The Supplementary Conditions contain additional provisions and modifications to the General Conditions. A few examples of some modifications made in the Supplementary Conditions include:

- AIA Form A201, General Conditions of the Contract, which specifies arbitration as the remedy to dispute resolution. Some owners, due to their governing rules and regulations, cannot use arbitration as a remedy but must seek relief in the courts, in litigation. The Supplementary Conditions, then, would modify that section of the General Conditions.
- AIA Form A201 requires property <u>insurance</u> on the project itself (Builder's Risk Insurance) to be carried by the owner. Many owners prefer that the contractor carry this insurance with a provision in the <u>Supplementary Conditions</u> modifying the General Conditions.
- AIA Form A201 does not contain specific monetary levels of insurance that the contractor is required to carry for the contract. The specific levels, such as "combined single limit, $1,000,000 per occurrence," are contained in the <u>Supplementary Conditions.</u>
- Prevailing wage rates, as determined by the public agency, also will be included in the <u>Supplementary Conditions</u>.

As the General Conditions and Supplementary Conditions are separate sections of the Project Manual, the project manager may want to annotate the General Conditions, indicating which articles and paragraphs have been modified by the Supplementary Conditions. Some contractors will "cut and paste" the provisions of the Supplementary Conditions into the General Conditions to provide a comprehensive set of Conditions of the Contract. By integrating these conditions, the contractor has a complete and quick reference for particular project situations.

Additional Information To Bidders

Many project manuals will include information to bidders that is not considered a contract document. The most common piece of information to bidders is the geotechnical or <u>soils report.</u> Because this report is prepared for use by the architect and engineer in determining the design of the foundations, and not specifically for the contractor's use, the soils report is normally furnished only as information, without designation as a contract document. This information, however, may be the only available subsurface information. The contractor is responsible for making decisions about the information reasonably available. Many Project Manuals reference the soil reports, relating that they are available at the architect and engineer's offices for examination.

Divisions 1–16, Technical Specification

The Technical Specification, which relates to labor, material, equipment, and procedures to accomplish the required construction work, is usually divided into sixteen divisions as per the master format published by the Construction Specifications Institute (CSI). This sixteen-division format is commonly used for most building construction specifications in the United States and Canada. There may be different organizations for specifications in heavy, highway, and industrial construction.

The sixteen divisions of the specification include:

Division 1 General Requirements
Division 2 Site Work
Division 3 Concrete
Division 4 Masonry
Division 5 Metals
Division 6 Wood and Plastics
Division 7 Thermal and Moisture Protection
Division 8 Doors and Windows
Division 9 Finishes
Division 10 Specialties
Division 11 Equipment
Division 12 Furnishings
Division 13 Special Construction
Division 14 Conveying Systems
Division 15 Mechanical
Division 16 Electrical

More detailed items, known as broad-scope and narrow-scope sections, are contained within each division. The CSI suggests a five-digit numbering system for these items. A typical broadscope item is 03300, Cast-in-place Concrete. The first two digits indicate the division, which in this case would be Division 3, Concrete. The other three numbers indicate the broad- and narrow-scope sections for the particular item. Under the broad-scope classification of 03300, Cast-in-place Concrete, the following narrow-scope sections are standard:

03310 Structural Concrete
03320 Concrete Topping
03330 Architectural Concrete
03340 Low-Density Concrete
03345 Concrete Finishing
03350 Special Concrete Finishes
03360 Specially Placed Concrete
03365 Post-Tensioned Concrete
03370 Concrete Curing

More specific topics under these headings can be added as necessary, such as 03311, Footings; 03312, Foundation Wall; and 03313, Above-grade Walls.

Knowledge of the organization of each specification will help project management personnel quickly and easily access appropriate information. For example, if the field engineer wants to find the specification on curing compound, he would look in Division 3, Concrete, Section 03300, Cast-in-place Concrete, with the material specification in section 03370, Concrete Curing. This organizational system strives to group items of similar properties, not to create work packages or subcontract packages. A typical subcontract may contain several broad-scope areas, for example, a floor covering subcontract might include the following broad-scope sections: 09550, Wood Flooring; 09650, Resilient Flooring, and 09680, Carpet.

Division 1, General Requirements

The General Requirements section specifies project-related overhead items for the particular project. The majority of these items are general items in that they relate to all phases of the work, rather than an individual activity or a subcontract. This section specifies the necessary project facilities that are anticipated to complete the work. This division will relate to meetings, submittal requirements, construction schedule requirements, testing services, supervision requirements, temporary facilities, cleaning, and project closeout and warranty requirements. The majority of the cost items in this section would be included as "direct overhead" or "general conditions costs" for the contractor.

Some of the items included in this section are:

- Insurance and bond requirements, including procedures for submittal and completion
- Allowances (specific amounts to be included in the bid for items of work)
- Supervision and coordination requirements
- Field engineering and layout requirements
- Project meeting requirements, including meeting types, frequency, and minutes responsibility
- Submittal requirements for shop drawings, material data, material samples, photographs, and schedules
- Construction schedule requirements, including schedule type and procurement schedule
- Quality control requirements, including field testing services
- Temporary facilities, including temporary utilities, heat, field offices, barricades, and other temporary construction elements
- Temporary controls, including traffic, noise, and security
- Cleaning, including progress and final cleaning
- Closeout procedures and requirements, including Operation and Maintenance Manuals, warranties, and certifications

This division is effectively used to prepare a checklist of required submittals early in the project and of closeout requirements at the end of the project. The contractor should assume that the owner intends that all of the procedures in this division will be completed. The information provided here provides a guide for the minimum overhead required for the project.

Divisions 2–16, Technical Specifications

The technical specification in Divisions 2–16 generally describes the products, equipment, and systems that are intended to be installed during the project. These specifications usually describe products that are shown on the drawings, however, there are some products that will be required and not shown.

The most common format for the technical specification is the "three-part" format consisting of Part 1, General; Part 2, Products; and Part 3, Execution. All three parts contain important information for the contractor, to ensure that the proper product and systems are installed.

> **Part 1, General:** This part contains several essential items that can help determine the scope of the product required. Included are the following components:
>
> Related work specified elsewhere: This describes other sections where related work is specified. For example, if the section being examined is section 03300, Cast-in-place Concrete, "related work specified elsewhere" would appear as:
>
> Section 03100, Concrete Formwork
> Section 03200, Concrete Reinforcement
> Section 05500, Metal Fabrications
>
> Description of Work: This normally is a short description of the work included under the section. For Cast-in-place Concrete, the "description of work" would appear as:
>
> Cast-in-place concrete footings, foundation walls, walls, structural slabs, slabs-on-grade, and exterior concrete in location and dimension as shown in the architectural and structural drawings.
>
> Quality Assurance: Several items can be included in this subsection, including Qualification of Manufacturer, Qualification of Workmen, Codes and Standards, and Inspection and Testing By Independent Laboratory. Not all products will require all of these items.
>
> Submittals: This is a guide for submittals for particular products. For Cast-in-place Concrete, the submittals required would appear as:
>
> Ready Mix Concrete: name of supplier; mix design for each type of concrete
> Admixtures: product data, including name and address of manufacturer and supplier
> Membranes: product data, including name and address for manufacturer and supplier

Product Delivery/Storage/Handling: This subsection would detail any special handling of the product.

Part 2, Products: This section details the products intended for the project. They may be generically described.

Curing and Sealing Coating: Clear, liquid acrylic based polymer compound for curing and sealing concrete slabs.

The product also may be specified using a brand name. To encourage competition, products often are specified with the brand name and the phrase "or approved equal." A procedure is usually specified in the Instructions to Bidders upon submitting a request for substitution prior to the bid. Using the brand name when specifying the product, the aforementioned specification would appear as:

Curing and Sealing Coating: "KURE-N-SEAL," manufactured by Sonneborn Building Products, Minneapolis, Minnesota, or approved equal.

Sufficient information should be provided in the specification to order the material. Care must be taken by the contractor to assure that the product desired is being furnished. Some products have true equal substitutions, and others do not. The architect's approval of a product helps ensure its acceptance, but the product still must perform as well as the specified standard.

Part 3, Execution: The construction documents rarely provide complete installation instructions for products and systems. The execution section may be as simple as "Follow manufacturer's instructions." In some instances, particularly with new or sensitive products or usage, there may be step-by-step instructions for installation. Generally, the contractor is assumed to have the expertise for installation of the material and the use of appropriate means and methods.

The execution phase can include information about suitable substrate, necessary preparatory work, tolerances of installation, quality of workmanship, type of finish, special techniques, weather and environmental limits, environmental constraints, curing and drying time, cleaning, and other necessary steps to achieve the desired product level.

The information contained in this section can be used as quality control checklists. The superintendent and project manager should carefully review this section for all subcontract work as well. Careful coordination of the work in progress to the specified requirements will substantially reduce punch lists at the end of the project.

Addenda

Addenda are bulletins of additional information that are germane to the contract documents issued during the bid period. This information is added to the contract documents after their completion and prior to the contract being awarded. Information contained in the addenda is considered part of the contract documents, and an acknowledgment of receipt of the addenda and inclusion of the information in the bid is required on the bid form.

The information contained in the addenda include the following types:

- Clarification or correction of information contained in the contract documents. These clarifications result from a review by the designer, owner, and bidders. Answers to questions asked during the bid period are clarified in the addenda to provide the same clarification to all bidders. These clarifications can relate to both the Project Manual and drawings. Written and graphical clarifications are frequently made to make the documents more understandable to bidders.
- Approval of substitutions. Approval of substitutions for specified materials, usually listed "or approved equal." Items approved in the addenda are acceptable for use in the project.
- Additional information. Additional information is often distributed in the addenda. This could include supplementary specifications, drawings, or any other details relating to the contract documents.

As the information in the addenda becomes part of the contract documents, care should be taken to incorporate it into the contract documents used on the jobsite. Many contractors "cut and paste" the items of the addenda into the appropriate locations in the Project Manual and drawings. Merely including the addenda with the Project Manual, rather than pasting the items into the documents in their relative locations, is usually not effective, as the addenda tend to be misplaced. All documents used on the jobsite should be up-to-date, with the addenda and change order items noted in the appropriate locations.

✳ The Drawings

The set of contract drawings has several sections, depending upon the nature of the project. Most commercial building projects will have, as a minimum, the following sections:

- Civil engineering and site survey documents, usually identified by a "C" designation
- Architectural drawings, usually identified by an "A" designation
- Structural engineering drawings, usually identified by an "S" designation
- Mechanical engineering drawings, usually identified by an "M" designation
- Electrical engineering drawings, usually identified by an "E" designation

Additional sections are added for special construction areas, when applicable, such as, landscape architecture, commercial kitchen equipment, industrial equipment, detention equipment, and many other special construction considerations.

The drawings and Project Manual are considered a full set of documents. If something is shown or described once, in any of the drawings or contract documents, it is assumed to be covered by the documents, even though it may not be referenced in relating areas. Thorough knowledge of all of the documents is essential for contractor personnel.

Civil Engineering and Site Survey.   These drawings are concerned with civil engineering concerns on the site, such as site drainage, site utilities, roads, bridges, and other related items. Plan views, contour drawings, cross-sections, and details are usually contained in this section. Surveys of the construction site also are included. An existing site survey and the survey information for the constructed site could be shown as separate plans, or could be included on the same plan drawing, with different designations for existing and new contours. These drawings normally are drawn in "engineer's scale," at 10, 20, 30, 40, 50, or 100 feet to the inch.

The information contained on these drawings usually is used for site layout, including location of site features and elevations. The contractor's surveyor, whether a licensed surveyor or the contractor's field engineer, will use information on these drawings to establish benchmarks and related elevations. Dimensions from these drawings often are used with larger-scale architectural drawings to establish the location of roads, improvements, and structures on the site.

Architectural Drawings.   The architectural drawings normally are prepared by the architectural firm, occasionally with the consultant's input. The architectural drawings primarily relate to the buildings and structures, describing the building's location, size, form, systems, and materials. These drawings are always indicated with an "A" prefix. Several different numerical systems are found, such as, A1, A2, and A1.1.2, A.2.1.2, as well as several other methods. Generally, architectural drawings include the following sections, usually in the sequence listed:

1. Index, symbols, abbreviations, notes, location map: The first one or two pages, after the title sheet, list this general information. The drawing index usually lists the drawings in the set and the type of information contained on the drawings. This index is convenient for individuals who are not particularly familiar with the specific project drawings. Most drawings list standard symbols and abbreviations, which usually do not list special symbols and abbreviations specific to the type project. The notes, however, should relate to the particular project. Some of the notes on the introductory sheets can contain information pertinent to the building permit for the project. Several project location maps may be included, with increasing detail.
2. Site plans and details: When civil engineering drawings are not necessary, the architectural drawings will include a site plan, locating the improvements and structures on the site. Contours, indicating

elevations, often are included in these plans. Irrigation and planting plans also may be included.

3. Demolition plans: Indications of demolition areas often are included in the architectural drawings. These plans may illustrate large demolition areas, or selective demolition, such as partition removal. For interior renovations, the removed partitions are often shown as dotted lines on the floor plan, with new partitions shown in heavier, solid lines.

4. Floor plans: It is customary to show every floor plan applicable to the project, even when the floors are similar. Floor plans should show: wall locations and types, window openings, door openings, dimensions, where necessary, cabinets, equipment, stairs, room numbers, detail and section locations, and various materials. Most material designations are usually shown on the sections and details.

5. Roof plans: These plans indicate drainage, equipment, penetrations, hatches, and so on.

6. Ceiling plans: These plans usually are reflected plans, as though the floor were a mirror and the ceiling an image in the mirror, coinciding in orientation with the floor plan. Ceiling plans indicate penetrations, light fixtures, grid configurations, and material types when there are a variety of materials within the same room.

7. Building sections: These sections are normally "cuts" through the building, indicated by symbols on the floor plan. They are usually fairly small-scale drawings, showing the full structure of the building. These drawings do not show details, but rather provide a view of the relationship of building elements. By studying the building sections and elevations, an individual can visualize the three dimensions of the structure, rather than the two dimensions of the floor plans.

8. Wall sections: These sections are larger-scale vertical cuts through the building, which show and name the materials in the walls, floors, and roofs. These sections are very important when determining what materials and systems will be used in the building and the relationships of the material and systems.

9. Exterior elevations: The exterior elevations are two-dimensional drawings of the exterior of the buildings. These drawings illustrate a vertical picture of the building, showing materials, windows, doors, control joints, dimensions, and other information pertinent to the construction of the building. These are not renderings of the building, but two-dimensional drawings. Occasionally, a three-dimensional rendering is available, either on the cover of the drawings or perhaps at the architect's office. Models of the building are occasionally built to convey to the owner the spatial aspects of the building and may be observed to obtain a three-dimensional perspective of the building. Occasionally, isometric drawings of specific areas will be included in the drawings to convey relationships not available from the customary two-dimensional drawings. Three-dimensional drawings, usually

produced by computer-aided drafting (CAD) occasionally are used to illustrate the volume of the building elements.

10. Schedules: Schedules, in tabular form, often are included in the drawings. Some common schedules include door, window, louver, and finish. The schedules relate each particular feature, such as the door, window, or wall, to the material, style, detail, and color. They are commonly found in the Project Manual, usually near the applicable item specification, as well as on the drawings.

11. Large-scale floor plans and accompanying sections and details: Some areas will have large-scale drawings to show the complexities of the particular area. These drawings are usually found for toilet rooms, stairways, escalators and elevators, lobbies, and special-use areas.

12. Exterior details: Details of materials and systems, particularly at connections, are referenced from the plans, sections, and elevations.

13. Interior elevations and details: Interior details will relate to equipment, millwork, cabinetwork, and finish work shown on the plans and sections. These elevations and details are used primarily for finish work and for coordination of the various elements in the finishes of the facility.

Structural Drawings.   The structural drawings represent the structural elements, such as foundations and structural frame. These normally are prepared by a structural engineer, either in-house with the architectural firm or on a consulting basis. The structural design is coordinated with the architectural aspects of the building. Structural drawings usually have definitive dimensions for the foundations and structural elements, both horizontally and vertically. These and the architectural drawings should be continuously coordinated, however, as both may contain information necessary for the location and construction of the building elements. The structural drawing is the primary document used in establishing the building's structural system and thus the basic dimensional attributes of the building.

The following elements usually are included in the structural drawings:

1. Structural notes: These notes are normally more specific in regard to the material strength and tolerances than are those in the Project Manual. The notes are often exactly related to the project. Typical details also are included in the notes. Many typical details are used in the structural drawings, relying on the shop drawings to coordinate conditions. The shop drawings are done by the contractor, subcontractor, or fabricator and are reviewed by the structural engineer.

2. Foundation plan: The foundation plan should indicate all subgrade elements, including piles, caissons, footings, and foundation walls. References are made to details.

3. Framing plans for floors and roof: The structural drawings will examine each floor structure, showing the necessary structure. In some structural drawings, the different system types, such as concrete, steel,

and wood, will be separated. Some drawings will, however, combine all structural systems on the same drawings.

4. Elevations and sections: These structural drawings provide the three-dimensional aspect with the necessary elevations.
5. Details: These drawings provide the details of the systems, particularly connections of materials and systems. Structural details are normally provided for stair systems.
6. Schedules: Tabular schedules of structural elements, such as joists, beams, deck, and other elements, are commonly shown in these drawings.

Mechanical Drawings.   Mechanical drawings normally include both plumbing and heating, ventilation, and air conditioning (HVAC) drawings. These drawings usually are accomplished through a mechanical engineering firm, which often acts as a consultant to the architect. A number of different subcontractors may use mechanical drawings, including the plumbing subcontractor, HVAC contractor, fire protection subcontractor, temperature control subcontractor, and possibly other subcontractors or third-tier subcontractors. Information about the other drawing sections that is pertinent to the mechanical drawings is available. Therefore, full sets of the construction documents should be distributed to all subcontractors, to facilitate information.

Although mechanical and electrical drawings use site and floor plans similar to architectural drawings, many mechanical and electrical drawings are much more diagrammatic than are architectural drawings. Plumbing riser diagrams, for instance, are not to scale or detail, but indicate the plumbing system in a manner consistent with the plumbing craft. Additionally, different symbols and abbreviations are used in mechanical drawings, rather than in other parts of the documents. A legend to the symbols is usually included in the mechanical drawings for clarification.

The following drawings are often included in mechanical drawings:

1. Mechanical site plans indicate the site utilities, such as water, sanitary sewer, storm sewer, and natural gas from the connection source to the building. Occasionally, the mechanical and electrical site plans will be combined.
2. Floor, roof, and reflected plans indicate the location of fixtures and piping.
3. Large-scale plans for mechanical and equipment rooms indicate equipment location, piping, and duct work.
4. Floor plans indicate duct work locations.
5. Diagrams, such as plumbing riser diagrams, indicate piping flow and control diagrams.
6. Details are indicated for connections and special situations.
7. Fire protection drawings indicate fire risers, piping, and head locations. Because fire sprinkler systems normally are designed by the fire

protection subcontractor, these drawings may be very diagrammatic and indicate the parameters of the system.

8. Schedules are indicated for equipment, piping, and fixtures. Depending upon the mechanical engineer, these schedules might be included in the Project Manual rather than on the drawings.

Electrical Drawings. The electrical drawings, like the mechanical drawings, are particularly diagrammatic. The former are also part of the entire set of documents and should not be separated from the documents. They also have their own symbols and abbreviations, which are usually defined within. Numerous systems are shown on the electrical drawings, such as fire alarm, communication, or security, which may be shown on separate drawings or integrated with the electrical drawings.

Electrical drawings will usually include the following:

1. Notes, symbols, abbreviations, and standard details
2. Site plan, indicating the utilities that connect to the building, such as power, TV cable, and telephone
3. Floor plans showing separate floor plans for power and lighting, however both may be included on the same drawing for smaller projects
4. Diagrams, such as power riser diagrams
5. Details
6. Schedules, such as panels, devices, special systems, and light fixtures
7. Special system drawings, such as intercom, security, centralized clock, TV cable, and data cable systems

Use of the Construction Documents

The construction professional uses the construction documents each day in managing the construction project. These documents are used for layout, clarification, direction, checklists, compliance inspections, communications, and verification in every stage of the construction process. The construction superintendent and field engineer need to know the documents in detail to be able to reference them quickly and efficiently. Some contractors will add reference tabs to the Project Manual to facilitate quick access to relevant information.

As previously mentioned, it is extremely important to have a complete and updated set of documents, such as the Project Manual, drawings, addenda, and change orders, at the jobsite, but other documents are essential for reference as well. Submittals and shop drawings should be complete, approved, and available at the jobsite. Product literature that is applicable to the project also is important. Current safety information, such as safety regulations, safety manuals, and Material Safety Data Sheets (MSDS) should also be available. Copies of the local building codes and regulations should be readily available for reference by field management personnel, as should copies of subcontract agreements, detailing the responsibility of each subcontractor.

There are, of course, thousands of applications of construction document use and even more variations, depending upon the contractor and the situations on the jobsite. The following section illustrates the use of construction documents during a building project.

Familiarization With a Project

Construction professionals often are concerned with more than one construction project. Quick familiarization with a project usually is necessary, providing a broad overview of the project, which enables one to intelligently interact with other workers. Obtaining detailed knowledge about the project requires considerable time and involvement. Steps for quick familiarization with a project, using the construction documents, include:

1. Review the site plan. Become familiar with the relationship of the structure to the rest of the site. Note the site limitations and access, as well as adjoining property and the relationship of the project to the adjoining property.
2. Review the floor plans. Determine a two-dimensional size for the building and its floors.
3. Review the building elevations. Determine a three-dimensional size (volume of the building). Become familiar with the intended look of the facility. Additional non-document information, such as renderings or models, may help gain a three-dimensional perspective on the building.
4. Review the building sections, relating them to the floor plans and elevations. When reviewing the building sections and largerscale wall sections, one should begin to understand the type of construction and structure for the building. Review the structural drawings, starting with the foundation plan, then examine the structural system of each floor and the building.
5. Review the mechanical and electrical drawings, determining an overview of the involvement of both areas in the construction.
6. Review the General Requirements section of the specification, determining the special circumstances for the project. Determine the completion date for the project.
7. Depending on the level of involvement with the project, review contractor documents, such as estimate, schedule, subcontracts, and work plan.

Preparing Crew Assignments

Many contractors will prepare "work packages" or "crew packages" to enable the crew to accomplish a task. This package information will vary, but could include:

- Drawings, or references to drawings, which are applicable to the particular task assigned to the crew. Because the document set is cumbersome to use in the field, many contractors condense the drawings into 8½" × 11" pages. These drawings may illustrate the task location, floor plans, sections, and details. Copies of the partial construction documents can be made and given to the crew. When CAD drawings are available, specific areas can be selected and printed on 8½" × 11" paper.
- Technical specification, including material and execution of the material, which can be copied and given to the crew.
- Material list, obtained from the drawings, according to material specification. Purchase order information is also normally included.
- List of necessary tools and equipment, usually determined by the superintendent for the assigned task.
- Information from the estimate and schedule that relates to the situation, such as man-hours available, work time frame, and other cost-related issues.

Problem Solving

Construction documents are used as a reference for solving problems that occur on the jobsite. They are continually referred to for everyday clarifications, particularly when confusion arises about the direction of the work. Usually, reviewing the documents clarifies the situation. Occasionally, further details are needed from the architect and engineer. The following section illustrates how essential construction documents are in problem solving.

Example 1.   During excavation for the building site, the excavation subcontractor encounters an existing large concrete footing about ten feet below the existing grade and about ten feet above the finish grade of the excavation. The subcontractor notifies the contractor's field engineer about the footing immediately upon discovery, explaining that this footing was not expected and probably would require additional work to remove it. Because the area surrounding the footing needs to be excavated, the field engineer directs the subcontractor to continue excavation to discover the size, all three dimensions, of the footing. The following steps could be taken to determine the extent of the problem and if additional payment is appropriate for this work.

1. The field engineer would immediately look at the construction drawings to examine the site, floor, and structural foundation plans to

find any reference to the existing footing. In this case, the field engineer does not find any reference. The field engineer would then look in the specifications, Section 020100, General Requirements—Sitework; Section 02100, Site Preparation and Demolition; and Section 02200, Earthwork. No specific mention of the existence or removal of the concrete footing is made here. The notes in the structural drawings also are examined, again with no reference to the specific condition.

2. The next step the field engineer takes is examining the geotechnical report, contained in the Project Manual as information only. The boring reports are reviewed to determine whether any concrete was encountered in subsurface exploration. Four borings, twenty feet in depth, were taken on the building site. All four borings indicated a well-graded gravel strata between 7.5 feet and 15.3 feet below the existing grade. No mention is made of concrete footing in the boring report, although none of the borings were made in the location of the concrete footing.

3. It is apparent to the contractor that this extra work is not covered in the construction documents. The field engineer then looks in the General Conditions and Supplementary Conditions to determine the procedures to follow for pursuing a change order. The field engineer examines the following clause:

> AIA form A201, General Conditions, 4.3.6: "If conditions are encountered at the site which are (1) subsurface or otherwise concealed physical conditions which differ materially from those indicated in the Contract Documents, or (2) unknown physical conditions of an unusual nature, which differ materially from those ordinarily found to exist and generally recognized as inherent in construction activities of the character provided for in the Contract Documents, then notice by the observing party shall be given to the other party promptly before conditions are disturbed and in no event later than 21 days after first observance of the conditions. The Architect will promptly investigate such conditions and, if they differ materially and cause an increase or decrease in the Contractor's cost of, time required for, performance of any part of the Work, will recommend an equitable adjustment in the Contract Sum or Contract time, or both. . . ."

The field engineer, upon direction from the superintendent, will then write a Request for Information (RFI) to the architect and engineer to notify them about the problem (see chapter 4 for a detailed explanation about Requests for Information). The field engineer could use the following description of the problem:

> On Monday June 12, 19xx, at 10:15 A.M. the excavation subcontractor encountered an existing concrete footing, 4'0" wide × 2'0" deep, at approximately grid 2-B at about 10'0" below existing grade. At this time, the extent of the footing is not known, and excavation is continuing to reveal the full extent of the footing affecting excavation at the building site.

We have examined the following documents, and find no mention of this footing: Site Plan, A-2; Basement Floor Plan, A-4; Structural Foundation Plan, S-2; Structural Notes, S-1; Sections 020100, 02100, and 02200 of the specification. We also have examined the Geotechnical Report, prepared by GO Engineering, Inc., and also find no indication of this footing.

We feel that this concrete footing is a concealed condition, with no available information to the contractor prior to bidding concerning this footing. Our excavation contractor has indicated to us that there will be additional costs incurred to remove this footing. We will notify you of the extra costs when the full extent of the footing is known.

As we feel that the removal of this concrete footing is additional work to our contract, we will not remove the footing until directed. Please respond as quickly as possible.

4. The field engineer sends the RFI to the architect via a fax. The architect then carefully reviews the construction documents and visits the site to verify the situation. After carefully examining the facts related to the problem, the architect discusses the problem with the owner's representative, and the owner decides to proceed with a change proposal from the contractor.

The architect faxes a reply to the contractor:

We have determined that the concrete footing mentioned in RFI-010, June 12, 19xx, must be removed from the building excavation location. Please prepare a change proposal for this work.

5. The field engineer refers to the General and Supplementary Conditions for information concerning preparation of a change proposal. Paragraph 7.3.3, under Construction Change Directives, AIA Form A201, discusses three types of pricing for a change proposal: an itemized lump sum proposal for the cost of the work; a unit-price proposal, when an unknown quantity of work needs to be accomplished; and a cost-plus-percentage arrangement, when unknown methods and quantities apply to the change. As the subcontractor has now discovered the length of the footing, 210 L. F., the field engineer decides to prepare an itemized lump sum proposal and requests the appropriate information from the excavation subcontractor. The Supplementary Conditions for this particular project state

In subparagraph 7.3.6, the allowances for the combined overhead and profit included in the total cost to the Owner shall be based on the following schedule:

For the Contractor, for work performed by the Contractor's own forces, ten percent (10%) of the cost.

For the Contractor, for work performed by the Contractor's subcontractor, five percent (5%) of the amount due the Subcontractor.

For each Subcontractor or Sub-subcontractor involved, for work performed by the Subcontractor's or Sub-subcontractor's own forces, ten percent (10%) of the cost.

The field engineer would then prepare a change proposal, on the appropriate form, as follows:

Remove and legally dispose of existing unknown concrete footing 4'0" wide, 2'0" deep, 210 L. F.

Excavating Subcontractor:

Labor	Operator	16 hrs.	@$28.90	$462.40
Laborer		16 hrs.	@$24.63	394.08
Equipment 235C		16 hrs.	@$120.00	1,920.00
Extra Trucking		6 hrs.	@$40.00	240.00
Tipping Fees		124 tons	@$50.00	6,200.00
Subcontractor Cost				9,216.48
Subcontractor Markup			10%	921.65
Subcontractor Total				9,938.13
Contractor Markup			5%	495.91
Total, Change Proposal				$10,434.04

As this work will delay the excavation work, which is on the critical path, we will also request a Two (2) calendar day extension to the contract.

6. The architect, upon receipt of the change proposal and after approval by the owner's representative, will issue a Contract Change Directive, in accordance to the General and Supplementary Conditions of the Contract.

Example 2.   In the project described under Example 1, exterior and interior partitions have been framed. The insulation subcontractor is on the site and completing the insulation for exterior walls and ceilings. The insulation subcontractor foreman, in speaking to the field engineer, has mentioned that his crew is preparing to leave the jobsite, as their work is complete. The field engineer asks about installation of the acoustical insulation in the interior partitions. The insulation subcontractor foreman replies that the acoustical insulation is normally supplied and installed by the drywall subcontractor, as the acoustical insulation is friction fit and needs one side of drywall complete to install it. The field engineer tells the insulation subcontractor foreman that he thinks the acoustical insulation is in the insulation contract, but he will verify it. At the jobsite office, the field engineer takes the following steps:

1. The field engineer telephones the project manager for the insulation subcontractor. The project manager confirms that they do not intend to furnish and install the acoustical insulation.
2. The field engineer checks the subcontract agreement with the insula-

tion subcontract. Under description of work, the subcontract agreement reads

> Furnish and install insulation, as per Section 07200, Building Insulation, of the Project Manual. Furnish rigid perimeter insulation, no installation. (The subcontract agreement references all applicable work shown on the complete set of construction documents.)

3. The field engineer checks the subcontract agreement regarding the drywall subcontractor. There is no mention of Section 07200 in the drywall subcontract.
4. The field engineer examines Section 07200, Building Insulation. The following clauses are applicable to the situation.

> Part 2, Products, Paragraph 2.04: Acoustical Batt Insulation: Fiberglass unfaced batts for wood stud walls where noted A. B. I. on drawings. Flame spread and smoke developed ratings of less than 25 and a noncombustible rating in accordance with ASTM E 136.
> Part 3, Execution, Paragraph 3.02.C: Friction fit acoustical batts into place between wood studs. Use multiple thickness as required to completely fill wall cavities. Batts shall be installed in walls indicated on drawings by A. B. I. designation.

5. The field engineer then examines the drawings to be sure that acoustical insulation is shown by the designation A. B. I. On drawings A2.1 and A2.2, the interior partitions on both the first and second floors are shown with a pattern, which in the legend on each page indicates A. B. I. for that wall type. The interior wall sections on A3.4.1 also indicate acoustical insulation.
6. As the drawings show acoustical insulation in definite locations, Section 07200 specifies the acoustical insulation, and Section 07200 is listed in the subcontract agreement with the insulation subcontractor without exception concerning acoustical insulation, and the field engineer is certain that the acoustical insulation is required under the insulation subcontract. The field engineer then writes a letter to the insulation subcontractor project manager, faxing it immediately. The letter cites the subcontract agreement, the specification, and the location on the drawings.
7. The field engineer telephones the insulation subcontractor project manager after the fax is sent. The insulation subcontractor, although not particularly happy, agrees to install the acoustical insulation.
8. The next morning, the field engineer observes the insulation subcontractor delivering and installing the acoustical insulation.

In this example, the field engineer was able to effectively reference the construction documents and solve a dispute that could have delayed the project and impacted on the cost to the contractor.

Summary

The construction documents, although not prepared by the contractor, are an essential reference to the contractor when constructing the project. A thorough knowledge of these documents is necessary for the construction professional. As construction documents are quite complex, the construction professional should know where to locate the answers to questions in the documents. Construction documents should contain the following components:

- Advertisement/Invitation To Bid
- Instructions To Bidders
- Bid Forms and Certifications
- General Conditions of the Construction Contract
- Supplementary Conditions of the Construction Contract
- Additional Information For Bidders
- Technical Specifications
- Addenda
- Drawings (including civil, architectural, structural, mechanical, electrical, and specialty)

The documents can be used for the construction and organization of the project, as a general reference for all questions, and as an aid in solving problems on the jobsite.

CHAPTER 3

SUBMITTALS, SAMPLES, AND SHOP DRAWINGS

CHAPTER OUTLINE

Types of Submittals

Requirements For Submittals, Shop Drawings, and Samples

Review of Submittals, Shop Drawings, and Samples

The Procurement Schedule

Submittal Review By The Contractor

The Use of Submittals During Construction

Summary

The construction documents, specifically the technical specifications, mentioned in chapter 2, require the contractor to submit product data, samples, and shop drawings to the architect and engineer for approval. This is one of the first steps that is taken by the contractor after execution of the construction contract and issuance of the Notice to Proceed. Shop drawings, material data, and samples generally are referred to as **submittals.** The submittal process is very important, as it directly relates to the quality, schedule, and ultimately the overall success of the project. The submittal process can be complex, because there are literally thousands of different materials, fabrications, and equipment used in a construction project.

Product data submittals, samples, and shop drawings are required primarily for the architect and engineer to verify that the correct products will be installed on the project. This process also gives the architect and subconsultants the opportunity to select colors, patterns, and types of material that were not chosen prior to completion of the construction drawings. This is not an occasion for the architect to select different materials than specified, but rather to clarify the selection within the quality level indicated in the specification. For materials requiring fabrication, such as reinforcing and structural steel, the architect and engineer need to verify details furnished by the fabricator. The contractor also uses this information in installation, using dimensions and installation data from the submittal.

Types of Submittals

Product Data Submittal

The **product data submittal** usually consists of the manufacturer's product information. The information that is necessary for such a submittal includes:

- Manufacturer, trade name, model or type number: This information is necessary to compare the submitted item with the specified products and acceptable products listed in the specification and addenda.
- Description of use and performance characteristics: Information should be furnished describing the normal use and expected performance of the product. The architect and contractor should both review this information to confirm that the product is appropriate for the intended use.
- Size and physical characteristics: The size and physical characteristics, such as adjustment capabilities, should be reviewed by both the contractor and architect. The contractor has the most available information for comparing adjoining materials and equipment. The contractor also needs to know the size and weight of the equipment for lifting and handling considerations.

- Finish characteristics: The architect should review the available finishes and select the appropriate finish, if the finish was not previously specified in the documents. The contractor should confirm that finish requirements in the specification are being met by the product.
- Specific request for jobsite dimensions: Some material is custom-fabricated to job conditions, requiring dimensions from the jobsite. These jobsite dimensions are provided by the contractor, prior to release of the product for manufacture.

Figure 3–1 is an example of a material data submittal.

Shop Drawings

A **shop drawing** is a drawing or set of drawings produced by the contractor, supplier, manufacturer, subcontractor, or fabricator. Shop drawings are not produced by architects and engineers under their contract with the owner. The shop drawing is the manufacturer's or contractor's drawn version of information shown in the construction documents. The shop drawing normally shows more detail than the construction documents. It is drawn to explain the fabrication of the items to the manufacturer's production crew. The style of the shop drawing is usually very different from that of the architect's drawing. The shop drawing's primary emphasis is on the particular product or installation and excludes notation concerning other products and installations, unless integration with the subject product is necessary.

Concrete reinforcing is one of the many items requiring specialized shop drawings for the fabrication of the material. Concrete reinforcing is custom-fabricated from 60-foot-long reinforcing bars. The reinforcing bars are cut to length and bent to specific configurations. The shop drawing and accompanying "cut sheet" list the quantity, sizes, lengths, and shapes of the reinforcing bar. This information is provided for review by the structural engineer to ensure that sufficient reinforcing is being supplied; fabrication of the bar by the supplier's shop; an inventory list for the contractor, upon delivery (the typical project has thousands of pieces of reinforcing steel that need to be organized for storage and installation); and placement by the ironworker. The Concrete Reinforcing Steel Institute (CRSI) has developed standard symbols, graphics, and formats for shop drawings and cut sheets that generally are used by reinforcing steel fabricators. Each fabricator, though, will have a particular style for shop drawings and cut sheets, depending on the draftspeople and computer-aided drafting systems.

Figure 3–2A and B illustrates a simplified, comparative look at the reinforcing shown on a structural engineering drawing and reinforcing steel shop drawing. Figure 3–2A, found on page 59, is illustrative of typical information contained in structural drawings of contract documents. Shown is a partial plan view of a concrete footing and foundation, with the referenced Section A-A illustrating the reinforcing that is required for

PRODUCT DATA SUBMITTAL

PROJECT: Downtown Center Office Building
SUBMITTED BY: ABC Construction Company
SECTION NO.:   03250
PRODUCT TYPE: Curing / Sealing Compound
PRODUCT SUBMITTED: Sonneborn "KURE-N-SEAL"
MANUFACTURER: Sonneborn Building Products, Minneapolis, MN
PRODUCT STATUS: Meets ASTM C-309; Approved as equal, Addendum 2, 5/12/97
PROJECT LOCATION: Cure/Seal on Concrete floors, Rooms 110, 111, 114, and 116

KURE-N-SEAL

A Clear, Polymeric Liquid Compound That Cures, Seals, and Dustproofs All In One Application. Surfaces To Be Treated Can Be Damp or Dry; Horizontal or Vertical; Interior or Exterior

USE

Kure-N-Seal is specifically designed for curing and sealing freshly placed and finished concrete.

In addition to newly placed concrete floors, Kure-N-Seal is effective on:

> Exposed Aggregate
> Brick Floors and Walls
> Terrazzo
> Older Concrete Surfaces

Kure-N-Seal dries to a tough glossy membrane, resistant to construction traffic and workman abuse, adhesion of mortar droppings and paint, and many chemicals and stains.

Kure-N-Seal is highly resistant to discoloration caused by the effects of sunlight. Surfaces retain their brightness and color for longer periods of time without ugly yellowing.

DESCRIPTION

A liquid acrylic based polymer. Kure-N-Seal contains no oils, saponifiable resins, waxes or chlorinated rubbers.

It is a superior concrete floor sealing compound which, by locking in the moisture in freshly placed concrete and tenaciously adhering to the surface, cures the concrete and creates conditions for achieving maximum hardness of the concrete slab.

Simultaneously, Kure-N-Seal functions as a surface dustproofing sealer — producing a hard film with excellent resistance to traffic abrasion, water spillage, mild acids and alkalies. Helps prevent construction stains on concrete in new buildings.

The degree of gloss to the finish depends upon the porosity of the surface and the number of coats applied.

Note: After thorough curing, floors treated in accordance with directions with a single coat of Kure-N-Seal may be finished with resilient flooring according to the tile manufacturer's recommendations for tile and adhesive.

ADVANTAGES

• Cures, dustproofs and seals in one application.

• Eliminates dirt and construction stains.

• Protects against alkalies and mild acids.

• Quick drying — easy to apply and maintain.

• Construction work continues rapidly and economically.

• For interior and exterior concrete and masonry floors and walls.

• From construction through occupancy, maintenance and cleanup greatly simplified — housekeeping costs greatly reduced.

APPLICATION

New Concrete: Surface is application-ready when it is damp but not wet and can no longer be marred by walking workmen.

Aged Concrete: Surface must be free of any dust, dirt, and other foreign matter. Use power tools and/or strippers to remove any incompatible sealers or coatings. Cleanse as required.

Apply so as to form a continuous, uniform film by spray, soft-bristle pushbroom, long-nap roller or lambswool applicator. Ordinary garden-type sprayers, using neoprene hose, are recommended for best results.

For curing only, first coat should be applied evenly and uniformly as soon as possible after final finishing. Second coat should be applied when all trades are completed and structure is ready for occupancy.

Reducer 990 is used primarily to thin first coats when spraying or applying to aged floors. It is also used for cleaning tools and equipment with which Kure-N-Seal has been applied.

To seal and dustproof, two coats are required. For sealing new concrete, both coats are applied full strength. On aged concrete, when renovating, dustproofing and sealing, the first coat should be thinned 10% to 15% with Reducer 990.

At normal temperature and humidity, Kure-N-Seal will dry for application of additional coats in approximately two hours. Allow to dry hard . . . for normal traffic overnight drying. For maximum hardness, drying time of seven days is required.

LIMITATIONS

Kure-N-Seal should not be used on surfaces to receive concrete overlays and/or additional toppings. It can be applied to colored concrete, but mottling may occur. May rubber burn and highlight imperfections of dry concrete surfaces. Not recommended as a release agent or where other sealers or treatments are to be later applied.

Not for use where Kure-N-Seal will be subjected to immersion. Should not be used for a coating or interior lining of concrete tanks and pools.

COMPLIANCES

Kure-N-Seal is recommended for use on Class 1, 2, 3 and 4 concrete floors as classified in Table 1.1, ACI Standard 302-69. Product also meets the requirements for ASTM C-309, Asphalt and Vinyl Asbestos Tile Institute and AASHTO M-148. Available in conformance to Federal Specification TT-C-800A pigmented gray, and 30% solids. USDA approved.

COVERAGE

200 to 600 sq. ft. per gallon depending upon surface and type of application.

COLOR

Transparent

PACKAGING

5 gallon pails (18.93 liters) and 55 gallon drums (208.18 liters).

FIGURE 3–1   Example Product Data Submittal (*Courtesy of Sonneborn, Division of ChemRex, Inc.*)

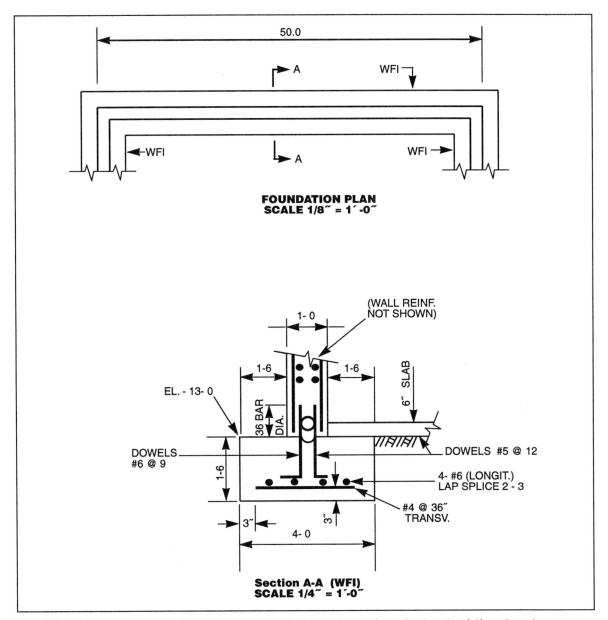

FIGURE 3–2A   Comparison of Structural Engineering Drawing and Reinforcing Steel Shop Drawing (*Courtesy of the Concrete Reinforcing Steel Institute*)

footing WF-1. Typically, dimensions and requirements for reinforcing are shown in the contract drawings.

Figure 3–2B, found on page 60, shows the steel reinforcing supplier's shop drawing. The required length of the reinforcing bars is illustrated, as well as the number and configuration of dowels, information needed for fabrication and placement of the reinforcing steel.

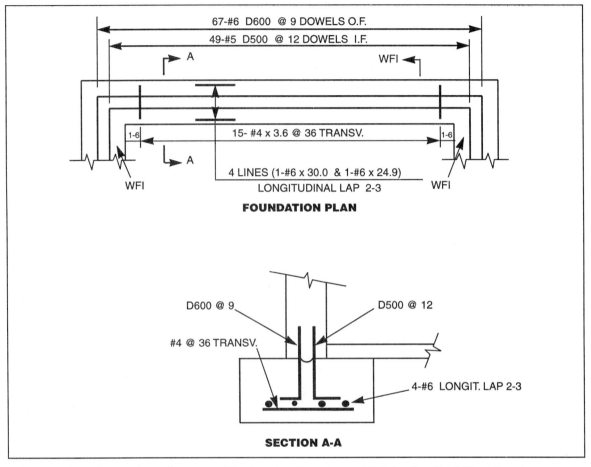

67-#6 D600 @ 9 DOWELS O.F.

49-#5 D500 @ 12 DOWELS I.F.

A   WFI

1-6   15- #4 x 3.6 @ 36 TRANSV.   1-6

A

4 LINES (1-#6 x 30.0 & 1-#6 x 24.9)

LONGITUDINAL LAP 2-3

WFI   WFI

FOUNDATION PLAN

D600 @ 9   D500 @ 12

#4 @ 36 TRANSV.

4-#6 LONGIT. LAP 2-3

SECTION A-A

FIGURE 3–2B   Comparison of Structural Engineering Drawing and Reinforcing Steel Shop Drawing (*Courtesy of the Concrete Reinforcing Steel Institute*)

Each supplier or manufacturer will have specific information that should be included on shop drawings. This information includes:

1. Comparison information for the architect and engineer: The shop drawings should include information for the architect and engineer to compare to the specifications and drawings. The shop drawing should address the appearance, performance, and prescriptive descriptions in the specifications and construction drawings. The shop drawing often is more detailed than the information shown in the construction documents to give the architect and engineer the opportunity to review the fabricator's version of the product, prior to fabrication. References to the construction documents, drawings, and specifications assist the architect and engineer in their review of the shop drawings. Attachment of manufacturer's material specifications, "catalog cut sheets," and other manufacturer's information may be

helpful to accompany these drawings. Because shop drawings facilitate the architect's and engineer's approval of the product, they should be as clear and complete as possible.

2. Notes of changes or alterations from the construction documents: Notes concerning changes or differences from the original documents should be made on the shop drawing for the architect's and engineer's approval. Ultimately, they are responsible for changes in these drawings and should have the opportunity to analyze any modifications. A dialogue should occur between the fabricator and the architect and engineer about any areas needing clarification. Successful installations are the result of collaboration between the designer, fabricator, and contractor.

3. Information needed to fabricate the product: Dimensions, manufacturing conventions, and special fabrication instructions should be included on the shop drawing. It should be clear to fabrication personnel what will be manufactured from the shop drawing alone. The construction documents are rarely used as a reference in fabrication, with the fabricators relying on the shop drawing for all information.

4. Indication of dimensions needing verification from the jobsite: Most jobsite dimensions, such as the dimensions between two surfaces on the jobsite, need to be verified. A dimension may be shown on the construction drawings, but the actual dimension may vary, from very small to large increments, depending on jobsite conditions. It is extremely important that the fabricated item arrive on the jobsite ready to be installed without field modification. Special care must be taken by the contractor to measure and verify dimensions. In new construction, plan dimensions usually are sufficient for ordering many fabricated items such as structural steel or precast concrete. In remodel and renovation work, it is essential that field dimensions be verified prior to fabrication. Some fabricators, such as cabinet and casework suppliers, prefer not to rely on the contractor's verification and will verify the dimensions with their own personnel.

5. Placement or installation information: Some fabricators and manufacturers will provide symbols, data, or instructions concerning installation. This can include a list of other materials, such as fasteners or adhesives, appropriate but not included for the product.

6. Samples: Some fabrications will require a sample submittal with the shop drawing, primarily for color and texture selection of finishes.

Shop drawings are required, in various forms, depending upon the practice of the architect and engineer. A specific number of copies may be required by the specification. An example distribution of the completed and corrected shop drawing may include the:

Owner—file or inspection copy
Architect—file copy
Architect—field or inspection copy
Consulting engineer—file copy

Consulting engineer—inspection copy
Contractor—file copy
Contractor—field copy
Supplier—original or one copy

(Total copies = 8)

Because writing comments on eight to ten copies is a tedious process and a waste of time for the architect and engineer, many times they will specify other methods for distributing their comments:

1. Submittal of one or two copies of the shop drawing. Corrections are made by the architect and engineer, and the shop drawing is corrected by the supplier, then the appropriate number of copies is distributed. This method can be time-consuming, as the shop drawing is not approved until the corrections are made on it.
2. Submittal of a copy that can be reproduced, such as vellum or sepia. The architect and engineer make comments on the reproducible, then copies are distributed. This method facilitates the timely approval and distribution of the shop drawing. Review comments usually are obvious on the reproducible copy. When sepia copies are used, the reproduction of the sepia often is not as clear as a normal blue-line print.
3. When the supplier and designer have compatible CAD software, the review can be made from a diskette or a modem-facilitated transfer. Comments can be made by the designer in a bold font or changes can be boxed for emphasis.

Quick review is essential during the approval process. Any method that facilitates this, while providing ample opportunity for comment and complete distribution, should be considered. Although a procedure may be specified in the contract drawings, most architects and engineers are open to suggestions and innovations that speed up the process.

Samples

Many products require submission of samples. A **sample** is a physical portion of the specified product. Some samples are full product samples, such as a brick or section of precast concrete, or a partial sample that indicates color or texture. The product sample is often required when several products are acceptable, to confirm the quality and aesthetic level of the material. The size or unit of sample material usually is specified. For some materials, a mock-up or sample panel is necessary. A common example of a sample panel is a brick panel in a large enough mock-up to demonstrate the full appearance of the material. The brick panel might be four feet wide by six feet in height, showing all of the brick colors and materials, if there is a required variation in color and size. The sample panel also shows the mortar color and type of joint and, in this case, provides a completed version of the look of the wall that is not available from the brick sample alone.

Samples usually are required for finish selection or approval. Color and textures in the actual product can vary considerably from the color and textures shown in printed material. The printed brochure gives an indication of available colors, but the colors are rendered in printer's ink, rather than in the actual material. A quality level may be specified, requiring a selection of color and/or texture from sample pieces of the material. Several acceptable manufacturers may be listed in the specification and a level of quality also may be specified. The contractor, subcontractor, or supplier may have a preference for one of these products, based on price, availability, quality, workability, or service. The contractor would then submit the color samples for the preferred product for the color selection the designer chooses.

Samples should be pursued as diligently as product submittals and shop drawings. They may take some time to obtain from the manufacturer. Most materials have substantial order, manufacture, and delivery periods that must be calculated along with the time spent to obtain samples. Sample processing should be accomplished as early as possible in the project, as delivery periods for construction materials can be considerable, due to production schedules and transportation.

Samples need to be stored at the jobsite and compared to the material delivered and installed. Confirmation of the correct material on the jobsite prior to installation of the product avoids costly delays. Comparison of samples with the product received is an important part of project quality control.

Requirements For Submittals, Shop Drawings, and Samples

The construction documents usually indicate which product data submittals, shop drawings, and samples are required by the architect for a particular project. The contractor can require these in purchase orders and subcontract agreements, whether or not they are specified in the construction documents. Submittals, shop drawings, and samples are used by the contractor and the architect as a reference for monitoring quality control.

The General Conditions of the Contract require product data submittals, shop drawings, and samples. AIA Form A201, General Conditions of the Construction Contract, requires the aforementioned in Section 4.12, which defines each. The Conditions also require that the contractor review these submittals prior to transmitting them to the architect. The contractor is required to verify that the materials to be used are the correct ones for the project. Article 4.12.8 requires that all submittals be approved prior to the commencement of work associated with the submittal. The Supplementary Conditions to the Construction

Contract may add provisions relating to the submittals required for the particular contract.

Specification Section 01300 in the General Requirements section of the Project Manual defines the information required in the product data submittals, shop drawings, and samples. This section also specifies the necessary format for project submittals. Many architects will list the required submittals for the project in this section, providing a handy checklist for the contractor. A checklist can be compiled by examining each product specification in the Project Manual. Part One of the material specification should list the submittal requirement for the product.

Review of Submittals, Shop Drawings, and Samples

Definite processing lines are required by most projects for approval of all submittals, shop drawings, and samples. The procedures can seem very cumbersome and time-consuming, however, there are substantial reasons for review steps by all parties. The designer is ultimately responsible for the design of the facility to meet occupancy needs and must ensure that the products being installed are suitable to meet these needs. Any change in material or fabrication needs to be reviewed for its acceptability with the original design. Both the architect and contractor need to be able to coordinate the product and installation of the product with other systems.

Figure 3–3 is a flow chart showing the typical process used in processing submittals for approval.

Each level must review, add information as necessary, and stamp that the submittal was examined and approved by that party. After the submittal reaches the primary reviewer, the subconsultant in the earlier example, it is returned through the same steps, which provides an opportunity for further comment and assures that each party is aware of the approval, partial approval, notes, or rejection. Obviously this approval process is cumbersome and time-consuming.

Typically, the architect will review the submittal for compliance to the requirements in the construction documents. Revisions may be noted on the submittal. Colors and other selection items will be made by the architect during this review. Sometimes the architect will reject the entire submittal and other times will request resubmittal of some of the items. The architect also will make corrections, which normally do not need to be resubmitted, but that do need to be applied to the product. A typical architect's submittal stamp is shown in Figure 3–4, on page 66, recognizing that each architect will have a slightly different stamp and wording.

The contractor should manage the submittal process just like any other process in the construction cycle. Careful planning is necessary to ensure that the products are ordered and delivered within the construction schedule, so as not to delay any activities.

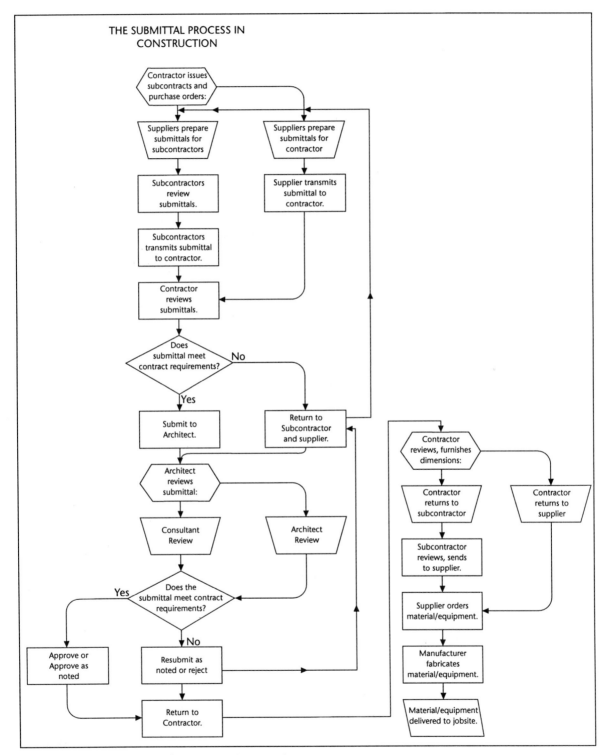

THE SUBMITTAL PROCESS IN
CONSTRUCTION

FIGURE 3–3   The Submittal Process in Construction

FIGURE 3–4
Example Architect's
Review Stamp

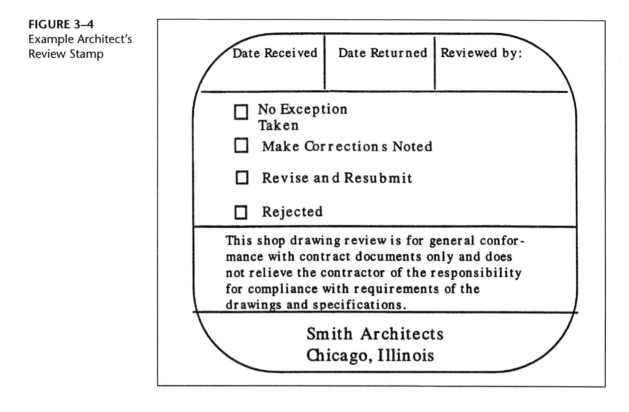

Date Received	Date Returned	Reviewed by:

☐ No Exception
Taken

☐ Make Corrections Noted

☐ Revise and Resubmit

☐ Rejected

This shop drawing review is for general conformance with contract documents only and does
not relieve the contractor of the responsibility
for compliance with requirements of the
drawings and specifications.

Smith Architects
Chicago, Illinois

The contractor must prioritize the submittal process, submitting and obtaining approval for materials needed for the first part of the project. Delivery and fabrication should be considered when establishing these priorities. Effective scheduling of the submittal process is facilitated by a dialogue with all of the parties involved, including the contractor, architect, subconsultants, subcontractors, and suppliers.

Occasionally, a contractor needs to divide a product submittal into packages, facilitating delivery of material early in the project. For example, reinforcing steel shop drawings frequently are separated into packages. Because the contractor needs reinforcing steel for the footings and foundations early in the project, the supplier will often produce a package of footing and foundation shop drawings immediately after receiving the purchase order, then submit this without the other packages of reinforcing steel. The footings and foundations package of reinforcing steel is then processed by the contractor, architect, and structural engineer approving the package for fabrication. The supplier will then immediately fabricate this package while still compiling the remaining shop drawings. The reinforcing steel for the footings and foundations, then, should be delivered to the jobsite in a timely manner, not impacting on the project's schedule.

The contractor must examine the construction schedule and determine the timing of the submittals and shop drawings so the material will be delivered to the jobsite prior to the scheduled installation activity.

The Procurement Schedule

The contractor will coordinate the submittal schedule with procurement and installation schedule requirements to ensure that the product will be delivered to the jobsite to coincide with the appropriate installation activity. AIA Form A201, General Conditions of the Contract, requires that the contractor provide a schedule of submittals to coordinate with the construction schedule.

> 3.10.2 The Contractor shall prepare and keep current, for the Architect's approval, a schedule of submittals which is coordinated with the Contractor's construction schedule and allows the Architect reasonable time to review submittals.
>
> (A201, General Conditions of the Contract, 1987 edition, *American Institute of Architects*)

The contractor needs to acquire data from the various parties concerning the durations used in a procurement schedule. The **procurement schedule** is a schedule of activities for particular products from submittal through delivery to the jobsite. The contractor must discuss the time that is necessary for the approval of submittals with the architect. Some of the factors that apply when determining the necessary time frame for these approvals include:

1. The size of the submittal package to be submitted. If a submittal is divided into packages, the approval time would be less per package than it would be for the entire submittal.
2. The complexity of the submittal has a large influence on the review and approval time.
3. The timing of the submittal. If all of the project submittals are given to the architect on the same date, it would take considerable time to process the submittal.
4. The number of consultants reviewing the submittal will influence the processing time. Occasionally, several consultants will review the submittal, which requires additional time for reviewing, handling, and mailing.
5. Some of the architect's consultants may live in other geographical areas, thus mailing time should be considered in the approval process.

The architect should have the opportunity to provide input and review the time frames allowed for approval of the submittals. The actual approval cycle will closely achieve the approval schedule if the contractor and architect first develop a realistic schedule.

The contractor also needs to obtain input from the supplier of the products or fabrications for the following information:

- amount of time necessary to produce the necessary submittal, shop drawing, or sample

- amount of time necessary to order the product after receipt of approved submittals
- amount of time necessary for manufacture and fabrication and delivery of the product to the jobsite

After receiving input from the architect and suppliers, the contractor can compile a procurement schedule. Figure 3–5 is a sample procurement

Procurement Schedule

Sect. No.	Item	To Contractor	To Architect	To Contractor	Order Date	Delivery Date
02150	Shoring Shop Drawing	1-May-97	2-May-97	9-May-97	9-May-97	15-May-97
02730	Drywell Submittal	1-May-97	2-May-97	9-May-97	10-May-97	24-May-97
02850	Irrigation Submittal & S.Dwg.	1-Jul-97	2-Jul-97	15-Jul-97	20-Jul-97	1-Mar-98
02900	Landscape Shop Drawing	1-Jul-97	2-Jul-97	15-Jul-97	20-Jul-97	1-Apr-98
03201	Rebar: Footings Shop Drawing	1-May-97	2-May-97	5-May-97	6-May-97	10-May-97
03202	Rebar: Found.Wall S.Dwg	4-May-97	5-May-97	10-May-97	11-May-97	20-May-97
03203	Rebar: Slabs Shop Drawing	10-May-97	11-May-97	15-May-97	16-May-97	26-May-97
03204	Rebar: T/U Panels, S.Dwg.	10-May-97	11-May-97	20-May-97	21-May-97	2-Jun-97
03300	Concrete Mix Design	1-May-97	2-May-97	3-May-97	4-May-97	10-May-97
03350	Cure/Seal Submittal	15-May-97	16-May-97	21-May-97	22-May-97	26-May-97
03400	Tilt-Up Panel Shop Drawings	15-May-97	16-May-97	22-May-97	23-May-97	2-Jun-97
04200	CMU Samples	15-May-97	16-May-97	18-May-97	19-May-97	1-Jun-97
04200	CMU Test Reports	15-May-97	16-May-97	18-May-97	19-May-97	1-Jun-97
05001	Steel: Embed, Bolts S.Dwg	1-May-97	2-May-97	12-May-97	13-May-97	20-May-97
05002	Steel: Columns Shop Drawing	15-May-97	16-May-97	20-May-97	21-May-97	28-May-97
05003	Steel: Beams Shop Drawing	1-Jun-97	2-Jun-97	15-Jun-97	17-Jun-97	10-Jul-97
05200	Steel Joist Shop Drawing	1-Jun-97	2-Jun-97	15-Jun-97	17-Jun-97	12-Jul-97
05300	Metal Deck Shop Drawing	1-Jun-97	2-Jun-97	15-Jun-97	17-Jun-97	12-Jul-97
05500	Steel Stair Shop Drawings	15-Jun-97	17-Jun-97	1-Jul-97	6-Jul-97	1-Oct-97
06400	Millwork Shop Drawings	15-Jun-97	17-Jun-97	1-Aug-97	15-Aug-97	1-Feb-98
07500	Roofing Submittal	15-May-97	16-May-97	1-Jun-97	2-Jun-97	1-Aug-97
07600	Flashing Submittal	15-May-97	16-May-97	1-Jun-97	2-Jun-97	1-Aug-97
07900	Joint Sealant Submittal	1-Jun-97	3-Jun-97	10-Jun-97	11-Jun-97	1-Aug-97
08200	HM Drs, Frs Shop Drawing	15-May-97	16-May-97	23-May-97	24-May-97	5-Jun-97
08300	Overhead Door Submittal	1-Jul-97	2-Jul-97	10-Jul-97	20-Jul-97	1-Nov-97
08500	Metal Window Shop Drawing	1-Jul-97	4-Jul-97	20-Jul-09	23-Jul-97	1-Oct-97
08700	Finish Hardware Submittal	1-Jun-97	2-Jun-97	15-Jun-97	16-Jun-97	1-Oct-97
09250	Drywall Submittal	1-Jul-97	2-Jul-97	10-Jul-97	15-Jul-97	15-Aug-97
09300	Ceramic Tile Samples	1-Jun-97	2-Jun-97	10-Jun-97	12-Jun-97	15-Sep-97
09500	Ceiling S. Drawings, Samples	1-Jun-97	2-Jun-97	10-Jun-97	12-Jun-97	1-Oct-97
09680	Carpet Samples	1-Jun-97	2-Jun-97	1-Jul-97	15-Jul-97	1-Jan-98
09900	Paint Submittal, Samples	1-Jun-97	2-Jun-97	1-Jul-97	5-Jul-97	1-Sep-97
10150	Toilet Partition Submittal	1-Jun-97	2-Jun-97	1-Jul-97	5-Jul-97	1-Dec-97
10800	Toilet Access. Submittal	1-Jun-97	2-Jun-97	1-Jul-97	5-Jul-97	1-Dec-97
15010	Plumbing R/I Submittal	1-May-97	2-May-97	9-May-97	10-May-97	11-May-97
15300	F.Sprinkler S. Dwgs.	15-Jun-97	16-Jun-97	1-Jul-97	2-Jul-97	15-Jul-97
15400	Plumbing Submittal	20-May-97	21-May-97	1-Jun-97	5-Jun-97	15-Jun-97
15500	HVAC Submittal	20-May-97	21-May-97	1-Jun-97	5-Jun-97	15-Jun-97
15950	Temp. Controls Submittal	15-Jul-97	16-Jul-97	1-Aug-97	2-Aug-97	1-Nov-97
16100	Electrical R/I Submittal	10-May-97	11-May-97	18-May-97	19-May-97	20-May-97
16200	Electrical Submittal	20-May-97	21-May-97	1-Jun-97	5-Jun-97	15-Jun-97
16700	Fire Alarm Shop Drawing	15-Jul-97	16-Jul-97	1-Aug-97	2-Aug-97	1-Nov-97

FIGURE 3–5   Example Procurement Schedule

schedule. Additional information can be added to this schedule, depending upon the contractor's needs and the use of the schedule. This information can include:

- vendor, supplier, fabricator, or subcontractor name and location
- additional steps in the process, such as approval by consultants
- actual dates achieved to be used for comparison with the scheduled dates
- correlation with the construction schedule activities, by number and description

The **submittal log** is used for tracking the actual progress of the submittal. Many contractors use this only for the dates the submittal was received by the appropriate parties. A more complete and useful submittal log would compare the projected dates with the actual dates, as shown in Figure 3–6. By comparing the actual and scheduled dates in a single form, variances are easily determined. The submittal log is used as documentation to indicate unnecessary delay in the submittal process. As the material and fabrications cannot be ordered until the approvals are received, any delay in the submittal process, whether by the subcontractor, supplier, contractor, or architect and engineer, can result in a delay of the delivery of the item to the jobsite. This delivery delay also can impact the construction schedule. The submittal log and procurement schedule are useful tools for the contractor's personnel, enabling them to track submittals at any given point in time and follow up with the individual who is responsible so the material will be ordered on schedule.

Submittal Review By The Contractor

Although the submittal process is intended primarily for the architect's review of material and fabrications, the contractor is quite involved in the submittal process. The contractor reviews submittals for

- compliance to specifications
- dimensional conformance to the construction assemblies
- interface with other materials and assemblies, avoiding duplication or omission of elements in the process
- the constructability of the assembly, recognizing the steps necessary in constructing the system

The contractor receives the submittal from the supplier or subcontractor prior to forwarding it to the architect and is expected to review the submittal and stamp it as having been reviewed prior to transmitting it to the architect. The contractor acts as a first-line filter with the submittals,

SUBMITTAL LOG

Sect. No.	Item	Scheduled: / Actual: To Contractor	ACTUAL	To Architect	ACTUAL	To Contractor	ACTUAL	Order Date	ACTUAL	Delivery Date	ACTUAL
02150	Shoring Shop Drawing	1-May-97	30-Apr-97	2-May-97	1-May-97	9-May-97	5-May-97	9-May-97	6-May-97	15-May-97	14-May-97
02730	Drywell Submittal	1-May-97	1-May-97	2-May-97	2-May-97	9-May-97	5-May-97	10-May-97	6-May-97	24-May-97	24-May-97
02850	Irrigation Submittal & S.Dwg.	1-Jul-97	15-Jun-97	2-Jul-97	16-Jun-97	15-Jul-97	25-Jun-97	20-Jul-97	28-Jun-97	1-Mar-98	1-Mar-98
02900	Landscape Shop Drawing	1-Jul-97	16-Jun-97	2-Jul-97	16-Jun-97	15-Jul-97	25-Jun-97	20-Jul-97	28-Jun-97	1-Apr-98	1-Apr-98
03201	Rebar: Footings Shop Drawing	1-May-97	2-May-97	2-May-97	4-May-98	5-May-97	6-May-98	6-May-97	7-May-98	10-May-98	10-May-98
03202	Rebar: Found.Wall S.Dwg	4-May-97	4-May-98	5-May-97	6-May-98	10-May-97	7-May-98	11-May-97	8-May-98	20-May-98	18-May-98
03203	Rebar: Slabs Shop Drawing	10-May-97	10-May-97	11-May-97	11-May-97	15-May-97	20-May-97	16-May-97	20-May-97	26-May-97	27-May-98
03204	Rebar: T/U Panels, S.Dwg.	10-May-97	10-May-97	11-May-97	11-May-97	20-May-97	20-May-97	21-May-97	20-May-97	2-Jun-97	27-May-98
03300	Concrete Mix Design	1-May-97	28-Apr-97	2-May-97	1-May-97	3-May-97	3-May-97	4-May-97	4-May-97	10-May-97	10-May-97
03350	Cure/Seal Submittal	15-May-97	1-May-97	16-May-97	5-May-97	21-May-97	7-May-97	22-May-97	10-May-97	2-Jun-97	24-May-97
03400	Tilt-Up Panel Shop Drawings	15-May-97		16-May-97	25-May-97	22-May-97	30-May-97	23-May-97	30-May-97	2-Jun-97	4-Jun-97
04200	CMU Samples	15-May-97	5-May-97	16-May-97	10-May-97	18-May-97	14-May-97	19-May-97	15-May-97	1-Jun-97	1-Jun-97
04200	CMU Test Reports	15-May-97	5-May-97	16-May-97	8-May-97	18-May-97	12-May-97	19-May-97	13-May-97	1-Jun-97	1-Jun-97
05001	Steel: Embed, Bolts S.Dwg	1-May-97	5-May-97	2-May-97	11-May-97	12-May-97	12-May-97	13-May-97	15-May-97	20-May-97	22-May-97
05002	Steel: Columns Shop Drawing	15-May-97	10-May-97	16-May-97	11-May-97	20-May-97	18-May-97	21-May-97	19-May-97	28-May-97	27-May-97
05003	Steel: Beams Shop Drawing	1-Jun-97	28-May-97	2-Jun-97	29-May-97	15-Jun-97	10-Jun-97	17-Jun-97	11-Jun-97	10-Jul-97	5-Jul-97
05200	Steel Joist Shop Drawing	1-Jun-97	28-May-97	2-Jun-97	29-May-97	15-Jun-97	10-Jun-97	17-Jun-97	11-Jun-97	12-Jul-97	5-Jul-97
05300	Metal Deck Shop Drawing	1-Jun-97	28-May-97	2-Jun-97	29-May-97	15-Jun-97	10-Jun-97	17-Jun-97	11-Jun-97	12-Jul-97	5-Jul-97
05500	Steel Stair Shop Drawings	15-Jun-97	30-Jun-97	17-Jun-97	3-Jun-97	1-Jul-97	17-Jun-97	6-Jul-97	29-Jul-97	1-Oct-97	28-Sep-97
06400	Millwork Shop Drawings	15-Jun-97	15-May-97	17-Jun-97	1-Jul-97	1-Aug-97	20-Jul-97	15-Aug-97	4-Oct-97	1-Feb-98	15-Feb-98
07500	Roofing Submittal	15-May-97	15-May-97	16-May-97	16-May-97	1-Jun-97	28-May-97	2-Jun-97	29-May-97	1-Aug-97	1-Aug-97
07600	Flashing Submittal	15-May-97	15-May-97	16-May-97	16-May-97	1-Jun-97	28-May-97	2-Jun-97	29-May-97	1-Aug-97	1-Aug-97
07900	Joint Sealant Submittal	15-May-97	18-May-97	3-Jun-97	17-Jun-97	10-Jun-97	30-Jun-97	11-Jun-97	1-Jul-97	1-Aug-97	1-Aug-97
08200	HM Drs, Frs Shop Drawing	15-May-97	18-May-97	18-May-97	18-Jun-97	23-May-97	23-May-97	24-May-97	24-May-97	5-Jun-97	10-Jun-97
08300	Overhead Door Submittal	1-Jul-97	15-May-97	2-Jul-97	17-May-97	10-Jul-97	25-May-97	20-Jul-97	27-May-97	1-Nov-97	25-Nov-97
08500	Metal Window Shop Drawing	1-Jul-97	10-Jun-97	5-Jul-97	10-Jun-97	20-Jul-09	24-Jun-97	23-Jul-97	28-Jun-97	1-Oct-97	1-Oct-97
08700	Finish Hardware Submittal	1-Jul-97	15-Jun-97	2-Jun-97	25-Jun-97	15-Jun-97	5-Jul-97	16-Jun-97	10-Jul-97	1-Oct-97	25-Sep-97
09250	Drywall Submittal	1-Jul-97	1-Aug-97	2-Jul-97	3-Aug-97	10-Jul-97	15-Aug-97	15-Jul-97	15-Aug-97	15-Aug-97	16-Aug-97
09300	Ceramic Tile Samples	1-Jun-97	1-Jun-97	2-Jun-97	3-Jun-97	10-Jun-97	10-Jun-97	12-Jun-97	12-Jun-97	15-Sep-97	15-Sep-97
09500	Ceiling S. Drawings, Samples	1-Jun-97	1-Jul-97	2-Jun-97	5-Aug-97	10-Jun-97	15-Aug-97	12-Jun-97	17-Aug-97	1-Oct-97	1-Oct-97
09680	Carpet Samples	1-Jun-97	1-Aug-97	2-Jun-97	3-Aug-97	1-Jul-97	3-Sep-97	15-Jul-97	10-Sep-97	1-Jan-98	4-Feb-97
09900	Paint Submittal, Samples	1-Jun-97	15-Jun-97	2-Jun-97	17-Jun-97	1-Jul-97	17-Jun-97	15-Jul-97	20-Jul-97	1-Sep-97	1-Sep-97
10150	Toilet Partition Submittal	1-Jun-97	15-May-97	2-Jun-97	10-Jun-97	1-Aug-97	3-Jun-97	5-Jul-97	5-Jun-97	1-Dec-97	1-Nov-97
10800	Toilet Access. Submittal	1-Jun-97	15-May-97	2-Jun-97	25-Jun-97	18-Aug-97	3-Jun-97	16-Jun-97	5-Jun-97	1-Dec-97	16-Nov-97
15010	Plumbing R/I Submittal	1-May-97	1-May-97	2-May-97	2-May-97	9-May-97	7-May-97	10-May-97	8-Jul-97	11-May-97	11-May-97
15300	F.Sprinkler S. Dwgs.	15-Jul-97	17-Jun-97	16-Jun-97	17-Jun-97	1-Jun-97	1-Jun-97	2-Jul-97	2-Jul-97	15-Jul-97	16-Jul-97
15400	Plumbing Submittal	20-May-97	15-May-97	21-May-97	16-May-97	1-Jun-97	29-May-97	5-Jun-97	1-Jun-97	15-Jun-97	15-Jun-97
15500	HVAC Submittal	15-Jul-97	15-May-97	21-May-97	16-May-97	1-Jun-97	29-May-97	5-Jun-97	1-Jun-97	15-Jun-97	15-Jun-97
15950	Temp. Controls Submittal	15-Jul-97	7-Jul-97	16-Jul-97	10-Jul-97	1-Aug-97	10-Aug-97	5-Jul-97	11-Aug-97	1-Nov-97	1-Dec-97
16100	Electrical R/I Submittal	10-May-97	8-May-97	11-May-97	9-May-97	18-May-97	15-May-97	19-May-97	15-May-97	20-May-97	20-May-97
16200	Electrical Submittal	20-May-97	20-May-97	21-May-97	21-May-97	1-Jun-97	7-Jun-97	5-Jun-97	8-Jun-97	15-Jun-97	15-Jun-97
16700	Fire Alarm Shop Drawing	15-Jul-97	1-Aug-97	16-Jul-97	10-Aug-97	1-Aug-97	30-Aug-97	2-Aug-97	1-Sep-97	1-Nov-97	5-Nov-97

FIGURE 3-6   Example Submittal Log

refusing products that do not comply to the project's specifications, and reviews the submittal for compliance with the contract documents. Many architects will not accept product submittals for products that have not been approved for substitution. Because the architect has selected products and approved substitutions, many contractors are reluctant to suggest further substitutions. If the product is specified as a primary product, listed as an approved equal, or approved in the addenda, it will comply with the contractual requirements. The contractor then needs to confirm that the product is acceptable for its intended use and compatible with the other products specified. If the product does not meet contract requirements, the contractor should immediately return the submittal to the supplier or subcontractor to avoid wasting valuable time in the submittal, order, and delivery processes. Some architects and owners will accept substitutions for products after the contract is awarded, but usually only with an appropriate deduction from the contract amount.

The contractor also is responsible for providing field dimensions for openings and items that are installed in instances where other construction elements determine the size of the material. Many materials have a very small tolerance in dimensional variance, requiring exact dimensions of the constructed elements, rather than relying on those in the construction documents. The field dimensions necessary may not be available at the time the shop drawings are submitted, requiring the contractor to hold the shop drawings until all are available, or provide them at a later date. Some suppliers and fabricators will measure the field dimensions themselves, rather than relying on the contractor for the dimensions. When fabricators visit the jobsite to obtain dimensions, they also are able to familiarize themselves with other aspects of the project that may affect their product's fabrication.

The contractor is able to determine the interface of the systems from the shop drawings and material submittals, occasionally becoming aware of duplication of material by more than one supplier. The contractor also may find that additional material is necessary to complete the assembly. In these cases, changes to the original purchase orders or subcontracts may be necessary.

By reviewing the shop drawings and submittals, the contractor will become aware of the means and methods of installing the construction materials, systems, and assemblies. When installing the material and equipment with their own forces, the contractor will use the submittals to determine the necessary equipment and workforce to construct the assembly. Although the contractor should have determined a plan for installation during the estimating phase, a detailed examination of the procedures is necessary during the construction phase. The information in the submittal should indicate the size, weight, configuration, and installation details of the material and equipment. It is possible that the particular material is not compatible with the construction methods intended for use on the project, requiring either a different material or different construction methods.

FIGURE 3–7
Example
Contractor's
Review Stamp

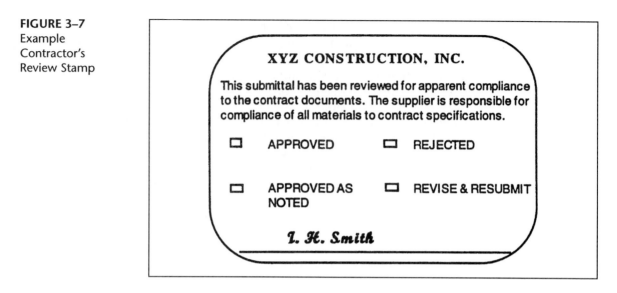

Review of shop drawings and submittals for the contractor should be done by an individual who thoroughly understands the project's requirements. The project manager often reviews the submittals on smaller projects, but on larger projects, the field engineer or office engineer reviews and processes the submittals. Because many submittals are received during the same period, a larger project will have several engineers reviewing submittals and shop drawings.

The contractor normally stamps the submittals, indicating that they have been reviewed. Some contractors prefer a brief statement on the stamp to avoid the transfer of liability. Some of these simple stamps may contain the following information:

- "Reviewed" statement
- Contractor
- Date reviewed
- Project
- Signature

Some contractors prefer to stamp the submittals with more information, as shown in Figure 3–7.

The Use of Submittals During Construction

The contractor will use submittals at the jobsite for several different applications.

- information for preparation of openings, support, adjoining assemblies, and general construction details
- quality control information to assure that the correct product has been supplied
- placement diagrams for installation of the material
- information relating to necessary handling and placement of equipment
- jobsite reference for architect and owner

The information in product data and shop drawings provides essential information that is usually beyond that of the contract documents, which facilitates the proper installation of the material and equipment. It is essential that this information be kept at the jobsite; normally it is kept in file cabinets and plan racks that are systematically organized by specification section number.

Summary

Submittal of product data, shop drawings, and samples in a complete and timely fashion is essential for the smooth execution of the construction contract. Submittals are used for approval for compliance to the contract and for quality control to ensure that the appropriate produce is being used in the constructed facility.

Complete information needs to be shown in the product submittal or shop drawing, regarding physical characteristics, finishes, and installation details or restraints. Samples and mock-ups are normally used to ensure that product and installation will meet the expectations of owners and designers.

Submittals are required in the following areas of contract documents:

- General Conditions of the Construction Contract
- Supplementary Conditions
- Section 01300 of General Requirements
- Individual Product Specification in the Technical Specification

The approval process for submittals involves several steps and can be very time-consuming. A procurement schedule, listing approval steps, order date, and delivery date, is required by some contracts and can be helpful in monitoring and managing the submittal process. Cooperation by all parties will aid in a smooth submittal process.

The contractor reviews submittals for several reasons:

- as the first-line review for compliance to the contract
- for interface with other materials and systems

- to determine the necessary equipment and procedures for installation
- to provide jobsite information, such as field dimensions

The contractor uses the submittals in the construction process for:

- placing details
- determining installation techniques
- inventory of material
- quality control

CHAPTER 4

DOCUMENTATION AND RECORD KEEPING AT THE JOBSITE

CHAPTER OUTLINE

Report Types and Content

Cost Documentation

Correspondence

Contractual Requirement Documentation

Meeting Minutes

Summary

As the contractor approaches the actual construction of the project, documentation will be needed for all activities. When preparing for the project, the contractor begins the documentation process by keeping logs and schedules for product submittals, as discussed in the previous chapter. This process continues throughout the construction project.

Accurate documentation of the construction project is essential in the current contracting climate. There was a time, within recent memory, when the contractor was able to pursue only the construction of the project, and the project itself spoke for its success. Today, however, the contractor operates within tight cost and time constraints. A number of activities beyond the control of the contractor can influence the contractor's performance, whether by actions of the owner, architect, subcontractor, supplier, building inspector, or other parties. An activity deviation may not appear to impact on the contract at the time it happens, however it may be felt at a later date. Because disputes can occur with the owner, architect, subcontractors, and code enforcement officials during the course of the project, clear documentation can assist in finding quick and equitable solutions.

All documentation should be clear and concise, as many individuals will be referring to it and using it as a basis for other actions. Documentation such as daily reports and meeting minutes may be the only contact some individuals, such as those in upper-tier management, have with the project, thus clear, objective, and complete reporting is necessary. Documentation should be objective and convey the facts truthfully, without bias. The personal diary may be slightly subjective, however, because it may be opinion-oriented, but this is a necessary part of the diary as it will provide a larger perspective regarding particular situations. Keeping this in mind, the individual should try to be as objective as possible when using this type of documentation.

Some keys to effective jobsite documentation and communication include:

- Objectivity and truthfulness—Provide a fair and honest assessment of the situation, without bias.
- Timeliness—Distribution of documentation, as required for the particular documentation tool, should be made as quickly as possible. Upper management needs current data to keep pace with the project.
- Retrievability—The documentation must be readily accessible and closely tied to the specific issue. A good, updated manual filing system or computerized search and sorting capabilities facilitate documentation.
- Appropriate distribution—Each piece of documentation has different distribution requirements, depending on the nature of the issue and the parties involved. It is important to distribute applicable information to the parties but to not send too much irrelevant information. Some parties may need weekly or monthly progress summaries rather than a daily report.

- Standard, uniform information—Each piece of documentation should contain certain information for quick retrieval. For example, if one needs to know how many roofing subcontractor employees worked on a project on a specific date, the daily reports would provide the information. There would be no need to search other documents for this information.

    Standard forms save time for all levels of management. They also ensure that company standard procedures are being followed at the jobsite. Standard information from project to project helps management evaluate conditions quickly, efficiently, and accurately.

- Completeness and comprehensiveness—The description of the event or issue should be as complete as possible at the time, without excessive elaboration. The entire documentation system for the construction project must include recording all facets of the project, without gaps in the information. The documentation must be continuous, from the beginning of the project to the end.

    When documenting the construction project, the primary areas covered include:

- Events—An event can be defined as an occurrence that should be documented during the project. These events can vary, from the number of employees on the project during a specific day to a catastrophic event such as a structural failure. Documentation of the events of the project might include: documentation of its progress, specific occurrences, and other information that might relate to the work.

    Documentation types—These consist of daily reports, weekly and monthly reports, diaries, accident reports, subcontractor logs, photographs, videos, time-lapse photography, progress schedules, schedule updates, and document control logs.

- Conversations—Included here are written records of conversations, including the parties involved, topics covered, and any directions or solutions. These records should be as objective as possible, but may also reflect a subjective view of the discourse. Conversations can occur by telephone or through direct personal contact. Formal meetings usually are covered by meeting minutes, but all conversations that concern the project should be documented.

    Documentation types—These consist of daily reports, diaries, telephone logs, and memos to the file.

- Costs—The contractor needs to track all costs on the jobsite and compare them to the estimated cost. All labor, material, and equipment used by the contractor needs to be reported and associated with work items. The cost data accumulated on the jobsite is used by the contractor's accounting department, as well as by the project management team. Cost information is accumulated during the work as a tool to determine the crews' productivity, which allows managers to alter the production factors during the process to control the cost within the budget that was established through the estimate.

Documentation types—These include the bill of materials, purchase order, receipt of delivery, delivery logs, time cards, labor reports, labor packages, labor cost control, equipment logs, and rental logs.

- Correspondence—Throughout the duration of the project, written correspondence is transmitted between the parties through hard copy via the mail, hard copy via a fax, or via E-mail. This correspondence should be sorted by party and issue. The contractor corresponds with the architect, the owner (usually through the architect), subcontractors, suppliers, building officials, and other parties active in the construction process. This correspondence ranges from a transmittal, indicating delivery of an item, to letters concerning serious contractual issues.

  Documentation types—These include transmittals, transmittal logs, submittal transmittals, submittal logs, Request For Information (RFI), letters, faxes, E-mail notes, speed letters/with responses, punch lists, payment requests, schedules, and schedule updates.

- Contractual requirements—Several items of documentation are required by the contract documents. Although these documents serve specific purposes, they also document certain events throughout the project. Specific forms will be specified for areas such as progress payment, change orders, and substantial completion.

  Documentation types—These include progress payment requests, schedules of values, contract change orders and proposals, certified payrolls, certificates of substantial completion, and photographs.

- Meeting minutes—Numerous meetings occur throughout the construction process. Some are regularly scheduled progress meetings and some may be special-topic meetings. A written record, or the minutes of the meeting, should always be saved and later distributed to meeting participants and other interested parties.

  Documentation types—These include meeting minutes for preconstruction, construction progress, special installation, subcontractor, and special issue meetings.

Report Types and Content

Event and Conversation Documentation

Complete and accurate documentation about what actually occurred on the jobsite during the construction process is perhaps the most important record keeping done. This documentation illustrates the actual sequence of the project, personnel on the job, materials delivered to the jobsite, equipment used on the jobsite, and many other factors. These reports are necessary to remain consistent in content and consecutive throughout the duration of the project, without time gaps between documentation. Although this record keeping seems burdensome at the time, it is extremely useful in recreating a realistic picture of the jobsite at a later date, as is often necessary in the settlement of disputes. These records can be transmitted to individuals who are not directly involved on a daily basis with the project, keeping management aware of jobsite activities. Much of the event documentation is recorded by the field engineer or superintendent on the jobsite.

Daily Reports

The daily report is the consecutive record of events on the jobsite. Its purpose is to provide a snapshot of the day's activities and conditions. The daily report records information objectively. The data is normally recorded on a preprinted form and contains consistent, daily information. The information and format will vary from contractor to contractor and even from project to project.

The contractor's daily report is, in most cases, an internal document, distributed only within the contractor's firm. All of the contracting entities at the jobsite, such as the owner's inspector, the architect's inspector, and the subcontractor's foremen, should keep daily reports for their own use as well. Although this seems like unnecessary duplication, each of the parties are particularly aware of the events and activities that the other parties may not be informed about. Most firms want to rely on information from their own representatives, rather than from another firm.

The daily report is a jobsite report, normally completed by the superintendent or field engineer. The daily report could be completed at all levels, that is, by the project manager, superintendent, and field engineer, with different contents for each report. Typical distribution of the daily reports would include:

- one copy on file at the jobsite
- one copy to the project manager; copy to remain in project file
- one copy, or a summary by week or month, to the company officer (partner, owner) in charge of the project

Information will be different for each contractor's daily report. Realistically, the report should be limited to one page. One of the objectives of any documentation system is to accurately record the information, but to minimize the amount of time spent by field personnel in documenting the events. Some forms have preprinted categories, such as types of subcontractors—roofing, drywall, painters, and so on, so only a number has to be added by the form preparer. Other forms require more information.

The daily report can be hard copy, with photocopies sent to the home office by mail, usually on a weekly basis. The daily report is commonly prepared on the jobsite computer and is transferred by modem or fax to the home office. Immediate transfer of this report, and others, helps home office personnel obtain an accurate picture of the current status of the project.

Typical information included in the daily report is:

- Date—The date reported and the date the report is written should be the same. Sequential numbering of the report also is required by some firms.
- Project name and number—The project may have several numbers. Because this is an internal document, the contractor's project number should be attached.
- Individual making the report.
- Weather information—Weather information is extremely important for project documentation. It indicates project conditions that can explain why work was not done on a specific date. This information should reflect the particulars of each work shift, such as the temperature at the start of the shift, at mid-shift, and at the end of the shift. It also could reflect the temperature during activities that are temperature-sensitive, such as concrete pours. Barometric readings help predict immediate changes in the weather. The weather section of the daily report should describe precipitation during a 24-hour period, including the period before the shift, such as overnight. The amount of precipitation, if known, and the time period, for instance 10 A.M. to 2 P.M., should be mentioned as well. The effect of the precipitation also should be noted, such as mud, ice, or the amount of snow on the ground. General wind conditions, such as speed and direction should be noted, as well as the time, duration, velocity, and direction of unusual wind conditions. Because radiated heat can affect some materials, particularly concrete, notes should indicate whether it was clear and sunny, cloudy, partly cloudy, and so on. If there is a disruption in work activities, including stopping work, extraordinary measures to protect work, or rescheduling of the activity due to weather conditions, the duration of the disruption should be noted.

Most of this data is available through observation at the jobsite. Additional information can be acquired through the U.S. Weather Service or several on-line databases. The contractor may wish to use a simple weather station on the jobsite, consisting of a thermometer, barometer, and rain gauge. An example of weather information that could be used in the daily report follows:

EXAMPLE

Weather: 42°F, 7 A.M.; 65°F, 12 P.M.; 70°F, 2 P.M. Concrete Pour
Precip: None; Month accum: 1.75"
Wind: Calm morning, 10 mph wind from south, afternoon

- Description of activities in progress—This includes a brief description of the work in progress on the jobsite during the report period (for the day or shift).

EXAMPLE

Carpenters, laborers, ironworkers, and electricians setting tilt-up panel forms 5 through 16 on Building A. Excavator digging foundation on Building B.

- Contractor's labor on the jobsite—This area should list the trades directly employed by the contractor on the jobsite and indicate the number of employees. This is not a time card or hourly listing, but provides an indication of the size of the crew on the jobsite.

EXAMPLE

Carpenter	5	Ironworker	2
Laborer	4	Truck driver	1

- Subcontractors on the jobsite—Some daily reports classify the type of craftsmen and/or the number of people working for the subcontractor. Actually, both pieces of information are valuable to obtain a full picture of the available productivity potential. This information is not meant for time cards or to track time and material accounts for subcontractors. It is used primarily to determine if subcontractors have adequately staffed the project. Personal observation of the subcontractors' crews is the most common method to determine crew size. The subcontractor's foreman also could be a source for obtaining the number of crew members. Occasionally, the contractor will require that daily reports be submitted from each subcontractor working on the site.

EXAMPLE

ABC Masonry, Inc.		FS Insulation	
Bricklayer	4	Carpenter	3
Hodcarrier	1	Laborer	1
City Plumbing, Inc.		State Electrical, Inc.	
Plumber	2	Electrician	3
Apprentice Plumber	1	Apprentice Electrician	1
Pipefitter	1		

- Equipment at the jobsite—The contractor should list the equipment used on the jobsite and indicate whether it is being used or it is idle. If a piece of equipment is not being used, its rental might not be applied to the cost item if the equipment is owned by the contractor. Equipment logs normally are used to establish hours for the contractor's use of the equipment. The list of equipment indicates the capabilities to perform the work at the time. Information from the equipment records can be used for cost control, for estimating historical data, and for availability costs. Idle subcontractor equipment should be noted because a record needs to be kept for equipment charges or cost-plus items. Normally, equipment listed in this area is larger and does not account for tools such as saws, drill motors, or hand tools.

EXAMPLE

Equipment	In Use (Hrs.)	Idle (Hrs.)
Our Equipment		
Concrete Pump	4	4
Forklift	8	
Tower Crane	8	
ABC Masonry, Inc.		
All-terrain Forklift	8	
Scaffolding: 200 frames	8	
Mixer	8	
City Plumbing, Inc.		
Hi-Lift	4	4

- Material deliveries to the jobsite—The material at the jobsite indicates the ability to pursue construction activities. Delivery of material also shows compliance with purchase orders and subcontract agreements (the delivery of material will be included in the purchase order records). Delivering large amounts of material to the jobsite can be disruptive, requiring the crew's time to unload and store material. Delivery of material also indicates the amount of crane time necessary for unloading material, which will detract from work activities.

EXAMPLE

Material Delivery

From	For	Time	Items
Valley Lumber	Our Firm (P.O. 293–124)	10–11 A.M.	2,000 BF 2 × 4
			3,000 BF 2 × 6
Smith Plmbg. Sply.	City Plumbing, Inc.	11–11:30 A.M.	1,000 LF 2″ PVC
			24 P-2 Fixtures
Iron City Steel	Our Firm (P.O. 294–003)	1–2 P.M.	4th shpt. Rebar

- Visitors to the jobsite—Verification of the dates for visits from the owner's representatives, building officials, architects, engineers and other interested parties is necessary in some disputes. Job security also is maintained by requiring visitors to register with the field office. A separate visitor's log, which records the visitor's signature, could be kept in addition to or in lieu of this activity item on the daily report. The contractor's purpose for this log is to monitor visitors to the jobsite in compliance with the safety program. It is important that the person be identified and that the name of the firm, the purpose, and the time and duration on the jobsite be recorded. Documentation of visitors' conversations normally is addressed in records such as the personal diary.

EXAMPLE

Visitor	Firm	Purpose	Time
A. J. Jones	OSHA	Safety Inspection	9–11 A.M.
M. Smith	ABC Architects	Construction Observation	1–2:30 P.M.
O. K. Anderson	Anderson Safety Sply.	Salesman, Safety Protect.	2–3 P.M.

- Occurrences on the jobsite—This section is used to describe occurrences that fall outside of the description of "work in progress." This area could indicate an action or a conflict on the jobsite, an accident, discovering an unknown condition, or another notable event. Most of these items also will be covered in other reports, such as accident reports, diary conversations, or a Request for Information to the architect. It is essential, however, that the occurrence be noted in the daily report. The date of occurrence is crucial to any potential change or claim when establishing the time frame for processing either. Specific information should be included here, including reference to other reports. Each occurrence item should include what happened, who was involved, when it happened, how it happened, where it happened, and the resolution, if any, that was reached.

EXAMPLE

Occurrences

10:15 A.M.: John Smith, AAA Excavation, notified us that they have encountered solid rock in the excavation of Building B, between grids A2 and G4 at about an elevation of 1567.0. Excavation is intended to go to 1560.5. Called M. Smith, ABC Architects. Faxed RFI 034.

2:35 P.M.: Jack Johnson, carpenter, cut hand with circular saw, in forming of building A. Art Anderson, foreman, administered first aid and took him to emergency room at City Hospital. Eight stitches necessary, bandaged, and back at the jobsite; able to perform tasks. See Accident Report 294–004.

3:15 P.M.: Conflict between plumbing rough-in and electrical panel in room A-103. RFI 035 faxed to architect.

- Signature and date—A signature and date usually are necessary to establish the report as an official document. Initialing the report also may indicate authorship by the preparer. Electronically produced documents do not provide the opportunity for a signature. A signed or an initialed hard copy in the job files should provide enough legal credence to the document.

As previously mentioned, contractors should develop their own forms containing information needed for project management. Blank and completed sample forms for the daily report are shown in Figures 4–1 and 4–2.

Weekly and Monthly Reports

Some contracting firms require weekly or monthly summaries of the project, usually to inform upper management of the project's progress. Daily reports are read by those who are involved on a day-to-day basis with the project, but the weekly or monthly report or summary keeps upper management aware of the project and enables them to assist when necessary. Occasionally, such reports also are required to be submitted to

```
FGH CONSTRUCTION COMPANY, INC.
                    DAILY REPORT
PROJECT:                      PROJECT NO.    DATE      NO.
REPORT BY:                    WEATHER:       TEMP.
                              WIND:      PRECIPITATION:
WORK IN PROGRESS:




LABOR: FGH EMPLOYEES
CARPENTERS_____      LABORERS_____    OPERATORS_____
TRUCKDRIVER_____     CEMENT MASONS_____ IRONWORKERS _____

LABOR:  SUBCONTRACTORS:



EQUIPMENT:
ITEM                       FIRM      HRS.IN USE     HRS. IDLE



MATERIAL  DELIVERY:
FROM            FOR              TIME          ITEMS




VISITORS
VISITOR                  FIRM           PURPOSE       TIME




OCCURRENCES:





SIGNATURE:_____
DATE:_____
```

FIGURE 4–1   Blank Daily Report Form

FGH CONSTRUCTION COMPANY, INC.
DAILY REPORT

PROJECT: New City Plaza, New City, CA   **PROJECT NO.** 2-98   **DATE** 5/24/98 **NO.** 29

REPORT BY: Bob Smith, Field Engineer   **WEATHER:TEMP.** 42 deg.F 8:00AM

75deg.F.2:00PM

WInd:10mphS   **PRECIPITATION:**None;.55"mtd

WORK IN PROGRESS:

Forming for tilt-ups, building A. Plumbers installed water service. Electricians installing conduit in tilt-up panels.

LABOR: FGH EMPLOYEES

CARPENTERS__4___   LABORERS __3__   OPERATORS   _0_

TRUCKDRIVER_ 0 ___   CEMENT MASONS _0_   IRONWORKERS   _2_

LABOR: SUBCONTRACTORS:

City Plumbing, Inc: Sparks Electric, Inc.

Plumbers: Journeyman   2   Electricians   2

          Apprentice   1

EQUIPMENT:

ITEM	FIRM	HRS.IN USE	HRS. IDLE
Case 580 Backhoe	FGH Const.	0	8

MATERIAL DELIVERY:

FROM	FOR	TIME	ITEMS
Iron City Steel	FGH Construction	10:00-11:00 AM	Rebar,3rdshpmt

VISITORS:

VISITOR	FIRM	PURPOSE	TIME
A.J. Jones	OSHA	Safety Inspection	9:00-11:00 AM

OCCURRENCES:

10:30 AM: Safety Inspector red tagged four ladders, 2 electrical cords, and table saw without guard. Citation Faxed to home office.

SIGNATURE: *Bob Smith*

DATE:___5/24/98_____

FIGURE 4–2   Completed Daily Report Form

the owner. Obviously, different types of information will be contained in a report to the owner than in a report intended for internal management.

The weekly or monthly report for internal management is normally written in narrative form, with some organizational parameters, rather than on a form such as the daily report. The organization of each report should be similar. The following areas examine some of the topics that could be included in a weekly or monthly report.

- Identification of the project—This includes the project name, location, and internal project number.
- Summary of activities—A summary of the activity progress since the last report. A discussion of site conditions, such as weather and the results on the site, also should be included here.
- Schedule analysis—This determines the relationship with the construction schedule, identifying the areas that are not meeting the schedule, what impact they will have on the overall schedule, and what can be done about those areas.
- Cost analysis—This briefly analyzes the profit picture for the project to date and clearly delineates any areas of concern and the steps that can be taken to minimize the impact.
- Subcontract and purchase order management—This involves any problems or project impact caused by subcontractors, vendors, or fabricators. Concerns about subcontractors could include the number of personnel on the project, competency of jobsite personnel, schedule compliance, and potential areas of dispute. Concerns about vendors and fabricators could include delivery date compliance, amounts of material delivered, and quality of material delivered.
- Change orders—Used to describe change order progress and problems that might result in change orders.
- Summary—Used to describe any additional problems or positive aspects relating to the project. Areas of interest or concern are noted, particularly unresolved issues that may become claims.
- Signature of preparer—The preparer should be the superintendent or project manager, as the information submitted requires management insight.

Longer reporting periods, such as monthly, can provide a forum for further analysis than can shorter term, or weekly, periods. A firm may want only weekly reports to keep abreast of progress, without the detail of daily reports, or a monthly report may be desired for problem-identification and problem-solving analysis. Some firms may want all three levels—daily, weekly, and monthly. Figures 4–3 and 4–4 show weekly and monthly reports, written by the superintendent to the project manager.

Diaries

Diaries are the personal records of conversations and occurrences that are kept by each management individual on the jobsite. Although this is a personal journal concerning the individual's contact and activities, it

WEEKLY REPORT
NEW CITY PLAZA, NEW CITY, CA
FGH PROJECT NO. 02-98
FOR THE PERIOD FROM JUNE 21 THROUGH JUNE 27, 1998

PREPARED BY: John Johnson, Superintendent

Summary of Activities:

We completed the subgrade work for the slab on Monday, and poured the slab for Building A on Tuesday and Wednesday. Curing compound was applied immediately after the slab was poured. Form material for the tilt-up panels was delivered Thursday AM. Tilt-up panels for the first lift have been formed on the east half of the slab as of Friday evening. Reinforcing steel and electrical rough-in have started on those panels. Excavation started on Wednesday, June 24th for the foundation at building B.

Schedule:

We still appear to be about a week ahead of schedule on Building A, due to time gained during the clearing and excavation for Building A. Forming of the tilt-ups is going well, and all material has been delivered for the tilt-ups, so probably will gain a few days. Excavation for Building B started on schedule, due to commitments of the excavation subcontractor.

Cost:

Forming and pouring of the foundation walls for Building A cost $176.90 / cubic yard, compared to $180.00 / cubic yard estimated.

Subcontractors:

City Plumbing has been hard to get on the job to do the underslab rough-in. They did finally got their work done, but started two days later than scheduled. We need to give them adequate notice and follow-up frequently to be sure they do their work. State Electrical has kept up well, doing good work. Fred's Excavating is right on schedule on Building B.

Change Orders:

No occurences or notice of changes this week.

Summary:

The project is progressing well. Architect seems to be processing submittals within schedule. We may be able to make some time in the next few weeks.

Respectfully Submitted,

John Johnson   June 29, 1998

FIGURE 4–3   Example Weekly Report

MONTHLY REPORT

NEW CITY PLAZA, NEW CITY, CA

FGH PROJECT NO. 02-98

FOR THE PERIOD FROM AUGUST 2 THROUGH AUGUST 29, 1998

PREPARED BY: John Johnson, Superintendent

Summary of Activities:

During this period, the exterior envelope of Building A has been completed. The metal joist and steel roof system, including steel roof deck, were installed by City Steel during the week of 8/16/98. The roofing crew were able to complete the roof insulation and roof membrane Friday August 28, 1998. Foundations, subgrade, and slab-on-grade have completed on Building B. Forming for tilt-up panels for building B are planned to start August 31, 1998.

Schedule Analysis:

We are currently 14 days ahead of schedule on building A and 28 calendar days behind schedule on Builiding B. As the critical path goes through the construction of building B, we are 28 days behind Schedule. The 28 days, however, relate to the unforeseen rock excavation. A request for an extension of 28 days is included in change proposal 08.

Cost Analysis:

Currently, we are under budget for the project. Our concrete work is about 75% complete with a predictable profit. Our major risk areas with our own labor are nearly complete.

Subcontractor Management:

We continue to have problems with City Plumbing. They come to the job three or four days later than scheduled, rarely finish their task, and are hard to get back on the job. We have called a meeting with Ed and George Roth on September 1st to get them to tend to their subcontract.

Change Orders:

Change Proposal 08 for the rock excavation has been approved and will be incorporated in Change 04 by September 1st. The architect has indicated that we will be able to be bill for the work in the August billing.

Summary:

With the time extension in Change Order 04, we will be back on schedule, although the project will complete a month later than originally intended. All work has been going well, as the weather has been good for the entire month. The only problem that we see in the future relates to the plumbers, and we are working on solving that problem currently.

Respectfully submitted,

John Johnson   August 31, 1998

FIGURE 4–4   Example Monthly Report

really is the property of the employer, as it concerns the record of the individual who is employed at the firm.

The diary should be as factual as possible in relating conversations and occurrences. Memory alone, in most cases, is not enough to recall specific conversations and events that took place weeks, months, or even years earlier. The diary is used primarily for disputes that involve specific events or conversations. As it is often used in arbitration, depositions, and court appearances, the writing should be clear and businesslike, and the facts should be accurate. A business diary or journal is a legal record and can be subpoenaed in court cases, therefore, the construction professional should be mindful of the importance of each diary entry.

Traditionally, an individual's daily diary should be:

• written in the individual's own hand, to prove authenticity,
• bound, so pages cannot be inserted,
• consecutively written with each day dated.

The form of this diary varies greatly, but usually conforms to the aforementioned guidelines. Some individuals use a twelve-month hard cover book, with consecutively dated pages. Several commercially produced diary-type books are available, including the spiral-bound type. Some firms use a bound book with duplicate copies, so one copy is kept in the bound book and the other copy is kept in the job file.

Because most writing is currently done on word processors and computers, it seems logical that this would be true for the diary as well. Some computer software programs are in journal format, which does not allow the writer to rewrite comments at a later date. However, even a running diary done in a word processing program can provide a historical context to a problem, enabling the writer to recollect conversations and events that relate to a particular issue. It is generally held that this type of documentation is valid, although it does not have the weight a traditional diary does.

There is a further benefit, not always obvious, to the constructor who keeps a daily diary. In writing, the individual concentrates on the events of the day and selects the most important items, thus establishing priorities when problem-solving. Solutions to problems often are reached in this way.

What follows is an example of the type of information contained in a superintendent's diary.

EXAMPLE

Wednesday, June 1, 1994
Our crew working to complete subgrade for slab, west part of building A. Three laborers backfilling and spreading gravel. Two carpenters setting screeds. Plumber not here to finish underslab rough-in. Two electricians completing underslab conduit.
9:10 A.M.: Called Ed at City Plumbing. Asked him where his crew was. He said the plumber and apprentice had an emergency repair and would be at the jobsite around 10 A.M. I emphasized that we needed the plumbing rough-in in the S. O. G.

done in the west end of building A today. Concrete pour tomorrow. Ed said they would be there.

9:30 A.M.: Call from Smith at the architect's office. Wanted to know when we are going to pour the slab, so he can arrange for the testing agency to be there. I told him tomorrow at 10 A.M.

10:30 A.M. Screeds set, sub-grade done, except in area of plumbing. Moved crew to building B. Finishers to be here tomorrow for slab pour.

11:00 A.M. Ordered 75 cu. yds. concrete 4,000 #, ¾" gravel, 4" slump—first truck to be here at 10:00 A.M.—rest to follow in 10-minute intervals.

11:30 A.M. Ordered concrete pumper from Pump City—ready to pump at 10 A.M. tomorrow, will arrive on job at about 9 A.M.

11:35 A.M.: Called Ed at City Plumbing. Not in, secretary said she'd call him on the mobile phone.

1:15 P.M. Ed from City Plumbing on jobsite. I asked him where the crew was. He thought they should be there. I explained that the rough-in had to be done today, concrete ordered for tomorrow. Ed said he'd take care of it.

2:00 P.M. Two plumbers from City Plumbing showed up on jobsite. They weren't familiar with the job, didn't know where the material was, left the job at 2:20 P.M. I tried to keep them there, but they snuck off.

2:20 P.M. Called Ed at City Plumbing. He said the repair job was bigger than anticipated. He was disappointed that his second crew left. He said that he wouldn't be able to get the rough-in done before the pour tomorrow. I told him we were going to back charge him for time lost. He said there was nothing he could do.

2:45 P.M. Notified Fred at home office of problem with City Plumbing. He will follow up. Agrees that I have to put off the pour.

3:00 P.M. Canceled concrete for tomorrow. Canceled inspection for pour.

3:10 P.M. Canceled order for pumper from Pump City. Guy at the desk said they will charge me a cancellation charge. I asked for Bud, but not in—he will call me.

4:45 P.M. Bud at Pump City called. He said he wouldn't charge cancellation charge, but went on for a while that I should get my act together.

5:05 P.M. Called Ed at City Plumbing: No answer.

Many firms and individuals prefer to keep a log of telephone conversations, separate from the diary, because the log entry would be made during or immediately after the phone call was made. This also can be accomplished with the diary, as most individuals write their diary entries at the end of the work day. The advantage of combining the diary and telephone conversation record is having a complete record of what occurred in one source.

Logs

Many logs can be kept on the jobsite to collect specific information. These logs are kept in addition to other reports. They can contain a mix of information in each form of document. Many firms will include all telephone call information in a telephone log, while some firms list the calls in a diary. A visitor's log often is kept separate from the daily report. Subcontractor information can be kept either in the daily report or daily diary. The method usually is chosen based on which method works best for the individuals who are recording the information. The method is not as important as providing complete, accurate, and up-to-date information about the activities and conversations at the jobsite.

Telephone Log.   The telephone log can be an accurate account of telephone conversations, or it can be a list that contains the time of the call, the person called, and the person calling. If used for recording the content of the call, information should be accurate and should include when the call was made, who called, who received the call, topics discussed, and promises made. Entries in the telephone log normally are made at the time of the call. When telephone discussions are recorded in the diary, they are included with other events that occur on the jobsite.

Visitors Log.   Many contractors maintain a notebook in the jobsite office for visitors to sign as they enter the area. This practice is probably not as effective as assigning a member of the jobsite team to track visitors and enter their names on the daily report.

Subcontractor Log.   Some contractors like to keep a close count of the subcontractors used on the jobsite. A subcontractor log can be organized to record on a daily basis the manpower for each subcontractor. The log can be organized by the date, listing the subcontractors and employees, or by the subcontractor, listing the date and number of employees. This is an important tool when determining whether subcontractor staffing is adequate for the project.

Document Control Log.   Construction documents often are changed and supplemented during the construction phase of the project. A document control log records these revisions, with dates and reference-associated documents, such as change orders. This log is very important when organizing large projects.

Accident Reports

Accident reports are specific reports written about particular incidents or accidents. These reports are part of the company's safety program, which is discussed in chapter 8. Safety activities, including occurrences and documentation of meetings and safety programs, are performed throughout the construction phase. (See chapter 8 for a discussion about documentation information and a description of safety activities.)

Progress Photographs

An essential part of jobsite documentation is to photograph the project. Still photographs should be taken at specific intervals throughout the project. Weekly photographs are probably adequate to record the project's progress. Photographs of a special condition, such as defective work, should be taken when the occurrence is encountered. Photographs are used to communicate certain situations both before and after the project.

If a dispute arises, a photograph can radically change a person's idea about events and conditions. The old adage that a "photograph doesn't lie" is mostly true, in that the photograph accurately portrays its subject.

Because a contractor takes thousands of photographs during a project, a primary concern is labeling and identifying photographs. The following information should be recorded when taking jobsite photographs.

- Date and time the photograph is taken—Some cameras are equipped with an automatic date and time stamp that appears on the negative and the actual photograph.
- Location of the project—The location of the photograph should be noted, as many jobsites look similar. Notation should be made regarding the direction of each photograph, such as "looking northwest from corner G-9."
- Subject matter—Notation should be made regarding the subject that was photographed.

There are numerous ways to record the aforementioned information. A photograph log can be kept, recording pertinent information, including film speed, lens aperture, and shutter speed, although these details are recorded by only the most serious photographers. Most photographs used are of the type that need developing. The instant photograph also is used on the jobsite, however, the quality is normally not as good and the photograph may not last as long as the developed one. However, instant photographs are convenient when one is needed at a remote location as quickly as possible, but one-hour film processing is available in most areas. Electronic (digital) photographs that can be transferred into the computer also are used on the jobsite. These photographs can be processed by a computer, inserted into reports or documents, and electronically transmitted to a remote location, such as the contractor's home office or the architect's office.

After the photograph is developed, it needs to be catalogued. Positive identification is ensured when information about the photograph is contained on the back. Because that information alone is not very retrievable, a computer database of information that has retrieving capabilities can facilitate the labeling of photographs and provide a database that can be sorted or searched for particular information. Typical data fields that might be made for a photograph database include:

- Photograph number
- Date, time, and photographer*
- Project, location*
- Subject*
- Issue
- Stage of completion
- Construction system
- Subcontract

* This information can be added to a label and attached to the back of the photograph.

Video Recordings

Video recordings are an excellent way to document project activity. Any activity that involves motion within a relatively short duration is a good subject for a video recording. Video recording does have some limitations in low-light conditions, however. Most recorders have a time and date feature that permanently stamps the video. Although video can be a good representation of some areas, it involves a time and equipment investment, whereas a few photographs may be all that is needed to quickly tell the story. Some areas where video can provide good records include:

- Discovering an unknown condition. The camera operator can record while he or she is describing the condition.
- Showing a process, such as an excavation, that can be used on the project. The video can show the equipment used and how long the process took. This type of video is useful for doing time and productivity studies. Real-time recordings should be made for activities of a fairly short duration, considering the cost of the recording media and the likelihood that anyone would watch hours and hours of these recordings.
- Recording existing conditions. Particularly in renovation or remodeling projects, the contract may require the contractor to restore an area to its original condition if it was damaged during construction. The contractor can perform a video walk-through, recording the original conditions. At the end of the project, this video can be reviewed to determine what damage was actually done.
- Videotaping installation techniques that can be used for reference by other superintendents on future projects.
- Videotaping special installations that can be referenced by maintenance personnel. Some special installations should be recorded to provide maintenance personnel with information about how the system was built.
- Videotaping operating and maintenance instructions during the closeout of the project. Videotaping these instructions provides not only a record of but a reference for maintenance personnel.

Like photographs, videotapes should be carefully labeled and stored where they will not be damaged by heat, light, or moisture. Indexing the videos, possibly in a computerized database, makes the information more accessible.

Time-lapse Photography

Time-lapse photography involves a series of still pictures taken at one- to four-second intervals by a special camera from a fixed location. This process provides an accelerated view of the construction process compared to real-time video recordings. This type of recording shows the flow of the process, but not the details. Time-lapse photography is primarily used for information in disputes, showing the sequence, equipment, and flow of the project, productivity studies, showing the effect of jobsite layout and facilities, and public relations for the contractor, showing how the facility was constructed.

Special attention should be paid to the initial placement of the camera. It should be placed in a location that will show the entire process for the duration of the project, or at least until the exterior work is complete, without the camera being moved during the project. Recording is automatic, but care must be taken to maintain the film and to check its operation.

Progress Schedules and Schedule Updates

The progress schedule can be used as a record of progress and sequence, as well as a plan for the project. Actual events, activities, and occurrences can be tracked and entered on the schedule, providing a record of what actually was done on the project. Chapter 11 provides a detailed description of the uses of the project schedule during construction.

Cost Documentation

Much of the documentation that is done in the field for a contractor relates to tracking cost information. Documentation of cost information has two primary purposes: to transmit information to the company's accounting system for disbursement of funds and to accumulate information to control the project's costs. Both types of cost information are usually collected and disseminated from accounting.

The objective of cost data reporting is to report the information in a manner that will permit cost control during the project and establish a cost database for future reference. Cost information should be assigned to cost codes that correspond with estimating activities, to enable the contractor to compare the actual cost with the estimated cost. If this comparison is done quickly and accurately, the contractor can use this information to control the cost during the activity, by taking positive action concerning labor, material, subcontractors, or compensation for the item.

Because a wide variety of methods of construction cost accounting exist, many methods of reporting are used. Time is of the essence in most cost reporting, as there are finite periods defined for payment of wages, materials, and subcontracts. Time also is important because problems need to be recognized immediately for activities where cost is exceeding estimate. Accounting systems that provide immediate compilation of data to the jobsite and to accounting are preferred by most contractors. Cost control information available to the jobsite after the activity is completed is not as valuable in project management as is current information.

The following section describes the types of cost documentation that are taken during the construction of a project.

Labor

The field needs to report labor hours to the accounting system. Because this information needs to be specific for payroll information, the accounting office needs to know the number of hours attributable to each employee for established payment periods. For the cost control system, the labor hours and classification need to be reported for specific construction activities. The following example shows information for payroll purposes, submitted weekly:

EXAMPLE

Name	Soc. Sec. #	Classification	Rate	Hours
Smith, Robert	547-09-8905	Carp 1	$18.57	40

This information is used by accounting to determine the amount of the payroll check, applicable fringe benefits, insurance premiums, and taxes for each employee.

The next example shows additional information that is needed for the cost control system.

EXAMPLE

Name	Hrly. Cost	Activity	Hours	Quantity
Smith, Robert	$25.40	03101-03	16	2,000 SFCA
	$25.40	03101-05	24	5,000 SFCA

Hourly cost is the cost to the contractor, including wages, fringe benefits, insurance, and taxes. The activity number is a cost code for a specific construction activity. In this case, the construction activity would be concrete forming, with two different form types or locations. The quantity would indicate the amount of production related to the hours worked.

Material

Prior to beginning project construction, the contractor normally prepares a bill of materials that lists the types of material, the amount, the vendor, and the cost. The bill of material also should relate the material to the specific construction activity. Purchase orders are generated from the bill of material. A purchase order is an agreement to purchase items from a vendor and may include several different materials and activity codes and multiple delivery dates. The purchase order is used for ordering material and for payment by accounting. The following example shows a bill of material for concrete formwork, footings.

EXAMPLE

Code	Material	Quantity	Vendor	Cost	P.O.#
03100-01	2 × 6 F&L, 2+	3,000 BF	Acme Lumber	$1,500	294-03
	1 × 2 stakes × 18″	500 ea	Acme Lumber	$ 100	294-03
	2″ Waterstop	1500 LF	U. S. Concrete Co.	$ 750	294-04

Some computerized estimating software produces a bill of materials that automatically combines all activity code items for the material classification, similar to the bill of material that follows:

EXAMPLE

Description	Takeoff Qty.	Order Qty.	Unit Price	Amount
Carpentry-Lumber:				
2 × 4 × 8 Standard & Better	45.00 each	.26 MBF	$475/MBF	$123.50
2 × 4 × 10 Standard & Better	21.00 each	.15 MBF	$475/MBF	$ 71.25
2 × 4 × 12 Standard & Better	34.00 each	.29 MBF	$475/MBF	$137.75

This information also can be sorted by vendor to provide information for purchase orders.

The purchase order, because it is a purchase agreement, details the terms of the agreement and references the construction documents as necessary. The purchase order also specifies the terms of payment for material. It cites the date of the bid or pricing and who made the offer. A list of the material, its cost, its anticipated delivery date, and the activity code for the contractor's code also should be provided. In addition to this information, other particulars may be included, such as the catalog or inventory number and the sales tax amount. Further information on the bill of material and purchase orders can be found in chapter 9. The following example shows a list of materials for a purchase order:

EXAMPLE

Code	Material	Quant.	Unit Price	Cost	Delivery Date
03100-01	2 × 6 F&L, 2+	3,000 BF	$500/MBF	$1,500.00	5/22/96
03100-01	1 × 2 stakes × 18″	500 EA	$ 20/C	$ 100.00	5/22/96
03100-04	2 × 4 F&L, 2+	10,000 BF	$450/MBF	$4,500.00	5/29/96
Total P. O. 294-03				$6,100.00	

Equipment

The contractor's owned equipment has certain rates associated with it, reflecting the computed cost of investment, fuel, repair costs, and miscellaneous costs, recognizing that the equipment does not work full-time. Reporting the hours of usage then relates to the amount of time the equipment is actually being used for its intended tasks. As described previously in the daily report, it is important to record the presence of the equipment and its immediate availability on the jobsite.

Owned equipment at the established internal rental rate is charged to the actual work activities. Operating labor is usually not included in the rate because the labor reporting should be done separately from the equipment reporting. The following is a weekly report on the activities of a Caterpillar D-7N bulldozer on the jobsite.

EXAMPLE

Act. Code	Description	Rate	Hours
02210-01	SITE GRADING, GRID A2-G4	$63.40	16
02210-02	SITE GRADING, GRID G5-J10	$63.40	16
02210-03	SITE GRADING, GRID J10-K12	$63.40	8

Some equipment is rented from outside sources. In this case, the rental cost and miscellaneous costs, such as fuel, should be invoiced to accounting for payment, with cost information going to cost control accounts.

Correspondence

Correspondence with other parties in the construction process is a major form of documentation of the construction project. Each construction contract and agreement requires that all requests, directions, and changes be in written form. Although legal agreements can be made orally and confirmed by a handshake, current construction customs dictate that all discussions be recorded, or written, to provide proof of the existence of such an agreement. The construction process is complex and can extend into several years, requiring reliance on written documents, rather than on human memory.

All correspondence should be well-crafted by the author. Written correspondence needs to accurately convey the message completely, while being as brief as possible. Care must be taken by the author to avoid misinterpretation of the intent or content of the message. The correspondence probably will be reviewed by people other than the addressee, thus a clear, objective statement of all of the information is required in the correspondence. Humor, emotional appeals, threats, and strong language are inappropriate for the majority of construction correspondence. These elements can be easily misconstrued by the reader who may attach an unintended meaning to the text. The writer must always be polite and professional, despite the emotion that may be generated by an incident.

Most construction correspondence, with the exception of transmittals, address specific issues or problems encountered during the project. The following guidelines can be used for construction correspondence.

1. **Reference:** Reference the project name and number. When writing to the architect and owner, the contractor should list both the official project number and the contractor's project number. Further reference may be added to a letter, RFI, particular issue, or other previous correspondence.

EXAMPLE

Re: Construction of New City Plaza; Project No. 98-001-NCP
     FGH Construction Project No: 2-98
     RFI from FGH Construction Co., dated 5/28/98
     Unsuitable Soil at Building B

2. **Description and Location:** For problem resolution, it is essential that the situation be described exactly, without confusion. Many times an incident can cause several problems that need to be solved independently. The exact location of the problem needs to be detailed as accurately as possible. It is not uncommon for the same problem to occur in more than one location on the project, which requires an accurate description of the location for each instance. Detailed information about a particular problem normally will expedite its solution. A description of the condition, references to drawings and specifications, quantities, and even pricing information are appropriate when explaining the problem. Withholding information as a strategy to avoid showing the firm's full position usually confuses and lengthens the discussion, turning it into a dispute.

EXAMPLE

As per our RFI dated 5/28/98, faxed to your office, our excavator located solid rock and rock described as solid rock, (larger than a $\frac{1}{2}$ yard bucket), in the rock clause, Section 02200, paragraph 4, page 2-2-1 of the Project Manual, on May 28, 1998. This solid rock appears at elevation 1556.0 at grid G-0 and appears to extend to about Grid E to the north and Grid 4 to the east, at Building B. The bottom-of-slab elevation in this area is intended to be 1550.0. Test pits in the area of Building B do not indicate solid rock. Elevation 1556.0 is below the bottom of the nearest test pits, B-11 and B-2. We estimate the rock to be 1,068 cubic yards in quantity, however, the exact amount of rock will vary from this amount, due to the irregular rock formation, which will be revealed during the rock removal.

3. **Objective of Correspondence:** Without clearly stating the objective of the correspondence, the description of an occurrence could be ambiguous. The writer needs to state, with certainty, his or her position and justification and what is desired. Some individuals try to avoid being direct, in an effort to be polite. Being direct is not impolite, but rather a route to resolution. If a monetary settlement is sought, the amount, or at least the method of payment, should be specified.

EXAMPLE

FGH Construction Company feels that the removal of this material was not required in the contract documents and that we should be compensated for the removal of the rock material, as per the specified rock removal rate, as per our proposal of 2/23/98, in the amount of $114/bank cubic yard. Using an approximate quantity of 1,068 bank cubic yards, the resulting addition to the contract would be $121,752 for the removal of this rock. We suggest that your on-site inspector accurately measure the exact quantity of the rock excavation, with our surveyor's assistance, as excavation progresses to provide an accurate estimate of the rock quantity.

4. **Summary:** The summary of the letter should clearly indicate what has happened or what is expected to happen. Most decisions or actions, within contractual confines, have further actions associated with them. If the construction firm feels they deserve extra compensation for additional work, they should request a change order. If the architectural firm rejects a request for a change order, they normally will request that the work be done at no additional cost. When further action is requested, usually a stipulated time restraint is added as well.

EXAMPLE

FGH Construction Company is hereby requesting a change order to the contract for the removal of the solid rock encountered in Building B. We feel that this contract change order should be compensated for the amount of rock removed, inspected and measured by your inspector and our surveyor, at a previously negotiated price of $114/bank cubic yard. Excavation work can continue for a brief period of time before we are delayed by a decision. Please notify us on June 1, 1998, of your decision for this additional work.

5. **Signature:** Every piece of correspondence should be signed by the originator of the document. Some correspondence, such as a request for a change to the contract, has significant contract implications and should be signed by the authorized party.

The construction contractor uses a number of vehicles to correspond with other parties in the construction process, depending on its purpose. Some of the most common forms of correspondence are described next.

Letter of Transmittal

The letter of transmittal is a dated record of when a particular document or item was sent to another party. It is necessary proof that a particular piece of information was sent from one party to another, but it is not conclusive proof that the item was received. To ensure receipt, the contractor should request that the document be sent by certified mail, with a return card showing the signature of the receiving party. A wide variety of couriers and delivery services that obtain signatures of receipt are available. Signed receipts usually are necessary for important documents, such as contracts; time sensitive documents, such as change order documents or payment documents; security items, such as keys, and valuable items.

The letter of transmittal usually contains the minimum following elements:

- name and address of sender
- name and address of intended receiver
- project number (both project numbers, if applicable)
- item sent (full description)
- number of copies of the item
- notes relating to the item

Preprinted forms are available for transmittals. Special forms are available for contractors, containing boxes for the most commonly sent items, such as submittals, shop drawings, contracts, change orders, and so on; what they are being sent for (review, recipient's use, etc.); and what should be done with them, such as "return four approved copies." Prepared from templates or form-making software are available for word processors and computers. These forms are convenient to fill out at the computer.

Fax transmittals are common in construction correspondence. As the fax message is instantly received rather than being received several days later, it is currently used for most construction correspondence. Fax transmittals also can be customized for the contract's specific needs. Many businesses send faxes directly from their computers, with a computer-generated fax transmittal sheet.

Request For Information (RFI)

The contractor has numerous questions throughout a project concerning documents, construction, materials, and numerous other items. Traditionally, a phone call or informal conversation with the architect solved the problem. It is now necessary, however, to document every request and reply. Extra costs or complications can arise during the process, requiring use of the RFI to substantiate a project's direction.

RFIs have a number of uses, several of which do not have cost implications. The request for clarification can be a non-change item and can merely require information about something shown in the documents. It can be a simple query to the architect regarding a particular in the construction document. Initially, the RFI normally just asks for clarification. It also can be a notice to the architect about an occurrence or knowledge of an occurrence, as well as a notice of latent, or unknown conditions.

Most RFIs are brief questions requesting clarification. It is important, however, that the RFI contain all of the necessary information and not be too brief. If the contractor is requesting a clarification, the architect needs to know exactly what is unclear. If multiple issues are contained in the RFI, then each one should be clearly delineated, requesting a reply for each one. When asking for clarification, all relative facts should be presented. Complete and honest representation of conditions and factors will result in timely and fair responses.

The guidelines for correspondence, mentioned earlier, apply to the RFI, although the RFI is usually considered a short communication. It is important that the RFI clearly state the problem and define what type of response is desired. Each RFI is numbered in the sequence issued, the number being used as a reference. Coordination among members of the jobsite team is necessary to avoid double-numbering or leaving a gap between actual RFIs. The RFI is a time-sensitive document that requires immediate information. The date the RFI was sent should be included, as a dated response is needed.

The RFI (sometimes called the Request for Clarification) also is available in preprinted form. Most contractors custom-prepare their forms and include specific information relevant to their business. Multiple copy forms, using carbon paper or NCR forms, can be used. A template form can be prepared for the computer, which facilitates communication. Some integrated document management programs, such as Expedition (discussed in chapter 12), include a form for RFIs. RFIs are commonly faxed to parties to speed communications. Most RFI forms use the top half of the sheet for the request and the bottom half for the reply. Figure 4–5 illustrates a sample blank RFI Form, with the completed information shown in Figure 4–6.

FIGURE 4–5
Example Request for
Information Form
(Blank)

FGH CONSTRUCTION CO.
REQUEST FOR INFORMATION

PROJECT: PROJECT NO.   RFI #   DATE

TO: FROM:
METHOD SENT: FAX___   MAIL____ COURIER ____
COPIES SENT TO:
INITIATED BY:
DESCRIPTION OF REQUEST:

ADDITIONAL SUPPORT DOCUMENTS ARE ATTACHED.
RESPONSE NEEDED BY:_____

_____
_____
RESPONSE:

BY: DATE:
FIRM:

FGH CONSTRUCTION CO.
REQUEST FOR INFORMATION

PROJECT: New City Plaza   **PROJECT NO.** 2-98 **RFI#**082   **DATE**August 21,1998
                                                            98-001-NCP

TO: R. G. Smith   **FROM:** F.W. Johnson

      Smith Architects   FGH Construction Co.

      P.O. Box 2334   P.O. Box 2356

      Atascadero, CA 93784   New City, CA 93209

      FAX: (805) 834-0987   FAX: (805) 342-7654

METHOD SENT: **FAX__x_**   **MAIL____**   **COURIER ____**

COPIES SENT TO: City Painting

INITIATED BY: F.W. Johnson

DESCRIPTION OF REQUEST:

The Finish Schedule shows the north wall of Room 245, Building A, to be painted Canary Yellow. The drawings, on sheet A-22 , detail 4, note that the color of the north wall in 245 is to be "Robin's Egg Blue". The painter will be in this area on 8/23/98, and we need to know the proper color to get the paint mixed.

ADDITIONAL SUPPORT DOCUMENTS ARE ATTACHED.
RESPONSE NEEDED BY: _____8/22/98 12:00 pm_____

RESPONSE:

The color of the North wall in Room 245, Building A, is intended to be "Robin's Egg Blue". Please submit a sample of this paint prior to application.

We are assuming that this is a no-cost clarification. Please change the notation in the Finish Schedule to Robin's Egg Blue.

BY:  R.G. Smith   **DATE:** August 21,1998 3:00 PM

FIRM: Smith Architects   Sent via FAX

FIGURE 4–6   Example Request For Information Form (Completed)

An increasing belief in the industry is that the contractor should be compensated for time spent processing RFIs, as the majority of them relate to errors, omissions, and conflicting information in the construction documents. Further, many contractors feel that the mere number of RFIs indicates incompetent completion of the documents that warrants an award for damages. Most of these cases need to be determined in the courts. The RFI should be used as a request for further information or clarification, rather than as a vehicle for a claim. Documentation of the clarification is necessary to eliminate unnecessary disputes.

Letters

Letters are used to provide and obtain information, request action, reply to a request for action, or present an explanation. They are used to encourage activity by parties in the construction process as well. The telephone call is the quickest form of communication, whereas the letter is probably the most powerful. A letter, as a valid piece of documentation, can make demands or state facts more strongly, thus having more of an impact than a telephone conversation. Both forms of communication should be used to obtain results, but obviously the letter is more time-consuming and expensive than the telephone.

The previously mentioned guidelines for correspondence also apply to letter writing. The letter should address a specific issue, using standard language in most cases. Inconsistencies and unnecessary language can result in the letter not being taken seriously by the reader. Form letters, although appropriate in some situations, always read as such and do not get the recipient's attention. There are situations, however, where mail-merging functions in the computer can personalize a form letter, avoiding the tedious task of writing individual letters. In the example that follows, the contractor is sending letters to all of his subcontractors requesting shop drawings or submittals. In the database, the contractor has included the subcontractors' names and addresses, the type of submittal they are required to submit, and the date the submittal is due.

EXAMPLE

Database Listing:

First/Last

Name	Firm	Address	City	Submittal	Date
A. C. Jones	AC Acoustical	P.O. Box 3456	New City, CA	Ceiling drawings	6/10/98
Fred Nelson	Nelson Tile	P.O. Box 3478	New City, CA	Ceramic tile sample	6/01/98
Dave Anders	Anders Plbg.	P.O. Box 1233	New City, CA	Plumbing submittal	5/14/98

A letter to all of the above subcontractors can be written using the mail-merge function in the word processor. The items inserted from the database are underlined in the sample letter shown in Figure 4–7.

Several types of letters are used in the construction industry, the most familiar being the standard letter form on company letterhead. Numerous other forms, such as speed letters with multiple copies and carbons and

FGH CONSTRUCTION COMPANY
P.O. BOX 3888
NEW CITY, CA 93478
(805) 444-9900
FAX: (805) 444-9901

May 1, 1998

A.C. Jones
AC Acoustical
P.O. Box 3456
New City, CA

RE: New City Plaza, New City, CA
FGH Project Number: 2-98

Dear Mr. Jones:

FGH Construction has now established our submittal schedule. Your firm is required to submit ceiling drawings in compliance with the shop drawing and submittal requirements for this contract. We need these submittals in our office on 6/10/98.

Please contact John Johnson at the jobsite, (805) 342-9387, if you have any questions on your requirements.

Sincerely,

O. E. Olsen

O.E. Olsen
Project Manager

FIGURE 4–7   Example Mail-merge Letter

space for a reply are frequently used for short notes. Regardless of the form, all correspondence needs to be filed and retained. The filing system may reflect the particulars of outgoing correspondence, such as who is writing the letter, or the issue that is involved. Whatever the filing method, it should be consistent for document retrieval, as needed. Integrated document systems allow immediate retrieval of the document, or parts of the document, by issue.

E-mail

E-Mail is becoming more popular for business correspondence, as well as for personal correspondence. It is instantly available and convenient for both the sender and the receiver. However, the same considerations for other types of communications apply to E-Mail. E-Mail often is a short, abbreviated message, but it can be a full letter as well, using the guidelines previously discussed. Software is available for sending a variety of documents, including schedules, via a modem. Due to the fragile nature of computer hardware, all electronic correspondence should be copied to avoid the risk of losing documents.

Contractual Requirement Documentation

Numerous pieces of documentation of the construction process are required by contract documents. Most follow a required standard format, whether on a form or not. These documents, while conveying a particular purpose within, provide a record of the activities of the project. Progress Payment Requests are specifically formatted documents the contractor uses to show the amount earned during the construction period. Progress Payment Requests also provide a record of the contractor's progress. A list of the types of contractually required documents follows.

Project Start-Up:
Construction Schedule
Schedule of Values
List of Subcontractors
Product Submittals, Shop Drawings, and Samples
Construction Progress:
Progress Payment Requests
Construction Schedule Updates
Certified Payrolls
Contract Change Orders: Proposals, Directives, Change Orders
Project Closeout:
Certificate of Substantial Completion
Lien Releases
Consent of Surety
Warranties

Operation and Maintenance Manuals
Spare Parts
Punch Lists

Meeting Minutes

Meetings are frequent occurrences during the construction process. A meeting provides several people the opportunity to discuss current matters of concern about the project, face-to-face. The construction meeting is formal, with an agenda, a leader, and some structure. Different types of meetings during the construction process include: partnering meetings and workshops, pre-construction meetings, progress meetings, subcontractor meetings, special installation meetings, schedule coordination meetings, safety meetings, and post-construction meetings. All have various purposes, and thus have different participants and a unique atmosphere. (See chapter 6 for a further discussion about these meetings.)

Because important discussion occurs in the aforementioned meetings, careful minutes should be taken to reflect the conversation and any resolution of issues. Minutes do not provide an exact record of the meeting, but an accurate summary of the major points and who made them. They are as objective and truthful as possible. Because topics are summarized, redundant discussion, trivial points, or long explanations are avoided.

The minutes-taker has a certain amount of control of the project. The primary goal when taking minutes is to be as objective as possible. Even the most objective minutes-taker will be biased to some extent. Most individuals who participate in project meetings view taking minutes as a burden, rather than an opportunity. Recording, editing, writing, and distributing the minutes does take extra effort but, as stated earlier, can result in subtle control advantages.

Occasionally the minutes-taker does not accurately report the action that occurred at the meeting. If this should happen, any objections should be sent to the recorder and discussed in subsequent meeting. When the minutes are consistently inaccurate, concerned individuals should keep their own notes and write and distribute an alternative set of minutes.

Progress meetings are the most common during the construction phase of the project. The elements that follow are key to progress meeting minutes, which also are similar for other meeting types.

Title.   The title should address the type of document, type of meeting, date of the meeting, and sequential number of the meeting. The time and location of the meeting also should be included.

EXAMPLE

Minutes of the Meeting
Construction Progress Meeting
Meeting #4
May 6, 1998, 9:00 A.M.
FGH Construction Co. Conference Room

Project Designation.   Project name, project number (may be several numbers, as in the other documents).

EXAMPLE

Project: New City Plaza, New City, CA Project No. FGH Const #2-98
Smith Architects # 98-001-NCP

Parties in Attendance.   There are two ways to record the parties in attendance at a meeting. The first method is to list the parties at the meeting. The minutes-taker should note who is present and the time individuals arrive or leave. The list of individuals also should include their firms. This method ensures that all parties are recorded, as long as the recorder is familiar with all meeting participants.

EXAMPLE

Parties in Attendance: Thomas Thompson, New City Plaza Development Corp.; R. G. Smith, Fred Stone, Smith Architects; O. E. Olsen, John Johnson, FGH Construction; George Roth, City Plumbing (arrived at 9:45 A.M.); Charlotte Smith, State Electric; A. C. Jones, AC Acoustical (left at 9:30 A.M.).

The second method is to provide an attendance roster or sign-up sheet for all meeting participants. This method documents attendance in the event that a dispute later arises. It does not, however, ensure that all attendees will sign the sheet, particularly if they arrive late. The minutes-taker should follow up to obtain the signatures of all parties listed on the roster.

EXAMPLE

Parties in Attendance: See Attached Attendance Roster (Figure 4–8).

Minutes From the Previous Meeting.   This section indicates any additions or corrections to the previous meeting. Usually there are none, which also should be noted. This section may seem like a waste of time, but it provides corrections that complete the validity of the minutes as a viable project document.

EXAMPLE A

Minutes from the Previous Meeting: No corrections or additions were made to the meeting of April 28, 1998.

ATTENDANCE ROSTER

PROJECT: NEW CITY PLAZA, NEW CITY, CA

MEETING DATE AND TIME: May 6, 1998, 9:00 AM

MEETING LOCATION: FGH Construction Conference Room

Parties in Attendance:

NAME	FIRM	PHONE #	FAX #
O.E. Olsen	FGH Const.	444-9900	444-9901
Bob Smith	FGH	456-0234	456-0235
R.G. Smith	Smith Architects	834-0986	834-0987
F.W. Johnson	FGH	456-0234	456-0235
A.C. Jones	A.C. Acoustical	325-9712	325-9900
Ed Roth	City Plumb.	225-0962	225-0960

FIGURE 4–8   Example Meeting Attendance Roster

EXAMPLE B

Minutes from the Previous Meeting: George Roth noted that he had stated that City Plumbing was complete with the underslab plumbing rough-in in Building A and just starting the rough-in in Building B, not "complete with the rough-in in Buildings A and B," as shown in the minutes.

No other corrections or additions were made.

Project Progress.   This section can be handled in a number of different ways, in a full section or in an items of business section. One of the most important elements of a construction progress meeting is the update of current construction progress. This is an item of interest to the parties attending the meeting and to those reading the minutes who are not able to attend the meeting, thus creating a record of the project's progress.

EXAMPLE

Project Progress:

John Johnson, FGH Construction Co., updated the progress of the project. The slab-on-grade for Building A was poured on May 4, 1998. They are now starting forming for the tilt-up panels. Tilt-up drawings have been approved for Building A. Tilt-up drawings for Building B will be submitted on May 10, 1998. All of the reinforcing steel for Building A has now been delivered to the jobsite. Excavation has started for Building B. Johnson's only concerns are about some of the submittals (next section).

George Roth, City Plumbing, said that they have completed the underslab rough-in and the water service to Building A. They are currently furnishing and installing blockouts and sleeves in tilt-up panels. The underslab rough-in in Building B has started and can't go much further until the excavation is complete. They have installed the water line and sewer line to Building B.

Charlotte Smith, State Electric, stated that they are currently installing the rough-in in the tilt-up panels. As there isn't much for their crew to do in connection with the building, the electricians will start work on the underground power service line to the complex tomorrow (5/7/98).

Submittals.   During the first half of the project, a primary concern is processing submittals, samples, and shop drawings. The timely processing of these items relates to the release for ordering and fabrication of the material and equipment. Discussion about submittals normally occurs in a construction progress meeting. In addition to the discussion on changes in submittals, a submittal log might be attached to the minutes to concerned parties to indicate the status of any submittals.

EXAMPLE

Submittals:

The mechanical submittal in total has been submitted, reviewed, and returned to FGH Construction. AC Acoustical submitted the ceiling drawings at this meeting to FGH Construction. They will be forwarded to Smith Architects immediately, following review by FGH Construction.

The Submittal Log Is Attached:

Section #	Item	To FGH	To Smith	To FGH
03200	Rebar Dwgs.	4/15/98	4/17/98	4/28/98
03350	Tilt-up Dwgs.	4/20/98	4/28/98	5/4/98
05100	Str. Stl. Dwgs.	5/1/98	5/6/98	
06400	Cabinet Dwgs.			
07400	Roofing Sub.			
08200	HM Doors	4/10/98	4/11/98	4/15/98
08700	Fin. Hdwe.	4/10/98	4/11/98	4/15/98
09300	Cer.Tile			
09500	Acoust. Clgs.	5/6/98		
09650	Carpet			
09900	Painting			
15000	Mechanical	4/10/98	4/11/98	5/4/98
16000	Electrical	4/28/98	5/1/98	

More information might be included in the submittal log, such as scheduled dates, delivery schedule, actual delivery date, activity relationship, and start of activity. (For a broader discussion about the submittal or procurement schedule, see chapter 3.)

Change Orders

The progress of change orders is a major project issue for both the contractor and subcontractor, who are interested in the authorization of the change order, so work can proceed and in the final approval of the change order, so work can be billed on the monthly progress payment request. The current status of change orders should be discussed in the progress meeting. A log of the current progress of change orders, attached or included in the minutes, prevents many problems when answering phone calls for subcontractors who are inquiring about their status. (See chapter 13 for a more complete discussion about change orders.)

EXAMPLE

Change Orders:
Change Proposal 06, "Additional floor hardener, Bldg B" has been submitted and approved, directive issued. It will be combined with other items in C. O. 2.
The change order log is attached:

C.P.#	Descrip.	Submit	Approve	Direct.	C.O.#	Date
01	Brick type	3/28/98	4/2/98	4/2/98	1	4/15/98
02	Drain Tile	4/1/98	4/3/98	4/3/98	1	4/15/98
03	Water PR Valve	5/5/98	Resubmit			
04	Sprinkler Cont.	4/5/98	4/16/98	4/18/98	2	5/6/98
05	Lt. Fixt. F-12	4/15/98	4/28/98		2	5/6/98
06	Hardener, B	4/28/98	5/5/98	5/5/98	2	5/6/98

The distribution of minutes goes beyond the contractual line among the owner, architect, and contractor. It should be noted that the amount of the change orders is not included in the minutes and is basically confidential information. Depending on the change order structure, other items may be included in the log as well.

Old Business

The old and new business areas relate to discussions about project issues. Each item of business should be numbered (with the meeting number and item number). This system indicates the age of the item in old business. All items that have not yet been resolved should be included here.

EXAMPLE

Old Business:
298-02-03:   Water Pressure Reducing Valve: The price on the pressure reducing valve included in Change Proposal 03 is unacceptable to the owner. The mechanical engineer has suggested pricing an "ACME" valve, #304, and resubmitting.

EXAMPLE
continued

298-03-01: Power Interruption: The owner still needs to know the date of the power interruption with the underground connection. Charlotte Smith, State Electric, stated that the connection should be made in the next two weeks and will establish a date with the maintenance staff.

New Business

All new business should contain the following information: who introduced the problem; a description of the problem; who will take action; what action will be taken; and when the action will be taken. Common practice at construction progress meetings is for any individual present to introduce items.

EXAMPLE

New Business:
298-04-01: R. G. Smith, Smith Architects, mentioned a concern about proper curing on the slab on grade. John Johnson stated that a curing compound was applied the day after the slab was poured. Smith agreed that the curing compound was adequate. No further action necessary.

298-04-02: John Johnson, FGH Construction, mentioned that the excavator found what appears to be solid rock in Building B. He will notify the architect by RFI if there is a significant rock problem, probably by 5/8/98.

298-04-03: A. C. Jones, AC Acoustical, asked if ¼" scale was adequate for the ceiling shop drawings. Smith indicated that it was adequate. Jones submitted the shop drawings to FGH Construction at the meeting.

Meeting Adjourned, Next Meeting

The minutes should indicate the time the meeting was adjourned and should announce the next regularly scheduled meeting. Including the meeting date in the minutes, assuming that the minutes are issued immediately, saves an additional meeting announcement.

EXAMPLE

Meeting adjourned at 10:45 A.M.
Next Meeting: May 14, 1998, 9:00 A.M., FGH Construction Conference Room

Meeting minutes are traditionally distributed to those in attendance and the three major participants in the project. Normally the architect is responsible for forwarding copies to subconsultants and the contractor for forwarding copies to subcontractors. The contractor should carefully review the contents of the minutes and send them to the affected subcontractors, if a blanket mailing is not standard.

Immediate distribution of the minutes minimizes confusion. Minutes should be read, noted, and corrected, if necessary, and kept on file for future reference. Meeting minutes are among the most important documentation, indicating the direction of the construction project.

Summary

Project documentation is an essential function of project administration. Several different areas need to be documented.

- Events, occurrences, and conversations—documented by daily reports, weekly reports, monthly reports, diaries, telephone and other logs, accident reports, photographs, time-lapse photography, videotape recordings, and schedules.
- Costs—documented by purchase orders, subcontracts, material reports, labor reports, and equipment reports.
- Correspondence—documented by transmittals, Request For Information (RFIs), and letters.
- Contractual requirements—documented by payment requests, schedules of values, change orders, punch lists, and certificates of substantial completion.
- Meetings—documented by meeting minutes.

Documentation should be honest, accurate, complete, and usually sequential. Careful documentation, considering all aspects of the events, helps the contractor solve problems on the project and facilitate budget and duration goals.

CHAPTER 5

JOBSITE LAYOUT AND CONTROL

Jobsite organization is essential for a productive construction project. Because estimated profit margins are small, the relative efficiency of the construction site can influence the profitability of the project. The jobsite layout affects the cost of material handling, labor, and the use of major equipment by the general contractor and the subcontractors working on the site. A well-organized jobsite has a positive effect on the productivity of the entire jobsite workforce. Productivity and worker morale are optimized by an effective jobsite layout. An efficient jobsite is a good indication of high-quality professional management by the contractor.

The **jobsite layout plan** is a plan for temporary facilities, material movement, material storage, and material handling equipment on the jobsite. It is similar to the construction plan and schedule and is a long-term system that considers all factors of the construction project. Although the jobsite layout plan is planned, funded, and implemented by the general contractor, it should consider the needs and requirements of all of the subcontractors working on the site. Some of the provisions of jobsite implementation, such as transportation of materials, are often the financial responsibility of subcontractors, but should be included in the overall plan established by the general contractor.

The jobsite layout plan includes the following aspects:

- Jobsite space allocation—areas on the jobsite for material delivery, material storage, temporary offices and facilities
- Jobsite access—access to and from the jobsite and to work areas within the jobsite, including haul roads
- Material handling—including material movement on the jobsite, both horizontally and vertically; lifting equipment, including fork lifts and cranes
- Worker transportation—personnel movement and access on the jobsite
- Temporary facilities—temporary offices, storage facilities, dry shacks, sanitary facilities, temporary water, power, and heat
- Jobsite security—temporary fencing, guard dogs, security patrols, electronic alarm systems, and watchmen
- Signage and barricades—protection of the public from construction hazards on the jobsite.

Based on the aforementioned aspects of the jobsite layout plan, the following four areas should be taken into consideration:

1. Material Handling
2. Labor Productivity
3. Equipment Constraints
4. Site Constraints

All of these areas have considerable impact on the decisions made concerning the jobsite layout plan. An optimum jobsite layout plan minimizes on-site labor in material movement, travel, and transportation, allowing workers to spend the maximum amount of time possible performing their construction activities. If a worker has more time to spend on the work activity, it follows that more production will be accomplished during the work shift. A well-organized jobsite provides a working environment that is conducive to completing work tasks for all jobsite personnel, including direct and subcontractor employees.

Material Handling

Efficient material handling is extremely important in optimizing the construction worker's productivity. Ideally, the necessary material should be within reach for the craftsperson as it is needed for the installation. Obviously, this rarely happens. Material could be stored on site in a remote location, it could be delivered as it is used, or it might not even be available on the jobsite. The primary concern is to have the craftsperson complete the construction activity as quickly as possible.

Guidelines for material handling on the jobsite follow.

- *Always move material with the least expensive labor possible.* Normally, the lowest-paid member of the crew, such as a laborer, apprentice, or helper, should be responsible for moving material. The craftsperson, such as the carpenter, is paid a higher wage to do a specific type of work. The crews should be organized to avoid using the journeyman craftsman for moving material.

EXAMPLE

A wood framing crew consists of two journeyman carpenters, a first-year carpenter apprentice, and a laborer. The laborer and carpenter apprentice will move the lumber and equipment where necessary and the two journeyman carpenters will do the framing. Because the laborer and carpenter apprentice are not both needed 100 percent of the time to move material and equipment, the carpenter apprentice can assist the carpenters with the framing when not moving lumber and equipment. The laborer can collect scrap material and clean the worksite when not moving material and equipment.

- *Deliver material as close as possible to the location of installation.* When possible, material should be stored near the installation location, making it ready for installation without much extra labor. This is a common practice when gypsum drywall is "stocked" on the jobsite. The drywall usually is initially distributed to the different floors or locations where installation will occur.
- *Deliver the material to its location with delivery people.* Material delivery to the jobsite and even to the location within the jobsite is most economically done by delivery people employed by the supplier of the material. Although there is normally a delivery charge, the material supplier's delivery crew probably will receive a lower wage than the

jobsite labor. The delivery crew knows how to handle material efficiently and usually is interested in the prompt delivery of material. The delivery truck often comes with a boom or fork attachment that assists in the placement of the material. For example, drywall is normally delivered and "stocked" by the delivery crew. The crew will, as directed, put portions of the order in specific rooms and on specific floors. They also can use the fork boom on the delivery truck to deliver drywall to a second floor opening, if desired.

Within unionized jurisdictions, however, delivery people usually will deliver to the jobsite, but not within the jobsite, as they usually are not union employees. Union laborers would claim material movement on the jobsite. Under union jurisdictions, only certain union craftspeople can move material or equipment on the jobsite. Usually, laborers will have the authority to move material, while operating engineers run the lifting equipment. Some trades, however, move their own materials, depending upon the language in the agreements. To avoid jurisdictional disputes, the contractor should realize which craft claims each of the tasks on the jobsite.

- *Deliver from the truck to the installation location, if possible.* There are some instances when delivery can be made from the delivery truck to the material installation location. This is often done with structural steel. The steel fabrication, such as a column assembly, is lifted by the tower crane directly to the installation location and immediately bolted into place. Large equipment, such as air handlers and fans, often is lifted directly from the truck to its ultimate destination. This method avoids on-site storage of material and is utilized extensively when little or no on-site storage is available. Extra coordination is needed, however, to ensure that the correct fabrication is available and shipped at the time needed and the installation location is ready for the material or equipment upon its arrival.

- *Avoid moving the material more than once.* When material is stored or stockpiled on site, it should go from the storage area to installation. Too often on construction sites a storage pile of material has to be moved to another location because of other storage needs, providing access to work or excavation, or for any other number of reasons. Material piles rarely need to be moved if prior planning is done. Anticipation of the work schedule and delivery of material can help create a storage plan that does not need to be changed during the project. Material that is placed haphazardly on the jobsite usually results in relocation of the material prior to its installation.

- *Anticipate equipment needs for the entire project and ensure that the proper equipment is available.* The contractor should anticipate the amount, size, and packaging method of the material and have the appropriate means of moving the material on the jobsite prior to delivery. Planning at the start of the project can provide adequate provisions for lifting equipment through the duration of the project. Permanent equipment should be selected for the majority of lifting and material handling. Some equipment needs may be temporary and require special machinery such as a mobile crane, for a few hours or days, due to the size, location, or amount of material delivered to the jobsite. Some lifting equipment will be the subcontractors' responsibility, such as the crane for erecting structural steel or for lifting a chiller unit to the roof of the building. Although the subcontractor is responsible for the cost of the lifting equipment, the general contractor should provide a location for the lifting equipment and for the delivery of the equipment. The contractor must consider all of the material that is to be installed on the project, including that furnished by subcontractors, when establishing a jobsite plan.

- *Select the optimum equipment for moving the material.* There are many ways to move and lift material into place. Most construction lifts and moves can be performed manually rather than by using a crane, forklift, or other method. Cost, of course, is the prime consideration, but a number of other factors should be considered. The following aspects should be considered when analyzing how to use equipment.
    1. Cost: The rental cost of the necessary equipment, including minimum rental (four hours, eight hours, etc.) time and travel time to and from the jobsite should be considered. All necessary accessories, such as rigging and chokers, should also be taken into account.
    2. Availability: The availability of lifting equipment within the community. A one-time lift may appear to require a piece of equipment that is not readily available within the community. Any piece of equipment can be shipped to a community, but the shipping costs may greatly exceed the cost of an alternative. Sometimes, for small lifts, a gin pole with a block and tackle can replace a crane that is not available in the area. For example, when setting a precast bridge section, a contractor needs a large mobile crane to set the precast section. The mobilization cost to bring the crane into the community is quite expensive. Thus, the contractor uses two smaller cranes, one on each side of the river, resulting in a quick, efficient pick for a much lower cost.
    3. Capacity of the equipment: The contractor needs to be sure the equipment has the appropriate capacity, including the safety factor to move and lift the equipment necessary. Cranes are selected according to the location of the pick and the designated target. Each piece of equipment has a certain capacity that should be taken into account with exact jobsite conditions. All of the qualities of the equipment need to be considered, including load capacity and the reach or size of the equipment.
    4. Safety: Safety is of paramount concern on the jobsite. Safety considerations always supersede cost issues. A more expensive method may have to be used to protect workers from a more risky one. A little more expense during the initial lift is a much better investment than a lift or move that endangers the lives or well-being of the workers.

5. Quantity of material: The relative quantity of material to be moved also is a factor when deciding what equipment to use. High mobilization costs can be absorbed if a large amount of material is to be moved. For small quantities, the contractor will probably use a labor-intensive method, rather than pay travel or setup fees for lifting or moving equipment.

6. Access to the point of use: Situations arise where equipment must be used because access is not close to the point of use. The most efficient and cost-effective way to pour concrete is to do it directly from the ready-mix truck into the concrete form. This is not always possible, as the form may be too far from the nearest truck access or above the pouring level of the truck, requiring alternative methods. Some of these methods might include a series of gravity chutes, a concrete pump, a bucket lifted by a crane, or possibly a conveyor. Having the space for delivery of material and sufficient room for trucks and vehicles to maneuver also should be considered. This analysis is made by considering the factors discussed in this section.

7. Provide adequate delivery routes. The route for delivery should be clearly marked from the main roads and streets to the jobsite entrance. Construction sites often are located in remote areas, but firm and well-maintained roads to the point of delivery are needed. The majority of delivery trucks to the construction site are semi-truck-trailer setups, requiring wide turns and standard roadway width. The roadway, even if it is temporary, needs to be firm and stable. (See Figure 5–1 for some indications of truck size and weight.) The jobsite entrance for deliveries should be clearly designated and internal site roads should be adequate for deliveries at the times they will be made.

8. Coordinate deliveries. Deliveries should be scheduled and coordinated, even though exact delivery times are not always available, particularly when the delivery is to be made from a long-distance hauler or common carrier. Because the typical jobsite will receive many deliveries during the day, scheduling the time of deliveries will ensure that equipment and personnel is available to promptly unload the truck. The contractor should prepare for the delivery by having available the appropriate amount and type of labor and equipment. Quick handling of the delivery facilities the optimum number of deliveries, as necessary, and also frees up labor and equipment to accomplish work tasks.

On jobsites with limited receiving facilities, schedules for the delivery area and lifting equipment should be established. Because a number of subcontractors require deliveries, certain times should be set aside for these. This schedule needs to be set up early in the project so the contractor and subcontractors can stipulate delivery restraints in their purchase orders. An example of a delivery schedule follows:

Delivery	Truck Type	Load Capacity	Truck Length	Truck Width	Truck Weight, loaded	Vertical Clearance	Remarks
Gravel	Dump Dump w/ pup	16 tons (8 to 9 CY) 30 tons (15 to 17 CY)	27 feet 63'-2"	8'-2" 8'-2"	53,180 lbs. 91,800 lbs.	9'-9" 9'-9"	"Pup" is trailer to dump truck
Concrete	Mixer truck	9 CY (36,458 lbs.)	40 feet	8'	66,000 lbs.	12'-0"	
Lumber	Tractor/trailer flatbed	48,000 lbs. 12 units lumber or 16 units plywood	16' tractor + 42' trailer = 58'	8'	80,000 lbs.	13' max.	Unload with forklift needs 16' clearance each side
Masonry	"Maxi" Semi w/ pup	CMU: 22 pallets, 1900 ea Brick: 14,000 ea	85 feet	8'	103,000 lbs.	13'-6" minimum	Forklift unload, 16' clearance each side Cannot back up (drive through)
Drywall	Flatbed w/boom	16,000 lbs. 160 sheets drywall 5/8" x 4' x 12'	30 feet	8'	48,000 lbs.	11'4"	Boom: 37' from roadway extended

FIGURE 5–1   Delivery Truck Size Data

EXAMPLE

7:00 A.M. to 10:00 A.M.:	General contractor deliveries
10:00 A.M. to 12:00 P.M.:	Mechanical contractor deliveries
1:00 P.M. to 2:00 P.M.:	Electrical deliveries
2:00 P.M. to 4:00 P.M.:	Miscellaneous subcontractor deliveries (varies, depending on the work being done)

A delivery schedule should be flexible to allow for special construction events and activities. Long-distance deliveries can arrive at any time and should be accommodated, if possible. The contractor should establish daily schedules and coordinate deliveries, in addition to the project delivery parameters. Because scheduling is a major item of concern on the jobsite, the superintendent or projected manager should make scheduling decisions and provide that information to their own forces and to the subcontractors.

- *Make arrangements for material or equipment to be placed in the installation location.* Certain materials or pieces of building equipment are too large to be taken through conventional routes to the area of installation at the time of delivery. The constructor should always try to place this equipment or material in the space provided prior to the installation of walls and doorways, which prevent the material from entering the space.

EXAMPLES

1. A large air handler, about 30 feet long, was designed for installation in the subgrade lower level (basement) of the building. Realizing the problem well before construction, the constructor pre-ordered the air handler and lifted it into place prior to the installation of the first floor slab. The air handler needed to be covered and protected during construction operations.

2. In a high rise building, transportation of gypsum drywall in the traditional 4' x 12' sheets is difficult after the installation of the exterior curtain wall. Moving large sheets by elevator also is difficult, requiring a large amount of labor, if that is even possible, depending upon the size of the elevator car. The easiest way to stock drywall on the floors is to use a tower crane or material lift prior to the installation of the exterior curtain wall. The drywall also needs to be covered and protected until its installation. The large size of the sheets may be in the way of some installations, such as ductwork, electrical rough-in, and suspended ceilings. Care should be taken when stocking drywall on elevated slabs to distribute the drywall to avoid overloading the slabs.

Equipment and material delivery cannot always be timed to arrive at the jobsite prior to the installation of constraining elements. When possible, the constructor can leave openings for the material or equipment to be moved into its assigned space. It also may be possible to have a piece of equipment broken down into smaller sections to enable it to be moved into the space.

EXAMPLES

1. The exterior precast concrete panels of a building have been installed prior to the delivery of a large exhaust fan on the upper floor. Provisions are made to omit a panel until the fan is delivered and lifted into place. The precast panel is then installed after the fan is lifted into place. Care must be taken to not interrupt the structural integrity of the exterior cladding. A structural relationship between the cladding panels may exist, requiring additional supports to accommodate a temporary opening.
2. An air handler is scheduled for a small upper floor mechanical room. The air handler, fully assembled, is too large to fit into the building's elevator and through the door into the mechanical room. The air handler is custom-made for the project, fully assembled and tested at the factory. It is scheduled to be delivered after the curtain wall is in place and the interior walls, including doors and frames, are installed. The air handler can be separated into sections that can be transported in the elevator car and through the door into the mechanical room. There is an additional charge from the manufacturer for breaking down the air handling unit. Extra cost is also incurred to reassemble the air handler in its place. An analysis and comparison of costs to create openings as well as to partially disassemble a piece of building equipment should be made to determine the most feasible alternative.

- *Storage of material on the jobsite should be systematic.* Most on-site storage areas are fairly small, requiring consolidation of material. The storage of material should be organized in a manner that is accessible. Deliveries are not necessarily made in sequential order for the material needed. An effective, well-organized storage area requires a knowledge of the construction schedule and of the materials used for construction.

    Unless well-organized, fabricated materials, such as reinforcing steel and structural steel, can become a serious storage and access problem at the jobsite. Literally thousands of pieces of reinforcing steel, even in fairly small projects, are delivered to the jobsite. The following steps can help organize the storage of material:

    1. *Coordinate shop drawings and fabrication with the supplier.* The contractor should work closely with the fabricator to establish a schedule of shop drawing production and fabrication that coincides with the construction activities. The sequence of the construction and the timing of the delivery of the material is not always obvious to the fabricator. For the fabricator, the ordering of production of fabricated items relates to the availability of raw material and economy of mass cutting, bending, and connecting, rather than the construction schedule. Delivery pertains to the completed fabrications, rather than the construction need. The constructor needs to clearly establish the progression of construction with the fabricator and ensure that the fabricator produces shop drawings and fabrications in a timely manner, that is, delivered to the jobsite prior to the construction activity.

EXAMPLE

Due to the physical location of the jobsite, the contractor decides to construct Building B, a warehouse building with a slab-on-grade, first, then Building A, an office building with a basement. This may not be the way in which the fabricator envisioned the project would be constructed. After completing the construction schedule, the constructor then establishes a sequence and timetable for the reinforcing steel fabrications that will be needed for the project. Specific information, rather than the entire construction schedule, will help the fabricator understand what is required to comply with the contractor's needs. A sample schedule of information the contractor can give to the fabricator is shown in Figure 5–2.

The fabricator can use this information to prepare a shop drawing and fabrication schedule to meet the required delivery dates.

2. *Establish areas for the delivery of each major material.* Establishing organized sub-areas in the storage yard for each major material, such as an area for reinforcing steel, a separate area for structural steel, and so forth, is essential. Mixing materials in the storage area causes chaos when identifying and retrieving items as the project progresses. A discussion with the supplier, concerning the timing of the delivery and the material being delivered, will help determine how much space to allocate for each material. The type of access needed when moving the material also needs to be examined when allocating space. If the material needs to be moved with a fork lift, wide enough aisles should exist between the materials to maneuver the fork lift. The area needed for the storage for each material will vary depending upon the stage of the project. During the first third of a building project, the area needed for storing reinforcing steel is quite large. In the subsequent two-thirds of the project, though, storage needs for reinforcing steel are considerably less, reduced to little or no space for the final third of the project. The construction schedule is a useful tool when determining the relative amounts of storage space needed during the project. The term **laydown area** is frequently used to indicate the storage area on the jobsite.

Building	Component	Schedule Date	Deliver By:
B	Footings	May 23, 1998	May 19, 1998
B	Foundation Wall	May 31, 1998	May 26, 1998
B	Slab-on -grade	June 13, 1998	June 9, 1998
A	Footings	June 13, 1998	June 9, 1998
B	Tilt-up Panels	June 20, 1998	June 16, 1998
A	Foundation Wall	June 20, 1998	June 16, 1998
A	Foundation Wall 2	June 27, 1998	June 23, 1998
A	Basement Columns	June 30, 1998	June 27, 1998
A	Slab-on-grade	July 8, 1998	July 5, 1998

FIGURE 5–2   Example Reinforcing Steel Delivery Schedule

3. *Sort the material upon its delivery to the site.* The material needs to be sorted as it is unloaded from the truck. Material that is needed first should be most accessible. The material pile should be ordered sequentially for materials that have a definite order in the installation. For example, the reinforcing steel previously discussed would be stored in this order: Building B footings, Building B foundation wall, Building B slab-on-grade, and so on. The material for Building A and Building B may be stored in separate areas, as well. Some material, such as lumber, may be used throughout the process, so organizing by size is probably the most efficient method to use, with 2 x 6s in one pile, 2 x 4s in another pile, and so on. Easy access should be provided to each stack of lumber, as it will be used throughout the process.

4. *Allow adequate space for sorting and storage of waste materials.* There was a time when all construction waste was accumulated and taken to a landfill. Sorting materials is now necessary to economically dispose of the construction waste. Materials should be sorted according to general types, such as wood, drywall, metal, and so forth, and then recycled. Adequate provisions should be made for sorting and accumulating in separate bins for major recyclable materials.

5. *Protect storage material, as necessary.* Many materials are sensitive to environmental conditions and should be adequately protected from rain, snow, wind, heat, freezing temperatures, and other adverse environmental elements. Most materials should be set on pallets or skids to protect them from ground moisture. Some materials can be covered by a tarp or polyethylene sheet, however, moisture has a tendency to condense in some of these conditions. Some materials are inappropriate for outside storage and should be stored either in the facility or warehouse. Portable storage trailers can be used to protect material on the jobsite, depending upon the space available.

Labor Productivity

Labor cost is the most unpredictable for the contractor. Although a fixed rate per hour is easy to determine, the number of hours to complete a task varies. The contractor estimates the number of hours each task will take at a certain rate under assumed conditions. Because the contractor is held to the estimate, the amount actually estimated for labor will be the budget for the item. If the actual labor equals the amount estimated, the contractor will achieve the estimated profit. If labor cost exceeds the estimated amount, the contractor will not achieve the estimated profit and could possibly lose money on the project. Concentration for labor then should focus on the work task, keeping all other activities minimal.

As previously mentioned, higher productivity rates will be achieved if time is spent installing the material, rather than moving it. The craftsperson should be instructed to perform the work for which they are trained and hired. A carpenter should be framing, installing doors and frames, or installing millwork. An ironworker should be installing reinforcing or structural steel. Part of the laborers' work, however, should be handling materials and equipment.

Current labor costs are quite high. The following example illustrates the cost of a carpenter to a contractor (labor rates will vary widely depending on the region and type of work performed).

EXAMPLE

Journeyman Carpenter (prevailing wage):

Wages	$18.85
Fringe benefits	$ 4.00
Taxes	$ 5.13
Total hourly cost to contractor	$27.98

(Total cost per minute ($27.98/hour/60 minutes/hour) = $ .47/ minute.)

These are direct costs for the carpenter's wages. Tools, equipment, support facilities, or supervision are not included.

At 47 cents per minute, the carpenter's time is very valuable. The contractor's goal is to be as labor-efficient as possible, utilizing the labor for the actual work task.

A major concern is travel time within the jobsite. The worker travels around the jobsite during the shift, possibly from the job shack to the location of the work task, from the work task to the restroom, or from the work task to lunch. Time spent traveling means work is not being accomplished. If it takes a carpenter fifteen minutes to travel from the entry gate to the work site, it will cost the contractor $7.05, without any production associated with that cost. The contractor does not expect the carpenter to run to his work, of course. Nonproductive time *is* expected during the course of the day, however, the jobsite facility should be arranged to keep nonproductive times to a minimum.

EXAMPLE

The site's sanitary facility (portable toilet) is located next to the job shack, which is convenient for the superintendent and visitors, but not for the carpenters. Four carpenters are working on the site. A trip to the sanitary facility takes fifteen minutes. Each crew member uses the sanitary facility four times during the shift. The labor cost for the trips to the facility is $112.80 per shift. If a sanitary facility is located near the work site, the average trip would be five minutes, with a cost of $37.60 per shift in labor cost. Assume that this work site would be active for the four carpenters for a month, or twenty working days. The cost of providing an additional sanitary facility near the work site would be $100 per month. By providing a sanitary facility near the work site, the contractor would save $1,404. This translates into fifty carpenter hours that could be applied to the work task.

When establishing the jobsite layout plan, the contractor should look at the areas of travel and nonproductive time for the crews and arrange the jobsite and facilities to minimize this time. The contractor must consider alternatives that will decrease labor time, economically, and within reason for the size and configuration of the project, in the following nonproductive time elements.

1. *Travel time from gate to work site*
   - Gate and parking should be close to work site (may possibly move as job progresses)
   - Provide transportation to work site: truck, bus, vertical man lift
   - In high-rise buildings, it may be possible to increase the speed of the man lift, with different equipment

2. *Travel time to sanitary facilities*
   - Provide sanitary facilities close to work site
   - Relocate sanitary facilities as job progresses

3. *Travel time to coffee break*
   - Provide tables and chairs in safe area near work site
   - Provide dry shack near work site
   - Discourage coffee break (may not be possible if in union agreement)

4. *Travel time to lunch*
   - Provide tables and chairs in safe area near work site
   - Provide dry shack near work site

5. *Travel time moving material*
   - Have material located close to work site
   - Discourage craftspeople from moving material

6. *Travel time to ask superintendent questions*
   - Project office should be near work site
   - Walkie-talkies for crew to communicate with office

Selective time studies can be made on alternatives, comparing the cost and cost savings of each. Care should be taken to include the extra costs, along with the savings in labor. For instance, the contractor may have to provide an extra dry shack, at an additional cost, near the work site. Some of the facility improvements, such as a manlift with increased speed, will benefit the subcontractors as well as the contractor, and such savings, if possible to realize, should also be included in the equation.

When optimizing transporting labor, particularly in high-rise buildings, delays need to be considered as well as the transportation from point A to point B. Vertical lifts rarely carry one person directly from the ground to the destination on an upper floor. Usually there are many stops for different people at different floors. The contractor must consider the average, quickest way to transport labor for the minimum amount of time. The following suggestions may optimize transportation.

- *Separate people and material.* Because moving material on vertical lifts takes much more time, particularly in loading and unloading, it is best to separate human transportation from material transportation. Material transportation can be limited to tower cranes or Chicago booms, with vertical transportation limited to human transportation. Certain parts of each hour could be divided for each for a vertical lift. For instance, worker transportation could be scheduled for ten minutes at the hour and ten minutes at the half hour, with the rest of the time (forty minutes of each hour) devoted to material transportation. Some high-rise buildings have two vertical lifts, one for worker and one for material transportation.
- *Schedule floor stops.* On taller high-rise buildings, the vertical lift for worker transportation may only stop on every other floor to limit the number of stops and to optimize the cycle time of the lift. If two lifts are used for worker transportation, one may stop more frequently than the other. The second lift might be the "express" lift.

Equipment Constraints

The jobsite layout can depend on the lifting and moving equipment that is used for the project. Once the equipment is determined, depending on its capabilities, cost, and other characteristics, the equipment puts some constraints on the site layout. Location of delivery points, site access for people and equipment, sequencing of events, and location of temporary facilities all affect what equipment is planned for the jobsite.

- *Point of delivery.* Once the equipment is decided upon, the point of delivery for material will relate to the location and capacity of the equipment. A tower crane, commonly used in building construction, has considerably more capacity for lifting near the mast than toward the end of the boom. Heavier loads then must be delivered fairly close to the mast and limited to a horizontal lift by the crane. If a fork lift is used to transport the material within the jobsite, the point of delivery should have solid road access from the point of delivery to the installation location. If lifting equipment is located some distance from the public road access, sufficient jobsite haul roads must be made and maintained to accommodate typical delivery trucks. If a permanent concrete pumping arrangement is established, whether a permanent concrete pump or permanent (during construction) pipeline, a location for concrete trucks during delivery, staging, and access also should be established, with a firm, solid roadway such as rolled gravel or asphalt paving.

- *Site access for people and material.* The gates and access into the site should be oriented to access the work site. Personnel entrance gates should be near personnel lifts, if they are selected for the project. Temporary facilities, such as field offices and storage sheds, should be accessible to the work site, via the site equipment.
- *Sequence of events.* The sequence of construction activities, particularly short-term, will be determined by the amount of equipment used on the project and equipment scheduling. The type of equipment used and designated traffic patterns on the jobsite are factors used when determining the schedule of underground utility lines.
- *Location of temporary facilities.* The location and properties of the equipment used on the jobsite will influence the location of temporary facilities. Material storage should have easy access, as well as the necessary equipment to facilitate its movement. If a tower crane is used for vertical transportation of material, movement would be facilitated by locating the storage within the reach and capacity of the tower crane. If this is not possible, as is often the case, additional equipment would be necessary for horizontal movement of the material on the site, such as a fork lift or a small mobile crane. Personnel facilities, such as tool storage, dry shacks, jobsite office, and sanitary facilities, should be close to personnel movement equipment.

Site Constraints

The site itself controls the layout and the equipment that is used probably more than any other factor. An urban site may consist of the property being the same size as the building footprint. In an extremely tight site such as an urban office building, the contractor needs a great deal of creativity to provide adequate storage, temporary facilities, and equipment. Additional property may have to be obtained on a temporary basis. Full or partial street vacation may be necessary to provide delivery and access to the site. In a very cramped site, the point of delivery becomes the focus of site layout. Some considerations for the point of delivery include:

- Traffic patterns—easy flow onto and off the site, avoiding congestion of traffic on city streets.
- Large enough space for all delivery vehicles—consider length; turning radius; height of truck, load, and on-board loading boom; and weight.
- Accessible for loading equipment.
- Queuing area for waiting trucks that will not disturb street traffic patterns—a concern particularly with concrete ready-mix trucks, where trucks need to follow each other closely to avoid interrupting the concrete pour.

When constructing on a tight site, different arrangements should be made for material storage. After the building is sufficiently constructed, the building can serve as a storage area for some material. An off-site storage yard, usually close to the construction site, can be used for this purpose. When an off-site yard is used, extra equipment, such as lifts and trucks, is necessary to provide efficient movement of material. In some urban areas, such as New York City, marshaling yards outside of the dense urban area are used for the receipt, storage, and disbursement of material to the jobsite on an as-needed basis. On the cramped site, material may have to be delivered daily, rather than stored on the jobsite. In dense urban areas, delivery of the material during the night may be the only feasible way. When a large amount of material is needed, for instance, for a large concrete pour, the construction activity may need to be scheduled on a Sunday or at night to avoid the congestion of city traffic. Large fabrications also may have to be transported at night or during the weekend. Large mobile cranes for a short duration lift that need full street access normally have to be set up, the pick accomplished, and broken down in off hours to avoid disrupting the normal traffic flow.

Large sites also have constraints that must be considered in the jobsite layout. A too-large site may be inefficient and costly. The following factors should be analyzed when laying out a large site:

- Temporary fencing—The cost of fencing can be a factor.
- Haul roads—The cost of constructing and maintaining haul roads on the site. Consider the climatic conditions during the project in relation to the soil, gravel, or other material used on the haul road. Mud, dust, irregular frozen ground, and snow are factors that prevent normal truck transportation from using haul roads. The haul roads need to be constructed and maintained to be passable when needed.
- Extra equipment needed—Distant storage areas may require additional lifting and transportation equipment.

Elements of the Jobsite Layout Plan

The jobsite layout plan has a number of elements, many of which have been discussed to some extent. The next section will describe the necessary elements the contractor should consider when making the jobsite layout plan.

Material Storage or Laydown Areas

Material storage was discussed quite extensively in the previous section. The following guidelines can be used as a checklist for establishing adequate storage areas.

- Estimate the amount of storage needed. Use the construction schedule to help determine the amount of materials stored at different intervals, so the storage area can be used for several different materials.
- Make sure adequate access is provided to the storage area for delivery and removal.
- Be sure the storage area is compact and solid. If an open-air storage, provide adequate drainage. Paved areas allow for good storage, as they are stable and easily accessed. Parking lots paved early in the project may provide good storage areas, however, some repair or a seal coat on the paving may be needed after use.
- Make sure all material is on skids or pallets to prevent moisture from entering the material from the ground surface.
- Protect weather-sensitive material from the elements. Tarps, temporary shelters, and other means may be necessary to protect the material from rain, snow, ice, wind, and sun.
- Parts of the structure may be used for storage of material. Storage within the structure is always limited, thus the material stored in the building should be that which will be used in the particular area, that which is weather sensitive, and that which requires higher security.
- Temporary storage units are often used on the jobsite. These include mobile home-type trailers designed for storage; 40-foot transit vans; containerized storage units; and job-built storage buildings.
- When determining material storage on the jobsite, always consider the needs of the subcontractors, as the majority of material is supplied by them.

Temporary Facilities

Jobsite Offices.   The jobsite office or "job shack" is essential for most commercial building projects. The jobsite office is the superintendent's headquarters, even when the superintendent is in the midst of the construction activities for most of the day. The jobsite office ranges from a clipboard in a pickup truck to multistory office facilities, depending upon the scope of the project. The jobsite office is often required in the General Requirements (Division 1) of the construction documents.

Functions of the Jobsite Office.   A list of the functions of the jobsite office includes:

- The superintendent's headquarters and office. This is where the superintendent and field management personnel, such as field engineers, offices are located.
- The place of business for the contractor. This is where visitors, salespeople, architects, owner's representatives, subcontractors, and direct labor enter the jobsite to establish contact with the appropriate management personnel.
- The location of telephone and fax facilities. The contractor's telephone and fax machine is usually in the jobsite office.

- The location of the full and current set of construction documents, including addenda and changes. Most contract documents require that the contractor keep a complete and updated set of documents at the jobsite. This set of documents is used as a reference for architects, owner's representatives, building inspectors, and subcontractors, as well as for the contractor's personnel.
- The location of the "record drawings" or "as-built" drawings, which show the actual dimensions and locations of the project as constructed. These "record drawings" are meant to be updated as the job progresses.
- The location of posters and information for employees. All employee information, such as safety posters, posters from state and federal agencies informing the employee of certain rights and obligations, and statements of company policy should be posted on a special bulletin board.
- The location of first aid information and equipment. Emergency phone numbers must be posted in a prominent location, with the jobsite address or location. First aid kits should be available at all work locations, but should also be available at the jobsite office. Major first aid equipment, such as stretchers, should be in the jobsite office, easily accessible.

Attributes of the Jobsite Office.   A list of the attributes and equipment commonly found in jobsite offices includes:

- It must be lockable and secure off-hours. Today's jobsite office contains valuable equipment, such as computers, printers, plotters, copiers, and fax machines. Other valuable equipment used on the jobsite, such as surveying equipment and construction lasers, are often stored in the jobsite office as well. Many contractors install security systems in the jobsite office, as this is a frequent target for burglaries.
- It must have adequate desk space for the permanent jobsite management staff. Considerations that might be made when determining the size of the jobsite office include:

Superintendent: Lockable desk; location for computer, including power outlets and telephone and data line; lockable file cabinet; plan table; telephone. If possible, the superintendent's office should be away from the public entrance to the jobsite office. It also should be out of sight of noncontractor personnel.

Jobsite clerk, office manager, or secretary: This person, when utilized on the jobsite, is the first point of contact at the jobsite and the office should be located at the entrance of the building. The desk should have a telephone, usually with a fax machine nearby. File cabinets also are necessary.

Field engineers: Each field engineer needs a desk. A common plan table can be used for more than one engineer. A great deal of time may be spent on the telephone with suppliers and subcontractors, so each engineer should have a telephone, if possible. Much of the field and office engineer's work is done on computers, so a space should be provided for a desktop model or laptop. A copy machine also is essential.

Foremen: Some jobsite facilities have a space that is jointly used by foremen for filling out time cards and reports.

- Space for a plan table with access for architects, owner's representatives, and subcontractors. A rack also should be included, with all of the pertinent contract and shop drawings. An area should be provided for the Project Manual and current approved submittals. Additional facilities for the architect or construction manager may be specified in Section 01500, Temporary Facilities, in the Project Manual.
- A window or windows with a direct view of the jobsite. This is not always possible, but field management personnel need to be in touch, even if only visually, with the project.
- Wall space for the project schedule. A schedule that is used by the contractor usually is very visible and accessible in the jobsite office. Updates and changes normally are posted on the schedule. The more readable and accessible the schedule is in the jobsite office, the more it is used as a tool to manage the project.
- Provide conference space, if necessary. The jobsite is often away from the owner's, architect's, and contractor's facilities, thus conference space should be available for project, safety, and foremen meetings, and all other gatherings at the jobsite. A conference room or another area that can be used as such often is included in the jobsite office facility.
- Storage space: Secure storage space often is included in the jobsite office for equipment, tools, and small, valuable material. The jobsite office normally will have lockable storage cabinets. It also may have rooms designated for storage of tools, equipment, and material.
- Adequate environmental features: Most jobsite offices need heat or air conditioning, depending on the climate of the jobsite locations.

Types of Jobsite Offices

There are as many types of jobsite offices as there are projects and contractors. Each jobsite office facility must fit the project and management of the project. In addition, several factors enter into the selection of the jobsite office, including:

- Cost—The cost of the jobsite facility must be within the budget established by the estimate. The project's needs are considered in the estimate when establishing a budget. The monthly cost of the office facility and the anticipated duration of the project are used to establish the budget amount. The contractor may use a trailer or portable building for minimal cost, or may rent or lease temporary facilities.
- Space at the jobsite—The space available at the jobsite will substantially affect the type of jobsite office. Tight space at the jobsite may force the contractor to find nearby offices. A small trailer may be the

most acceptable office, due to jobsite limitations, despite the fact that more office space would be desirable.

- Availability—The availability of adequate jobsite office facilities in the locality also will help determine the type of office facility used. If jobsite space is tight or unavailable, the contractor may wish to rent a nearby building, warehouse, or office facility. If these are not available, the contractor must consider utilizing the structure itself in the same way as the jobsite office. When a contractor uses a fleet of trailers or portable buildings as jobsites, the best ones that are available at the start of the project are generally used. This may not be the optimum jobsite facility for the project, but the price is usually right. Hauling costs also may limit rental facilities.

Common types of jobsite office facilities include:

- Existing buildings—Buildings near the jobsite can make good jobsite offices. Warehouses, office buildings, houses, and other types of structures can be used. These off-site offices must be adjacent or very close to the jobsite to avoid wasted travel time. Usually these facilities are rented on a short-term basis. Normally some cost is involved in remodeling the facilities for jobsite offices, which also needs to be considered when comparing rates.
- Modular office units—Numerous modular, portable units that are manufactured specifically for the temporary office are available in every size, configuration, and feature. There is, of course, a wide quality range also available. These modular units can be used as single units or several combined units. The contractor can purchase or lease them. When considering modular units, the following items must be added into the cost calculation:

Hauling costs to and from the jobsite. The larger modular units have to be transported like a mobile home, which requires special transportation beyond the contractor's normal equipment. When renting or leasing units, this expense may be part of the quoted cost. Because transportation can be costly, confirm that this price is covered in the quoted rental costs.

Set-up and take-down of the temporary assembly, as necessary.

Additional facilities to be constructed on the jobsite for modular units. These items could include stairs and ramps; wood decks between units; and floor and roof structures between the units.

Wiring and plumbing of the modular units. Additional exterior security lighting is necessary.

Additional furniture. Plan tables, plan racks, chairs, desks, file cabinets, and storage cabinets, beyond those furnished with modular units.

Monthly utility costs for heat, electricity, water, and power for the jobsite office facility.

Security system or special security provisions for the jobsite office.

- Trailers—Many contractors use trailer-mounted jobsite offices, which can be towed by a contractor's vehicle. These are usually smaller units, used on small- to medium-sized projects. The obvious advantage is the cost savings in hauling to and from the jobsite. These smaller, more mobile units can be accommodated on the small site more easily than mobile-home type units. Contractors use different types of trailers: manufactured temporary office trailers; converted travel trailers; and custom-built trailers. Another relatively mobile small office unit is the wood-framed building on skids, normally transported on a flatbed truck or low-boy trailer. Examples of some common office trailer configurations are shown in Figure 5–3A and Figure 5–3B.

  Figure 5–3A illustrates some office configurations available in prefabricated mobile offices. The first unit shown is a small office to be used only with the superintendent on the project. An additional storage cabinet can be added for survey equipment. This office can accommodate about four people. The second unit shown is a small office with storage space for tools, equipment, and some materials. A double door provides access to the storage area. Most contractors build bins on each side of the storage room to hold tools. The third unit shown is an office trailer, which utilizes the 32-foot-long module. This office can accommodate a superintendent and field engineer.

  Figure 5–3B illustrates larger units that are available in both the 48-foot and 60-foot lengths. These units have provisions for internal rest rooms. Some sites may have water and sewer hook-ups. The first unit shown is the 48-foot, the second the 60-foot. These larger units have a great deal of flexibility and can accommodate a staff of several individuals. For larger projects, a series of prefabricated units can be arranged to provide office space for the entire on-site management team. Decks can be built between the units to avoid unnecessary stair climbing.

- Site-built jobsite offices—Many contractors will build the jobsite office on the actual work site. They may use a prefabricated system, with panels and roof sections, hauled flat on a truck and erected on the jobsite. The contractor also may opt to build a wood-frame structure on the jobsite. When using a warehouse or the underground parking floors of a building, the contractor can build a wood-frame structure and finish it like an office, with carpeting and suspended acoustical ceilings.

  The photograph in Figure 5–4 shows a typical jobsite office. Note the copy machine, fax machine, and computer, adjacent to the superintendent's desk.

  Some subcontractors may also furnish jobsite offices. The size and type of the office will depend primarily on the size of the subcontractor's jobsite staff. Mechanical and electrical subcontractors, in particular, will have jobsite offices, as will other long-term subcontractors. Some subcontractors will combine their jobsite office with jobsite storage.

FIGURE 5–3A
Example Office
Trailer Configura-
tions

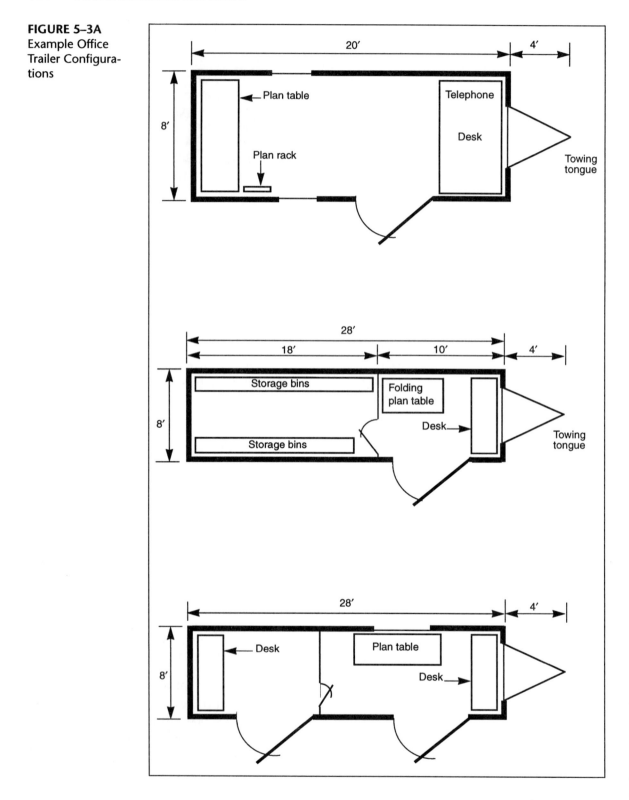

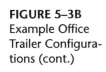

FIGURE 5–3B
Example Office
Trailer Configura-
tions (cont.)

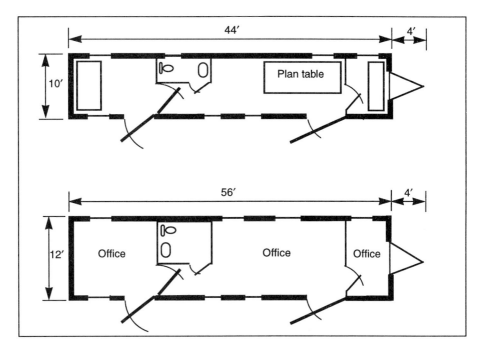

FIGURE 5–4
Jobsite Office Interior

Dry Shacks. Traditionally, the contractor provides a dry place for construction workers to eat their lunch and a place to change their clothes. This is commonly called a **dry shack.** Union agreements often have a provision that requires a dry shack to be furnished. The dry shack, like all other facilities on the jobsite, has an infinite number of types and variations. A few types include a wood-framed small building, with a table and a place to store lunches; a modular unit with one room; a room in the job trailer; or an area in the building, near the work activities. Although dry shacks do not need to be elaborate, they should be clean, have adequate light, and be relatively comfortable for employees.

Tool Storage. The contractor normally furnishes power tools and larger tools for the work to be accomplished. These tools can include circular saws, drill motors, hammer drills, demolition hammers, laser levels, ladders, scaffolding, concrete vibrators, pumps, surveying equipment, and many others to facilitate construction activities. However, these tools are subject to theft, due to their mobility and high value. Tool storage, then, must be extremely secure. Because many individuals on the jobsite use the tools they should be in bins and organized so they can be quickly accessed.

A separate tool shed also could be used on the jobsite, preferably close to the work site. It often is located near the jobsite office, for security reasons. The tool storage area needs to be secure as well, with strong locks and impenetrable window openings. This area often is combined with the jobsite office.

Sanitary Facilities. The contractor needs to provide sanitary facilities for its employees on the jobsite. Normally, the general contractor provides these facilities for all individuals on the work site, including subcontractors. The contractor will want to provide drinking water, washing water, and toilet facilities for jobsite employees, as they are necessities. Further, current safety requirements specifically outline what facilities must be provided on the jobsite.

Drinking Water. Fresh, potable drinking water must be available on the jobsite. To avoid wasted time, the water should be available at the crew's work site. Most contractors use five-gallon insulated containers, equipped with a tap. Paper cups are provided with the container, usually attached to it. Safety regulations prohibit the use of a common cup for drinking water. A receptacle adjacent to the water container should be provided, to prevent paper cups from being littered over the entire site. If potable water is not available on the jobsite, containers must be filled elsewhere and transported to the site. The containers should be rinsed and refilled daily and must be thoroughly washed weekly. Some union agreements may have additional requirements for drinking water, such as availability and adding ice to the water. During hot weather, adding ice to the containers maintains a cooler water temperature throughout the shift, preventing warm water in the afternoon.

Washing Water.   Safety standards require that clean, tepid wash water between 70 and 100 degrees be provided at the construction jobsite. The standards also require that individual hand towels, such as paper towels, be provided with an appropriate receptacle. Until permanent toilet room facilities are established in the building, this requirement may be difficult to meet. A container of warmer water, marked "For Hand Washing Only," could be provided, preferably near the toilet facility or dry shack.

Toilets.   Obviously, toilets are a necessity on the jobsite. They should be placed adjacent to the work area, to avoid unnecessary travel time. Safety regulations require that toilets be provided at the work site. An example of the minimum requirement for toilets on the jobsite from a state safety standard is shown in Figure 5–5.

These requirements apply to *all* employees on the jobsite, including subcontractors. The toilet facilities may be located as per the work site, which may have fewer employees, but better facilitates work. The safety requirements also address the location of the toilet facility:

1. At all sites, toilet facilities shall be located within 200 feet, horizontally, of all employees.
2. On multi-story structures, toilet facilities shall be furnished on every third floor.

Depending on the diversity of the workforce on the jobsite, company policy, and local, state, and federal regulations, the contractor may wish to provide additional and separate toilet facilities for each gender.

Prior to the installation of permanent sewer and water facilities, most contractors use portable chemical toilets. In most localities, these are rented and/or leased to the contractor by a business that also services them, usually on a weekly basis. The service includes emptying the tank and washing and disinfecting the unit. There is some variation in the size of these units, but usually they are about 4′ x 4′, with adequate ventilation.

Number of Employees	Toilets Required
1 through 10	1
11 through 25	2
26 through 40	3
41 through 60	4
61 through 80	5
Over 80	One additional toilet for each additional twenty employees

FIGURE 5–5   Quantity of Toilets Required For Number of Employees

When plumbing facilities are available, the contractor must also clean and disinfect the toilet rooms weekly.

Construction Waste Facilities.   For new construction or remodeling, a considerable amount of waste needs to be collected, transported within the site, sorted, and transported off of the site. Vertical chutes often are used to transport the waste from upper stories to ground level sorting areas or collection points. Separate bins are necessary for different types of waste to facilitate recycling. These locations, along with the general garbage collection "dumpster" should have good access for loading and transporting off-site.

Temporary Utilities.   Because building utilities, such as power, water, and heat, are not available until the building is about two-thirds complete, temporary utilities are needed for construction operations. The contractor must arrange for the installation of these utilities, their locations, and their cost until the owner assumes beneficial occupancy. A discussion of temporary utilities to be considered in jobsite layout follows.

Temporary Power.   The contractor will normally contract with the local power authority for a temporary electrical service for the jobsite. In many areas of the country, connection of this temporary service may have to be scheduled as much as three to six months in advance. Some companies can facilitate the connection two to three weeks from the order date. If construction activities need to commence prior to the connection of the temporary power, temporary generators normally are used. These usually are not a desirable alternative, due to the cost of the generator rental and the cost of fuel. Several small generators normally are needed for several pieces of electrical-powered equipment.

The temporary power service must be placed using the following parameters:

- Locate as centrally as possible to avoid long power cords.
- Locate where service will not be moved until disconnected, when permanent power is installed.
- Ensure that power lines will not interfere with the structure or any delivery equipment.
- Locates lines, if underground power lines are being used, where they will not be dug up by construction operations.

The utility company and municipality have codes for the safe installation of a temporary power service. The temporary panel should be installed on an approved mast or pole, with breakers and should be grounded. Both the utility company and municipality must inspect the panel, mast, and grounding prior to allowing the utility company to connect power to the temporary service.

Most construction tools run on 110 volts. Some, such as the masonry saw and electric-powered compressors (for pneumatic tools), run on 220-volt power. The contractor must provide a temporary panel that will meet the needs of the construction activities and will often subcontract the installation of the temporary service and panel to an electrical contractor. It should be noted that the electrical subcontractor on the project will not usually include this temporary service in the project bid, unless it is included specifically in the price quotation. Some contractors already have several temporary panels and masts and can furnish the service themselves.

The contractor must provide safe and code-compliant distribution of power from the panel. Distribution boxes should have ground-fault protection. All electrical power cords must be grounded and have a three-prong configuration. All extension cords must be in good repair and properly sized for their use.

Temporary Water.   The permanent water connection may not be made until several weeks after the project has started. Pipes and fittings have to be approved, ordered, and delivered before the connection can be made. The tap into the water utility's main line must be scheduled, similar to the power connection. The contractor may have to investigate other means for temporary water, such as a temporary line or making arrangements for water delivery to the site.

Water is used in several construction activities. It is used in earthwork operations for compaction purposes and dust control; in concrete operations, in wetting the surfaces and cleaning the tools; in masonry operations, as it is added to the mortar on the jobsite; and in fire protection, as a temporary fire hydrant may be necessary during the construction activities.

The temporary water connection, like the temporary power connection, has several parameters that should be considered:

- Provide water to the most central point possible. Consider the effect of hoses to access routes.
- Avoid water distribution in an area that will be susceptible to breakage by construction equipment and delivery trucks.
- Coordinate the location of the underground temporary water line to avoid breakage during future excavation.
- Install the temporary line according to local codes. Provide adequate cover over underground water lines to avoid freezing.

Careful coordination with the water utility company is necessary. Unfortunately, the company furnishing the water will probably be a different firm than the one providing the electric, depending on the locality. Some water utilities are owned and operated by the municipality, some are private utility companies, and some are water districts, separate from the municipal organization.

Temporary Weather Protection and Heat.   Although temporary weather protection and heat are not necessarily jobsite layout considerations, they should be discussed briefly as they are important elements during the construction process.

Weather protection may need to be installed during the winter weather to allow the construction process to continue. A variety of methods of covering the work are available, usually depending upon the severity of the weather and the type of work that remains to be done under the protection. A temporary frame structure can be erected, with a reinforced polyethylene fabric installed on the surface. Heat can be added with the temporary structure. Caution must be used in these types of structures, because there is a great deal of heat loss, resulting in extremely expensive heating fuel costs. Caution also must be exercised to use heating equipment that will not cause fire damage. Ventilation of fumes from the heating equipment is a consideration as well.

Temporary heat within the structure under construction also needs to be planned. Many contractors use electric heat, due to fire and ventilation considerations. This, however, consumes a large amount of power, and the temporary service should be sized for what heat is needed. All temporary heat is expensive, and each method has its advantages and disadvantages.

Jobsite Security

The jobsite needs to be secured, both for public safety and for the security of the installation and equipment. A jobsite, whether it is an excavated hole in the ground, or a steel-framed high-rise building in construction, is considered an attractive nuisance. It is the contractor's responsibility to prevent the general public from entering the jobsite. If an individual is injured while entering an unsecured site, the contractor is generally held liable for that injury. The construction jobsite, while safe for trained and equipped construction workers, is not necessarily safe for the general public.

Vandalism and theft at the jobsite are major concerns for the contractor. Any construction site, in any geographical area of the country, is subject to vandalism and theft. Theft of materials, tools, and equipment is a frequent occurrence. Even theft of large pieces of equipment, such as backhoes and bulldozers, happens throughout the country.

The jobsite must prevent nonauthorized personnel from entering its active areas. Further provisions also are needed on the site to secure materials, tools, and equipment from theft. The following list of provisions will help maintain a secure jobsite.

- A secure perimeter fence; secure and lock gates at night
- Bright yard lights at night
- Notify local police of the existing jobsite
- Secure or lock any equipment in concealed storage, if possible
- Do not leave keys in vehicles or on the jobsite at night
- If necessary, provide additional security at night, such as a watchman, security patrol, and so on

Perimeter Fencing

Construction documents may require a specific location for perimeter fencing. Usually the contractor is left to determine the amount and type of security fencing necessary around the project. Several options are available to the contractor.

- Rental fence—In most localities, businesses rent temporary fencing, usually by the month or year. The fence rental includes delivery, installation, and removal. A rental fence consists of galvanized chain link fabric, with temporary vertical pipes, usually at ten-foot centers. A rental fence usually does not have horizontal top or bottom pipe rails. This type of fence is economical, however, because it provides a mostly visual barrier and is fairly easy to purposely penetrate. This fence is popular for large sites, because it is relatively inexpensive.
- Permanent fencing—Permanent fencing, depending upon the specification, usually is more secure than temporary fencing. Many contractors will install permanent fencing early in the project, using it both for temporary and permanent fencing. Permanent fencing, however, may not cover the extent of the area necessary during the construction phase. In these cases, the contractor might use a combination of the two. Often, security fencing is used at the very start of the project. Permanent fencing may be not as immediately available, as the product submittal needs to be processed and the material needs to be ordered.
- Wood-frame and panel fencing—Solid material, such as plywood, can provide a good barrier around the site, while providing a visual barrier as well. This can be combined with sidewalk protection on urban sites. The municipality will have strict guidelines for the protection of pedestrians and sidewalks around the site, including overhead protection.

It is important to consider the size, type, amount, and location of the gates in the perimeter fencing. Most construction gates are swinging gates, but there may be site restrictions requiring sliding gates. The size of the delivery gates should be adequate for all delivery vehicles entering the site. Personnel gates also will be needed for access between employee parking lots and the worksite. In cases where both union and non-union employees are working on the same site, separate, marked gates are required for both. Separate delivery gates also may be necessary in these situations. Adequate, secure locks are a must for all access gates. Strict key control should be maintained, with only necessary parties receiving keys.

Additional Security Measures.   Many jobsites require more general security than perimeter fencing. Some of these measures might include:

- Night watchman—a permanent night watchman might be employed at the jobsite. The night watchman would need access to telephone and other communications devices. The night watchman might also be combined with a guard dog.
- Guard dog—A guard dog might be used in the construction area to prevent intruders from coming into the area. Notice of a guard dog may need to be posted, depending on local regulations. The contractor should always consult with local law enforcement authorities when establishing a jobsite security plan.
- Security patrol—Private security patrols can be contracted to provide periodic patrol of the jobsite.
- Intrusion alarms—Gates, job shacks, tool sheds, and storage buildings can be attached to security alarms that will notify a security service of intrusion. The security service, in turn, will call the police. Alarm bells at the jobsite also can be hooked to an intrusion detection system.

The general layout of the jobsite should consider security aspects. Exterior lighting should be provided, particularly at gates and other vulnerable points. The site should be observable from adjacent streets and roads, with occasional police patrols. The contractor might consult with local law enforcement authorities when formulating the jobsite layout and obtain their advice on security measures for the site, regarding local and neighborhood conditions.

Access Roads

Proper planning of access roads to the site and within the site is essential to the efficiency of the project. The following goals should be achieved by all of the access roads involved in the project:

- The roads must provide direct access to all points needing access. Such points would include storage areas, installation areas, and entries to the work sites.
- The access roads must be built solidly, to withstand the loads, the traffic, and the weather conditions during the period of use.
- The access roads must be built to accommodate the type of vehicle and load that will use it, which relates to the width of the roadway, the radius at the corners, and the slope or grade of the roadway. When con-

structing haul roads for excavation work, using trucks or scrapers, the contractor should always remember that the strength of the road surface, equated by "rolling resistance," and the slope of the roadway, or "grade resistance," affect the duration of the hauling operation, thus affect the cost of the hauling operation.

- Access roads must be placed during a time and in a location where they will not be moved or replaced. Underground utilities can disrupt the traffic on haul roads. Timing the underground utility construction prior to the construction of the haul road can avoid disruption of the traffic pattern. If the haul road does need to be disrupted during its use by the construction of underground utilities, then plans should be made for alternate temporary routes. Timing of these disruptions should be carefully coordinated with the construction schedule.
- Access roads should be appropriately constructed to the type of traffic that will be using them. A roadway for public vehicle traffic has different requirements than that for a haul road for scrapers.
- Haul roads should cause a minimal amount of dust. If roads are constructed of earth materials, a plan of dust control, such as periodic watering, should be made.

Two basic types of construction roads exist: access to and access within the construction site. Both road types affect the jobsite's operation. Both types of roads also have some distinctions, as discussed next.

Access to the Site.   The construction site is not always immediately accessible by public roads and streets. Roads often need to be constructed from the nearest public roads to the new facility. Usually the construction of these roads will be included in the construction contract. Roads need to be constructed according to the construction documents. As most access roads become public streets and roads, they must meet the road standards for the local municipality. If new access roads are in the contract, their construction should take first priority, if possible. Ideally, construction of the new permanent access roads should be done prior to the project, but this is not always possible. Many times, construction of the roadways is done by a separate contract, with a schedule totally unrelated to the project construction schedule. Temporary access roads often are needed until permanent access roads are available for use. In these cases, the access road should be built to accommodate the necessary traffic during its use, and should be maintained to that condition.

The access road to the site should be as direct as possible from the existing public roads. Because the construction project will require numerous deliveries, access should be identified and easy to describe to delivery companies. The access road should be constructed directly to the delivery area of the project, as the majority of construction traffic relates to deliveries. Adequate parking areas for construction employees need to be established and need to be close to access roads.

Access Within the Site. The access roads within the site are rarely designed or defined by the construction documents, because they are the contractor's responsibility. Some occasions allow permanent internal roadways to be used during the construction phase, but these usually are constructed after the buildings and structures on the site. The contractor should determine the location of these roads, how they will be used, and what type of roadway will be necessary. Compacted natural earth roadways can be good construction roads if the soil is gravelly, with good drainage. If the natural soil is clay or fine silt, the material drastically changes form upon the introduction of water and probably is not appropriate for an access roadway. Occasionally, the use of a geotextile material that provides stability to the soil is necessary, with the addition of gravel or crushed stone for the roadway surface.

Site Drainage and Water Control. The contractor needs to control site drainage within the area to prevent disruption of work. Drainage and possible pumping should be used to keep excavated areas clear of water while work is being done there. The contractor also will be responsible for controlling all site drainage, to avoid environmental damage or influence. Many states and municipalities require the contractor to submit a plan on site storm water drainage and containment.

Signs and Barricades

Signs and barricades are part of the entire jobsite layout plan. They are part of the safety plan of a jobsite, but also can identify correct routes to and within the area.

Signs. Signs can be used to direct traffic to and from the jobsite. Some of the signs that can be used in jobsite organization include:

- Project sign—Normally a specified sign listing information about the project, such as owner, architect, contractor, and major subcontractors.
- Location signs—Miscellaneous signage on roads to indicate the route to the jobsite for delivery trucks.
- Gate signs—Some gates are restrictive regarding which company's personnel can enter. The gate signs normally list the firms that are required to use the gate. This provision is used when both union and non-union contractors are working on the same project.
- Directional signs—Signs indicating location of deliveries, one-way roadways, and locations of contractors' offices.
- Safety information signs—Danger signs, Hard Hat area signs, and other signage indicating any safety precaution. (See chapter 8, for a further discussion.)

Barricades.   Barricades can be used for safety, security, and definition of traffic patterns. They are used in the following applications:

- In a safety situation, to prevent individuals from proceeding to an area.
- In dealing with the public, the barricades need to be identifiable and substantial enough to physically prevent passage into the area.
- To define the site area and protect the public, such as walkway protection.
- To divert traffic into certain areas. This could be used to direct concrete trucks to different areas other than the previous delivery location.

Organizing Jobsite Layout

The process of laying out the jobsite considers all factors regarding the project and utilizes all methods for its efficiency. Jobsite layout is probably best done by an individual who completely understands the project and construction plan for the project. Some references used for determining jobsite layout include:

1. Construction documents, specifically:

   - Division 1, General Requirements of the Project Manual—Information should be contained in this division concerning specific project instructions, such as parking facilities, use of the site, and required temporary facilities.
   - Site plan, architectural drawings—An existing site plan and the constructed site plan may be available, which will provide a base for the layout. Layout of the site plan can be facilitated by using a CAD, perhaps from the architect.
   - Mechanical and electrical site plans—These site plans often are separate from the architectural site plan. Location of underground utilities and mechanical and electrical site work is needed in jobsite layout for planning access roads and planning the location of temporary facilities.
   - Elevations and building sections—These drawings help determine the reach and length of lifting equipment.

2. Construction schedule—This schedule will help determine the length of certain construction operations and the time needed for use of equipment.
3. Capacity charts and technical data concerning cranes, lifting, and conveyance equipment.
4. Local codes for use of public roads, temporary vacation, and safety provisions required to protect the public.

5. Safety standards for safety-related requirements, with temporary facilities and lifting.
6. Data on the size of the larger lifting loads that will be encountered.
7. Information from subcontractors on storage needs for their materials.
8. Budgets established by the estimate that determine maximum cost for the elements of the jobsite layout.

A drawing should be made of the site, including all items that add a constraint to the jobsite layout. Some of these items include:

1. The existing building and structure location
2. The new building and structure location
3. The existing roadways and streets adjacent to the jobsite
4. The underground and overhead utility locations—both existing and new

   • sewer, both sanitary and storm, including drywells
   • water
   • natural gas
   • electrical power
   • steam, if provided by a central steam plant
   • telephone/data

5. The paved surfaces, both existing and new

All of this information probably is not on a single sheet of the construction drawings. Additional information could be drawn onto the site plan, but usually there is more information than is needed on the site plan. A tracing could be made of the site plan, for additional information. A CAD can be useful in these situations. The utility lines can be added in "layers," with designated colors. A CAD can be used without the original drawings, using outlines for structures and buildings.

The constructor, after having all of the constraining factors on a drawing, can then start laying out the roadways, access, lifting facilities, and temporary facilities. A CAD system can provide several different scenarios, quickly and easily. Many different solutions are available, and the constructor will be looking for the solution that best optimizes time and cost in completing the work.

Figures 5–6 and 5–7 illustrate a jobsite layout drawing. Figure 5–6 indicates information shown on the site plan in the construction drawings. Figure 5–7 shows site access, site offices, and other information determined during jobsite layout planning.

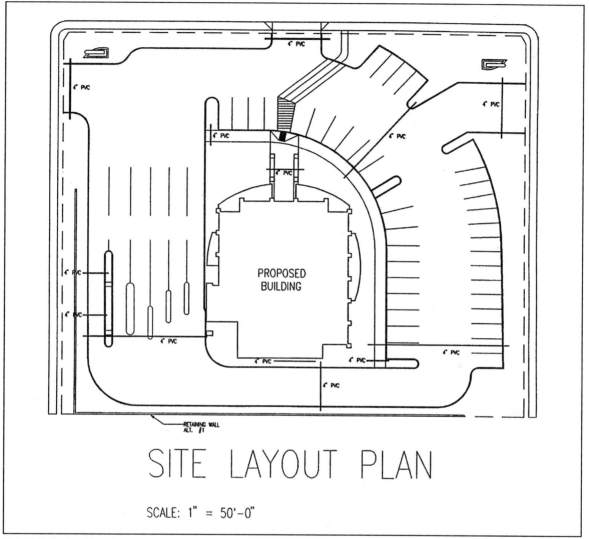

SITE LAYOUT PLAN

SCALE: 1" = 50'-0"

FIGURE 5–6 Site Plan From Construction Drawings

Summary

Jobsite organization is a factor in the profitability of a construction project. It has a distinct impact on the productivity of every craftsperson on the site. The contractor should carefully consider all aspects of the project when laying out the jobsite, such as:

- Delivery access
- Material storage

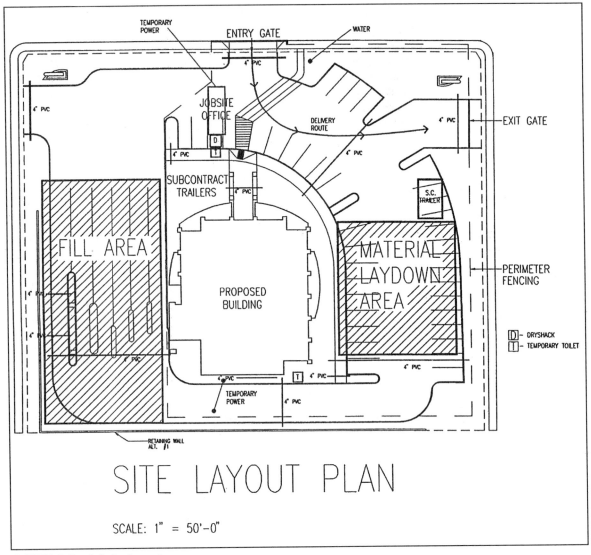

FIGURE 5–7   Jobsite Layout

- Access roads
- Material handling and lifting equipment
- Personnel movement on the jobsite
- Temporary facilities related to project requirements

The contractor must plan organization of the jobsite as the construction of the project would be planned. After data is gathered and the method is established, the contractor has to implement it, and as with any other plan in construction, it must be flexible and accommodate changes in the construction process.

CHAPTER 6

MEETINGS, NEGOTIATIONS, AND DISPUTE RESOLUTION

CHAPTER OUTLINE

Today's constructor must negotiate in one form or another, from signing the contract until the end of the project. Once a project has been sold to an owner, writing subcontracts, settling changes, applying for payments, and finishing the project punch list all require some form of negotiation with another party. Sometimes negotiations are with the owner, other times with the architect of subcontractor. Negotiations on the construction site occur at all levels and in all positions. Projects in construction are always complicated, individual trades are very specialized, and the potential for misunderstanding and confusion is ever-present.

As complexity increases throughout the project, additional formal lines of communication are required. **Meetings** are a convenient means of facilitating communication. They should be formally scheduled and used to inform individuals of progress, to discuss problems, and to propose and seek solutions. Jobsite-related meetings can be classified as:

- partnering meeting and workshop session (project team)
- contractor's pre-construction planning and organization meeting
- pre-construction meeting with subcontractors
- pre-construction meeting (owner, architect, and contractor)
- project meeting (weekly or biweekly)
- subcontractor coordination meetings (contractor and subcontractors)
- schedule meetings
- safety meetings (discussed in chapter 8)
- staff meetings
- specialized meetings
- project closeout meeting
- post-project review and evaluation

Construction meetings should follow the basic format as in other industries. **Agendas** should be developed for all meetings. Meetings are held to inform people and to find solutions to problems, not to socialize. Time is precious for all of the professionals who are active in a construction project. When possible, the agenda and list of attendees should be distributed before the meeting. Agendas for some meetings can be somewhat standardized, even though the details of all meetings differ from week to week and job to job. Specific tasks and ideas parties will be expected to discuss are placed on the agenda. Meetings should be held at a scheduled time during the project, using minutes or other types of logs and letters so individuals or companies can receive clarification when needed.

Not all meetings must be formal, but written details of the discussions that occur must be kept. **Meeting minutes** are an effective way to record the content of meetings and to hold individuals accountable for their statements. Meeting minutes should be a fair depiction of what actually transpires during the meeting. Many times the note taker will slant the minutes toward his or her company's favor. This form of deception should be avoided, as it breaks down any trust that might exist. After distribution, the minutes will be reviewed by all attendees. At the next meeting, the previous minutes will be approved and, if not accepted, corrections will be made. The question of who should take the notes and who should receive copies are important. Construction team minutes should be distributed to all attendees as early as possible, allowing enough time for review before the next scheduled meeting. Minutes can help clarify pending problems, place responsibility for solutions on individuals, and act as reminders before the next meeting. A list of action items leads to further discussion. Items are added and other areas are deleted. Any active project that has good communication will utilize an ongoing list, much like a "To Do" list that some people use to organize their day. It generally is preferred by the general contractor that minutes be taken by contractor personnel to maintain control of the meeting and the project. Though recording, formalizing, and distributing the minutes takes considerable effort, they are an important means of project communication. Many architects and owner's representatives do not have the additional manpower to produce and distribute timely minutes.

Figure 6–1, on page 152, is an example of one form of meeting minutes. A more detailed discussion about the elements of meeting minutes is contained in chapter 4.

Note that in Figure 6–1 areas exist for listing attendees and their companies. Discussion topics are itemized and numbered to facilitate later referral to a specific item or question. Note that on the right side of the form, on the same line as the item number, the column "responsibility" is given.

Partnering Meeting and Workshop Session

A partnering meeting or workshop of the complete project team is becoming a viable method for preparing project participants for the construction process. Partnering is a team-building process that starts with establishing trust between major project participants. The initial partnering workshop involves owner's representatives, architects and engineers, construction managers, contractors, and occasionally subcontractors. A partnering session generally will have a facilitator, a professional who conducts these type of meetings without any vested interest in the contracts to lead the group through the workshop. This workshop is free-flowing, but with a definite underlying structure and direction. The key element to partnering is the acceptance by all parties to reach an agreement and commit to making it a reality. A typical agenda could include:

- an introduction of construction team members
- a discussion about changing the "work as usual" attitude
- a discussion of the roles and positions for people involved with the project, and their importance

Meeting Minutes

FGH CONSTRUCTION COMPANY, INC. Date _____

Sheet _____ of _____

Minutes of Meeting _____   Project Name:

_____

On _____   Project Number:

_____

Subject: _____   Location:

_____

Attendee:

Name	Company
_____	_____
_____	_____
_____	_____
_____	_____
_____	_____
_____	_____

Item No.	Description	Status	Due Date	Responsibility
1	Discussed outstanding paint color submittal	New	6/6/95	Jim R. - FGH
2				

_____

Meeting Notes Taken By:

_____

FIGURE 6–1   Example Meeting Minutes

- a discussion of how problems can be solved at the lowest levels
- creating a contract or agreement of understanding to be signed at the end of the session
- finalizing a working agreement that sets specific project goals and methods to create and maintain cooperative relationships

The primary goals of partnering meetings are to break down adversarial relationships that have built up over the years; to create understanding between different organizations; and to set a common goal among all

FIGURE 6–2
Example of a Project
Partnering Agree-
ment

Project Mission Statement:

The mission of the project team, which consists of all the design organizations, the owner and the owner's personnel, and the construction group and its subcontractors, is the successful completion of the Atascadero High School Project. The success is defined as meeting the owner's expectations for building that is within budget, on time, and to a quality specified by the project documents.

Objectives:

1. No employee lost time due to accidents
2. Create an environment that all employee strive for quality
3. Commit to finishing the project a month early
4. Maintain total change order costs below 5% of total project budget
5. ....
6. ....

Signed by:

FIGURE 6–2 Example of a Project Partnering Agreement

parties. Partnering only works when the group attempts to create solutions to the anticipated problems that will be acceptable to all parties; each party values each team member; it creates an open and trusting relationship; and where innovation and resourcefulness are accepted as a group process.

This workshop can point out potential problems and attitudes, and create a better understanding between the project participants. A more thorough discussion of partnering is contained in Chapter 10, Project Quality Management. Figure 6–2 is a sample agreement reached and signed by all participants at the end of the initial partnering meeting.

The Contractor's Preconstruction Planning and Organization Meeting

The contractor's preconstruction planning and organization meeting is necessary to accomplish a successful project. A well-planned project from its inception has positive long-term returns. Prior planning solves problems before the project starts, enabling the management team to concentrate on project start-up. After the Notice to Proceed, the project management team focuses on implementing, rather than planning the project. Time is not unlimited, and with each extra day the contractor incurs indirect jobsite costs. The pre-job planning session can be broken into five or more distinct parts. Some firms have developed checklists for each of the phases, to expedite the planning and organization of the project. Pre-job planning involves the following five organizational sections:

1. Review of staffing requirements based on Project Manual requirements and any special project requirements, that is, environmental, training, hazardous waste, and so on. The complexity of the project also should be discussed. Organizational charts, listing names and responsibilities of each person should be developed. Descriptions of each position should accompany the organization chart.

2. A review of the estimate and estimator's intent should be made. A review of the subcontractors, long lead items, areas of risks, and special equipment needs also would be discussed by the estimator. The estimate must be revised into a project budget, which may require a major change in the estimate items to ensure that they are compatible with the cost control system. The contractor should begin writing major equipment purchase orders and any subcontract agreements.

3. Development of the construction plan and preliminary schedule. Decisions about how this particular project will be built are made. This schedule is the basis for the final construction schedule and should have a great deal of input and thought behind it. A detailed ninety-day schedule also should be completed.

4. Project safety, fire protection, environmental considerations, and other specification regulations requiring specific plans should be outlined and tasks should be assigned to complete them.

5. Review the external and internal administrative requirements of procuring required bonds, insurance, permits, minority/woman-owned business enterprises (M/WBE) requirements, and other miscellaneous items.

Figure 6–3 presents an example of a preconstruction planning and organization checklist, which can be used to set the agenda for a preconstruction planning meeting.

Preconstruction Meeting With Subcontractors

A preconstruction meeting with subcontractors is not always held but has some merit for more complicated projects. A meeting of all subcontractors is not possible until subcontractor scope, Disadvantaged Business Enterprises (DBE) selections, subcontractor negotiations, and other issues are resolved. Many problems can slow the selection of subcontractors, thus making it impossible to hold a pre-job meeting with all of the subcontractors involved. If relatively few problems are anticipated with sub-

FGH Construction, Inc.
Pre-Construction Planning and Organization Checklist

Project General Description:
Local agencies:
  License Requirements
  Building permits
  Any special air & water
  requirements
  Demolition permits
  Power Companies Name & Phone
  Water Companies Name & Phone
  Waste Disposal Name & Phone
  Local ordinances:
    Noise
    Street cleaning
    Special use permit - alleys &
    Streets
    Bond require. for
    Street excavation
Communications
  Telephone Comp. Name & Phone
  Local Phone Book
  Jobsite Mailing Address
  Postal Service
  Federal Express Information
Project Services
  Transportation:
    Air Transportation
    Local Maps
    Access Roads
  Weather
  Special Conditions
Local Labor
  Non-Union
  Hiring Halls
  Subcontract Language
Staffing and Responsibilities
Personnel
  Project Manager
  Project Engineer
  Office Engineer
  Field Engineers
  Superintendent
  Assistant Superintendent
  Carpenter Foreman
  Labor Foreman
  Other Foreman
  Crane Operator
  Secretary

Rental equipment procedures
  Office Requirements
    Facilities:
      Trailers - Contractor
      Trailers - Owner/Architect
      Storage
      Testing
      Other
    Office Equipment
      Computers
      Faxes
      Telephones
      Copy Machine
      blueprint Machine
      Radios
Budget and Costs
  Estimate
  Labor Recap
  Budget Recast
Construction Plan & Preliminary Schedule
  90 day schedule
  preliminary schedule
  submittal log
  equipment procurement & expediting
log
Safety, Fire Protection, Environmental Considerations, Special Requirements
  MSDS program
  Tailgate Meeting Schedule
  Local Hospital:
    Phone Number
    Time to and from Jobsite
  Other emergency number
  Job site documentation requirements
External Administrative Requirements
  Payment schedule
    Billing form
  Substitution Procedures
Internal Administrative Requirements
  Project Reports
  Payroll processes
  Use of Jobsite checks
  Procurement procedure

FIGURE 6–3   Preconstruction Planning and Organization Checklist

contractor selection, this meeting is helpful in completing project pre-planning. Material problems, work sequence, access and subcontractor storage, and hoisting arrangements are among the issues that can be discussed. The meeting also provides time for the contractor to discuss the contractor's administrative processes for submittals, payment, and jobsite safety. Figure 6–4 illustrates a sample agenda for a preconstruction meeting with subcontractors.

Project Preconstruction Meeting

The project preconstruction meeting is the meeting that "kicks off" the project. It is the meeting that first establishes the relationships that will continue to develop as the project is built. The preconstruction meeting's agenda will discuss and clarify critical areas that impact on project

FIGURE 6–4
Agenda, Precon-struction Meeting With Subcontractors

FGH CONSTRUCTION COMPANY, INC.

Pre-Construction Meeting with Subcontractors

Agenda

Date : 5/16/95
Time: 10:00
Location: Contractor Main Office - Conference Room

1. Introduction of general contractor's staff and the subcontractor's staff

2. General review of general contractor's preliminary schedule

3. General input from each major subcontract
   a. Subcontract work sequence
   b. Manpower requirement
   c. Long lead items and other material problems
   d. Special equipment requirements
   e. Access and storage of materials

4. Review of general contractors administrative requirement
   a. Pay estimates
   b. Submittals
   c. RFI's
   d. Changes
   e. Use of general contractors equipment

5. Review of any special project requirements

6. General contractors safety requirements
   a. Hazardous Communication (MSDS) forms
   b. Toolbox talk
   c. Other safety issues

personnel's ability to work and solve problems. A suggested minimum agenda for the preconstruction meeting should contain the following elements:

- introduction of all of the project team members and their responsibilities
- review of the contract
- discussion and agreement to a Notice to Proceed date
- layout and field control issues
- review of preliminary schedule and sequence of work
- owner's site restrictions (for example, traffic, noise, hour of work, etc.)
- temporary facilities to be placed on site, field offices for the owner's representative, and other facilities to be occupied by testing and other inspectors
- quality control and other areas of coordination and cooperation
- a general discussion on the weekly or biweekly project meeting to be held when work commences
- administrative procedures and processes to be used on the project, that is, payment requests, change orders, change directives, authority, RFIs, field orders, submittals and shop drawing process
- special certifications and paperwork requirements, such as certified payrolls and other documentation

The preconstruction meeting should be attended by the owner's representative, the architect and his consultants, the contractor's superintendent and project manager, major subcontractors, and major equipment suppliers.

Project Meetings

Project meetings are generally scheduled on a weekly or biweekly basis. These meetings are where questions are asked and answered, and provide a forum for owner's representatives, architects, contractors, and subcontractors to discuss their concerns. It also is an opportunity for the contractor to keep the designer and owner up to date on progress, any changes affecting work, future problems that may arise, and other issues. The meeting should be held at the jobsite, if possible, and should present a positive opportunity to move the project forward.

A typical agenda for this type of meeting would contain the following items:

- review of the last meeting minutes (As minutes are sent to all parties prior to the meeting, only changes or additions to the minutes are discussed at this meeting.)
- review of the submittal status, using a submittal log as a reference

- review of the schedule, including, work completed to date; a three-week preview of work to be accomplished; and a discussion of problems related to the schedule and project duration
- review of the Request For Information (RFIs), using the RFI log as a reference
- review of substitutions
- review of changes, both new and pending, using the change order log as a reference
- discussion of new issues, such as jobsite problems, safety inspections, and other current areas of concern
- field observations

Figure 6–5 is a sample agenda for a weekly or biweekly project meeting.

FGH CONSTRUCTION COMPANY, INC.
<u>Agenda for Project Meeting</u>

Date : 6/6/95
Time: 10:00
Location: Owner Trailer - Conference Room

1. Introduction of Attendees

2. Review of last meeting minutes

3. Submittal Status Review

4. Schedule Status Review

5. RFI's Review

6. Change Order Status

7. New Business Discussion

Note: All attendees should have available to them copies of all logs, schedules, past minutes, etc.   to use during the meeting for reference and clarification.

FIGURE 6–5   Project Meeting Agenda

Construction Phase Subcontractor Meetings

Many general contractors schedule a weekly subcontractor meeting the day before weekly project meetings. This coordination meeting's goals are similar to the project meeting's. Current problems are verbalized and an opportunity is provided to discuss and solve anticipated problems. Coordination among subcontractors is one of the major responsibilities of the construction's project team. If done properly, the owner and designer are unaware of the process. The agenda for this meeting should be as formal as the other types of meetings previously discussed. Without a written agenda, no organization will exist, resulting in little or no clarification or accountability of existing problems to complete the meeting. The attendees will range from the subcontractors' office people to the on-site foremen of different subcontractors. It is important that the superintendent be well-informed of the activities scheduled for the next period, so this information can be used to set the agenda and arrange for the appropriate subcontractors to attend the meetings. This may take some coordination and management skill on the superintendent's part. Quite often, if the work scheduled does not relate to a particular subcontractor, the subcontractor probably will not attend, even if requested. If these meetings do not solve coordination problems among subcontractors, the superintendent will be involved in a number of individual meetings with subcontractors, which makes coordination more complicated, not to mention disrupting the flow of the project.

Construction Staff Meetings

Construction staff meetings are held for larger projects to coordinate different areas of work. The superintendent and project manager will provide the leadership for these meetings, with all project staff present, to discuss problems and find solutions for any existing or future problems that may arise. The superintendent, assistant superintendents, area superintendents, project manager, field engineers, office engineers, project engineer, foreman, and other key craftspersons, for example, the crane operator, should attend. Staff meetings are held strictly for contractor's personnel, but major individual subcontractors may be asked to attend if a topic of discussion might involve them. The status of submittals, RFIs, and delivery of material and equipment is reviewed at these meetings. New RFIs from the superintendents present are discussed and formalized for transmittal to the architect. The status of pending changes also should be discussed and new potential changes should be written for presentation to the owner's representative. Quality control and jobsite safety should be discussed at these meetings, including accidents that may have occurred during the week and any plans for future safety prevention. A review of weekly "toolbox" talks and jobsite safety meetings, should be performed as well. Any potential hazardous areas must be brought to the attention of all project personnel.

A three-week construction schedule is a good planning tool for proposed work. The schedule normally would examine in detail the previous week and two weeks in advance. This schedule provides direction for all project construction staff for a two-week period. A ninety-day summary schedule also is helpful for determining longer-lead items that may be delayed in the submittal process or that may have RFIs attached to them. This meeting differs from other meetings discussed in that it is an internal management meeting. Specific tasks will be assigned and accomplishments will be expected by project leaders. For instance, the field engineer may be directed to implement a specific layout; the project engineer may be requested to locate a submittal that is needed to start the purchasing process; or the project manager may be asked to solve a certain change order problem.

Specialized Meetings

Specialized meetings, coordination meetings, or specific-issue meetings can occur throughout the construction process. These are often single-issue meetings, focused on a specific topic. Some typical topics would include:

- Installation procedures for a system or material prior to installation. The concerned parties, such as the superintendent, relevant subcontractors, and material supplier meet to discuss the sequence and special concerns of installation. The architect and the owner's representative also may attend these meetings for guidance concerning the intended installation, depending upon the circumstances.
- Problem-solving for a specific conflict or area of concern on the project. Affected parties would attend this meeting.
- Meeting with an outside representative, such as a fire marshal, safety inspector, building inspector, or other related individual. Attendance at these meetings could include all field management and active subcontractors, or just specific parties, depending upon the issue.
- Payment processing, if not included in the project meeting.

Because these meetings can be time-consuming, they should be limited regarding topic and attendance. They also should promptly address the topic, come to a resolution, and adjourn.

Project Closeout Meeting

The project closeout meeting also is based on an ongoing process. This meeting provides the opportunity to define closeout procedures and for the owner to further specify any additional items needed in the operation of the facility. This meeting should be held early enough to discuss the completion of the pending punchlist and to review the status of substantial completion for the project. Discussion can concern the record drawings and operating and maintenance manuals, relating primarily to schedule, content, and presentation. Project closeout meetings can help the facilities management team assume the responsibility for maintenance and operation of the building. Another topic of importance, depending upon when the meeting is held, is the keys and keying schedule. The keying requirement usually is spelled out clearly in the specifications, but facilities management personnel usually are the recipients of the keys.

A project closeout meeting can be a last chance to adjust outstanding changes, review allowances not completely used, and discuss procedures for final billings. A list of final clean-up, window cleaning, and a demobilization schedule should be given to the architect, owner, and the owner's facilities management group.

A project closeout checklist should be developed for use by the project manager. The list should include items discussed in chapter 15. This checklist also can be used as the agenda for the project closeout meeting.

Post-project Review and Evaluation

The final act of the field management team should be an in-house project review. This review should include a formal presentation of the project from start to finish. It is the last chance to save, for historical purposes, the positive and negative attributes from the project. A review of the cost reports, productivity charts, and other important estimating data should be examined with the estimating staff. In addition to the data, a description of all of the items that might have impacted on or influenced productivity is considered. The type and size of equipment used, how the concrete was placed, the quality of the craftspeople, the weather, and the management of the process all help explain the actual unit prices developed for historical estimating data.

Part of the post-project review should contain a formal review of the subcontractor's performance. This should be in the form of a written report, detailing how the subcontractor performed, who the foreman was, what problems occurred, and how these problems were resolved. Both the quality of the subcontractor's work and the safety practices of the crews should be included on the form and in the discussion. The subcontractor's evaluations can be used for selecting subcontractors for future projects.

Negotiations

A construction project is a fluid process, requiring negotiation and compromise by most parties. Often, areas of disagreement arise that are not completely defined in the contract documents. The team represented by the owner, architect, and contractor must be in constant communication with one another to accomplish the common goal of producing a facility that meets the owner's needs, the owner's budget and time frame, and the level of quality specified by the documents. The three major contractual participants in a construction project must learn to apply effective methods for problem-solving. The solution to most problems will require some compromise by all involved parties. The goal is not to "win" every situation, but to solve each problem with minimal impact to all parties. Solving the problem quickly and effectively should be the goal in all negotiations.

The process of negotiating is meant to resolve an issue, not to position one party against another party for possible financial gain. This process may be influenced by a stronger position by one of the parties. Position strength, however, changes throughout the project and changes with different issues. At times, the architect may be in a stronger position while other times the contractor may. All negotiations should be done in good faith, with each party remaining honest. Since negotiations continue throughout the project, a continuous honesty and directness is required by all parties.

Negotiations can commence at the onset of the project, including negotiations for project price, terms, and contract duration, and can continue throughout the project. Honest and complete negotiations should be handled fairly by all parties at each point.

When the negotiating process breaks down, all parties involved move to other forms of dispute settlement. Many owners are adding formal nonbinding mediation clauses in their contracts along with the standard arbitration clauses. Most contractors will not sign contracts without the standard arbitration clause, but have no problem when an additional form of dispute resolution is introduced. Arbitration has played a key role in construction dispute resolution. The American Arbitration Association is the leading dispute resolution group in the country. Many contracts state this organization's name in the contract's arbitration clause. Section 4.5.1 of Form A201, General Conditions of the Contract from The American Institute of Architects (AIA) is shown in Figure 6–6.

The notification times and the specific flow of a dispute or claim is fairly concise. A flow chart of the project can act as a reminder of what needs to be done if arbitration occurs. The flow chart in Figure 6–7, on page 165, shows the claims process from the AIA A201 General Conditions of the Contract.

Mediation is the process of bringing together disputing parties with the aid of a trained mediator (facilitator) to discuss the problems and find

4.5   Arbitration

4.5.1 Controversies and Claims Subject to Arbitration. Any controversy or Claim arising out of or related to the Contract, or the breach thereof, shall be settled by arbitration in accordance with the Construction Industry Arbitration Rules of the American Arbitration Association, and judgment upon the award rendered by the arbitrator or arbitrators may be entered in any court having jurisdiction thereof, except controversies or Claims relating to aesthetic effect and except those waived as provided for in Subparagraph 4.3.5. Such controversies or Claims upon which the Architect has given notice and rendered a decision as provided in Subparagraph 4.4.4 shall be subject to arbitration upon written demand of either party. Aribtration may be commenced when 45 days have passed after a Claim has been referred to the Architect as provided in Paragraph 4.3 and no decision has been rendered.

4.5.2 Rules and Notices for Arbitration. Claims between the Owner and Contractor not resolved under Paragraph 4.4 shall, if subject to arbitration under Subparagraph 4.5.1, be decided by arbitration in accordance with the Construction Industry Arbitration Rules of the American Arbitration Association currently in effect, unless the parties mutually agree otherwise. Notice of demand for arbitration shall be filed in writing with the other party to the Agreement between the Owner and Contractor and with the American Arbitration Association, and a copy shall be filed with the Architect.

4.5.3 Contract Performance During Arbitration. During arbitration proceedings, the Owner and Contractor shall comply with Subparagraph 4.3.4.

4.5.4 When Arbitration May Be Demanded. Demand for arbitration of any Claim may not be made until the earlier of (1) the date on which the Architect has rendered a final written decision on the Claim, (2) the tenth day after the parties have presented evidence to the Architect or have been given reasonable opportunity to do so, if the Architect has not rendered a final written decision by that date, or (3) any of the five events described in Subparagraph 4.3.2.

4.5.4.1 When a written decision of the Architect states that (1) the decision is final but subject to arbitration and (2) a demand for arbitration of a Claim covered by such decision must be made within 30 days after the date on which the party making the demand receives the final written decision, then failure to demand arbitration within said 30 days' period shall result in the Architect's decision becoming final and binding upon the Owner and Contractor. If the Architect renders a decision after arbitration proceedings have been initiated, such decision may be entered as evidence, but shall not supersede arbitration proceedings unless the decision is acceptable to all parties concerned.

4.5.4.2 A demand for arbitration shall be made within the time limits specified in Subparagraphs 4.5.1 and 4.5.4 and Clause 4.5.4.1 as applicable, and in other cases within a reasonable time after the Claim has arisen, and in no event shall it be made after the date when institution of legal or equitable proceedings based on such Claim would be barred by the applicable statute of limitations as determined pursuant to Paragraph 13.7.

4.5.5 Limitation on Consolidation or Joinder. No arbitration arising out of or relating to the Contract Documents shall include, by consolidation or joinder or in any other manner, the Architect, the Architect's employees or consultants,

Figure 6–6   Arbitration Section, AIA Document A201, General Conditions of the Contract

except by written consent containing specific reference to the Agreement and signed by the Architect, Owner, Contractor and any other person or entity sought to be joined. No arbitration shall include, by consolidation or joinder or in any other manner, parties other than the Owner, Contractor, a separate contractor as described in Article 6 and other persons substantially involved in a common question of fact or law whose presence is required if complete relief is to be accorded in arbitration. No person or entity other than the Owner, Contractor or a separate contractor as described in Article 6 shall be included as an original third party or additional third party to an arbitration whose interest or responsibility is insubstantial. Consent to arbitration involving an additional person or entity shall not constitute consent to arbitration of a dispute not described therein or with a person or entity not named or described therein. The foregoing agreement to arbitrate and other agreements to arbitrate with an additional person or entity duly consented to by parties to the Agreement shall be specifically enforceable under applicable law in any court having jurisdiction thereof.

4.5.6 Claims and Timely Assertion of Claims. A party who files a notice of demand for arbitration must assert in the demand all Claims then known to that party on which arbitration is permitted to be demanded. When a party fails to include a Claim through oversight, inadvertence or excusable neglect, or when a Claim has matured or been acquired subsequently, the arbitrator or arbitrators may permit amendment.

4.5.7 Judgment on Final Award. The award rendered by the arbitrator or arbitrators shall be final, and judgment may be entered upon it in accordance with applicable law in any court having jurisdiction thereof.

Figure 6–6   Arbitration Section, AIA Document A201, General Conditions of the Contract continued

a middle ground or solution to which both parties can agree. Mediation is a nonbinding process, so problems are not always resolved, but this should still be the first form of alternate dispute resolution used. The American Arbitration Association's contract language for mediation that follows can be used in conjunction with a standard arbitration provision.

> If a dispute arises out of or relates to this contract, or the breach thereof, and if the dispute cannot be settled through negotiation, the parties agree first to try in good faith to settle the dispute by mediation administered by the American Arbitration Association under its Construction Industry Mediation Rules before resorting to arbitration, litigation, or some other dispute-resolution procedure.

Mediation is first used when negotiations fail to resolve a particular dispute. An outside neutral player is used as a facilitator and advisor for each party trying to resolve the dispute. The facilitator, who does not impose solutions, will make suggestions in an attempt to move the parties closer to a resolution, guiding all members involved to resolve the dispute among themselves. Arbitration, another form of dispute resolution, pro-

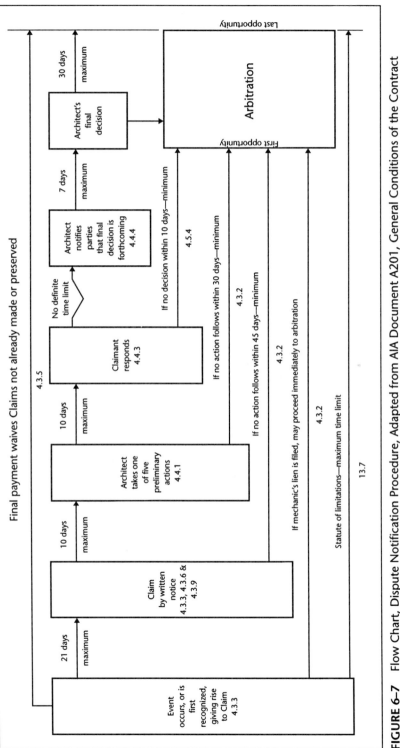

FIGURE 6–7   Flow Chart, Dispute Notification Procedure, Adapted from AIA Document A201, General Conditions of the Contract

vides an outside neutral party who not only hears the problem but will make a binding ruling. In contrast to mediations and arbitrations, negotiations do not need a neutral party to achieve an acceptable outcome. Other contractors have used the idea of dispute avoidance and resolution to include partnering strategies in their job contracts to make project members aware of and anticipate where sources of future disputes may arise. The key element to all alternate dispute avoidance or resolution strategies is to keep project members communicating and participating in the daily process of resolving the problems at hand.

Arbitration, as stated earlier, can be, and generally is, binding. Both parties select and agree upon an arbitrator or arbitrators from a list of professionals. Arbitration was developed to save time by eliminating the court litigation process. Generally, at this stage both parties will have an attorney present, but will not be represented by the attorneys during arbitration, as this process is not conducted like a court of law. Depositions will be taken before the arbitration, and attorneys' fees will not be considered part of the settlement, which makes arbitration costly for both parties.

Under the American Arbitration Associations' Construction Industry Arbitration Rules, three tracks are available. "Fast Track" rules involve claims of less than $50,000, "Regular Track" between $50,000 to $1,000,000, and "Large, Complex Construction Case Track" is for cases over $1,000,000. Highlights for these three tracks, according to the American Arbitration Association, include:

"Fast Track"

- a 60-day "time standard" for case completion
- establishment of a special pool of arbitrators who are prequalified to serve on an expedited basis
- an expedited arbitrator appointment process, with party input
- presumption that cases involving less than $10,000 will be heard on documents only
- requirement of a hearing within 30 days of the arbitrator's appointment
- a single-day hearing, in most cases
- an award no more than seven days after the hearing

"Regular Track"

- more party input into the AAA's preparation of lists of proposed arbitrators
- express arbitrator authority to control the discovery process
- broad arbitrator authority to control the hearing
- written breakdowns of the award and, if requested in a timely manner by all parties or in the discretion of the arbitrator, a written explanation of the award

- arbitrator compensation starting with the first day of service (with the AAA to provide the arbitrator's compensation policy with the biographical information sent to the parties)
- a revised demand form and a new answer form, both of which seek more information from the parties to assist the AAA in better serving the parties

"Large, Complex Construction Case Track"

- mandatory use of the procedures in cases involving claims of more than $1 million
- a highly qualified, trained Roster of Neutrals, compensated at their customary rate
- a mandatory preliminary hearing with the arbitrators, which may be conducted by teleconference
- broad arbitrator authority to order and control discovery, including depositions
- presumption that hearings will proceed on a consecutive or block basis.

Arbitration can result in a compromised solution for all parties, but most often this is not the case, and unfortunately, this can take a long time to complete. The wise contractor, architect, engineer, owner, and subcontractor will fairly and with sensible compromise, solve and implement solutions to problems before, during, and after the completion of a project.

If all members of the construction team are professional and informed participants, all changes and claims can be negotiated without mediation, arbitration, and litigation. Most problems that arise in construction are not black-and-white, but if all parties are looking for solutions, results are much easier to accomplish.

Summary

Meetings are a major form of communication during the construction project. Some types of meetings held for a construction project include:

- contractor's preconstruction planning and organization meeting
- preconstruction meeting with subcontractors
- project preconstruction meeting
- project meetings
- subcontractor meetings
- partnering meetings
- contractor staff meetings

- project closeout meetings
- post-project evaluation meeting

Agendas prepared prior to the meetings are essential to meeting organization. Minutes of all meetings are necessary to provide a record for reference of the meeting and discussions.

Negotiations are an important part of the construction project, from start to finish. Negotiations need to be made in good faith, with all parties expecting compromises. When negotiations fail, mediation and arbitration can settle project disputes.

CHAPTER 7

JOBSITE LABOR RELATIONS AND CONTROL

The term **labor** describes craftspeople who actually perform construction work, from the foreman to the individual tradesperson. The term "labor" indicates direct labor, or craftspeople employed by the contractor at the jobsite. The contractor pays the wages, fringe benefits, taxes, and workmen's compensation for these employees and is responsible for managing the direct labor on the jobsite, through the superintendent and through field management. The contractor provides supervision, task direction, material, and equipment for direct-labor employees. Because of this responsibility, the contractor also controls the labor and has flexibility in moving employees from task to task; an effect on employees' productivity; control of duration of activities; and control in fluctuating the crew size, as necessary. The contractor does not have direct control of subcontractor's labor, as these employees are controlled by the subcontractor's management. The contractor's use of direct labor depends on project requirements, labor force available, subcontractors available, local customs, company policy, and the expertise of the contractor's field management.

The contractor employs a variety of different craftspeople, including laborers, carpenters, operating engineers, and ironworkers. These employees are assumed to possess the knowledge and skill necessary for their crafts. Some contractors require that employees perform a variety of craft-related tasks, however, it is assumed that they are skilled in each task. The contractor must depend on a certain level of productivity of employee labor for each craft, so reliable cost estimates, crew compositions, crew assignments, and schedules can be prepared.

This chapter concerns the management of jobsite labor. An assumption is made that if the employee is classified in a certain craft, that is, carpentry, then they are skilled in that craft and are fairly equal in productivity potential. Differences exist in crafts, productivity rates, and customs in various geographical areas.

Labor Productivity

Labor productivity can be defined as the rate at which tasks are produced, especially the output per unit of labor. The goal in managing construction tasks is to produce the optimum output per labor hour. The output desired naturally complies with project requirements for materials, construction techniques, appearance, performance, and workmanship. Productivity rates should be established in the estimate regarding perception of conditions and project requirements. The productivity rates used in the estimate are generated from historical data, both from the contractor's own records and from a variety of available references.

Many different factors can affect labor productivity on a project. Project supervisors should be familiar with the most common factors affecting labor productivity, which include:

- lack of supervision or poor supervision
- lack of coordination of subcontractors with work activities
- improper or insufficient material available for tasks
- poor jobsite layout
- lack of proper tools for work activities
- congested work areas
- poor housekeeping
- accidents and unsafe conditions on the jobsite
- excessive moving of craftspeople from project to project
- adverse weather conditions
- poor lighting in the work area
- inadequate heat or ventilation in the work area
- tardiness of excessive absenteeism
- uncontrolled starting time, quitting time, coffee breaks, and lunch and breaks
- shortage and location of close parking, changing rooms, rest rooms, and drinking water
- high employee turnover
- use of improperly or poorly trained craftspeople
- supervisors not making timely decisions
- poor attitude among employees
- poor use of multiple shifts or overtime
- construction mistakes caused by complexity, poor drawings, or lack of communication
- impact of changes on production work

Most superintendents and foremen would admit that these are not revelations. Unfortunately, supervisors are aware of these problem areas in a broad sense, but sometimes are not cognizant of them on specific projects with their particular crews. The most serious problems are the ones that impact upon productivity without the superintendent's knowledge. Solutions are available for all productivity problems, but problems must be communicated to the superintendent with sufficient time to successfully alter conditions. Serious problems arise for many reasons. A discussion of these reasons follows.

Impact of Changes

Momentum or flow of work can be disrupted by changes in the work. If an error occurs in the construction drawings or if a latent condition is discovered during a work activity, a delay will occur. Changes will often consume field personnel, resulting in neglect in continuing other areas of work. Frequent changes produce confusion in crews, which reduces productivity. Crews often have a different attitude to work changes, consequently working without the same motivation to complete activities quickly and efficiently. The removal of recently completed installations can demoralize crews and slow productivity.

Although changes during construction are often beyond the control of the contractor, the field staff can minimize the impact of changes to construction productivity. Knowledgeable examination of the construction documents can reveal problems prior to crew assignments. It is desirable to solve problems before tasks are assigned. When changes are implemented, crew directions should indicate these changes. Construction documents must be notated to show changes accurately. The field management staff should work carefully with the crew to keep the momentum going, both in the area of change and in other areas of work.

Poor Weather

Bad weather often is not adequately anticipated, forcing changes in schedules, production, and damage to completed work. Productivity decreases in poor weather to varying degrees, depending upon the severity of the weather and the work tasks. Weather can affect some construction materials, such as concrete and mortar, as well as the efficiency of the labor. When protective clothing, such as rain gear or cold-weather gear is necessary, labor is impeded.

Initial project planning should consider seasonal weather conditions. Anticipating bad weather and planning weather-sensitive activities around it can reduce some of the impact. Flexibility should be built into the work schedule to allow for down time during inclement weather. When cold or rainy weather is predicted during construction activities, an analysis is necessary to compare the productivity loss with the cost of temporary weather protection, such as a tent and temporary heat. It often is desirable to protect the work area and continue with activities during these periods.

Special considerations are necessary for crews working in non-ideal weather conditions. During hot weather, enough cold drinking water must be furnished at the installation location. Coordination of work activities during cooler times of the day may be advisable. Shaded areas are often needed for work areas, as are wind breaks to control wind and dust effects on the construction installation.

Rainy weather requires protection of the work area, maintenance of haul roads and jobsite access, and dry shack provisions for changing clothes and eating lunch. In fine-grained soils, rain will produce mud, which slows work considerably. Preparation for these conditions can be made, with provisions such as preparing haul roads and jobsite access with a gravel surface and adequate drainage.

Material Problems

Late deliveries require crews to move to other work areas, halting the production at one area and requiring startup of new work activities. Also, shortages can stop crews from working, forcing workers to be laid off for short periods of time. If crews are required to carry materials long distances before work can occur, productivity is affected. Double-handling of materials will result in damage to materials, increased waste, and lost time.

Jobsite management is responsible for ensuring that the correct amount of material is available for installation and for appropriate jobsite storage of material. Careful attention to the construction documents, site conditions, work assignment packages, and progress of the project is necessary to order and schedule delivery of materials. Coordination of the procurement schedule, as discussed in chapter 3, is extremely helpful when material is being delivered to the jobsite to coincide with construction activities.

The material supplier may be responsible for incomplete deliveries, improper material delivered, defective material, and late delivery of material. Confirmation with the supplier prior to delivery is helpful in receiving the proper material for the project. Some material contracts and purchase orders have provisions for backcharging the supplier for damages, however, this is often not feasible.

High Labor Turnover

High labor turnover may be an indication of poor planning, general unrest, and lack of leadership by the foreman and/or superintendent. New training may be required each time this situation occurs and high levels of unrest, or low morale, can lead to a slowdown of work.

Construction labor is fairly mobile from contractor to contractor. The contractor may be using employees who are not familiar with the type of installation needed or the methods. Training may not be consistent in the available labor force.

Jobsite management must employ the correct labor for the project. It is important that jobsite management be experienced in the type of construction on the project. Many contractors have teams consisting of a superintendent, field engineers, and foremen, and a few craftspeople who will remain from project to project. This consistency helps screen appropriate labor for the project. Despite the fact that foremen are technically considered labor rather than management, they are an important bridge between management and labor, leading the crew in techniques, methods, and productivity. Trained foremen in labor techniques, safety procedures, crew management, productivity control, and cost control are essential for effective labor management. Every project has a selection period for its labor crews, which involves hiring and firing, but astute labor managers will keep this period to a minimum, selecting the proper crews quickly. The size and quality of the labor force in the community at the time of the project has a major effect on this type of labor problem, but active management can optimize each situation.

Accidents and Unsafe Conditions

All work will quickly come to a halt when an accident occurs on a jobsite. Employees who do not feel safe tend to be overly cautious when performing their work tasks, thus noticeably slowing production. A clean and safe jobsite is conducive to obtaining maximum productivity from labor crews.

Many construction companies realize the importance of safety on jobsite production. Contractors are required by state and federal regulations to have in place an active construction safety program, but the degree of implementation on the jobsite is a matter of commitment by the construction company. The impact of a safe and clean jobsite on productivity cannot be understated. The cost of keeping a clean jobsite is considerably less than the cost of lost productivity, accidents, and safety fines associated with poor housekeeping.

Working Overtime

Working longer days or adding additional days of work on a prolonged basis will result in increased injuries and safety problems. As workers become tired from longer periods of work, they begin to adjust their pace or slow their productivity to avoid fatigue. When forced to work overtime, an employee may become disgruntled, causing low morale among other employees. In the same respect, if overtime is offered to one employee and not another, jealousy may become a problem among employees.

Overtime is expensive to the contractor, both in additional wages and lost productivity. Although conditions exist when overtime is necessary, this can usually be avoided by proper project planning and crew size. Careful analysis of work tasks can reveal opportunities to use multiple or larger crews to facilitate work during regular shifts. Using the construction schedule to determine the time frame available for the tasks related to crew output should help determine the number and type of crews. Particular job conditions must be considered in this determination. Labor supply largely controls these strategies. Care must be taken to maintain effective leadership of added crews, with capable foremen and field management personnel.

When increased production is required for a long period of time, more shifts may be added. In conditions where extra crews are not efficient, the second shift can add production. Although there are additional costs associated with a second shift, such as a labor premium and additional supervision, the costs of a double shift are usually much less than overtime premiums. The double shift is efficient when all applicable trades participate.

Projects in Existing Facilities or Congested Areas

Areas where finishes are in place or where the client's employees are working will cause a slowdown in construction productivity. Either the craftspeople will damage completed finishes, requiring replacement or touchups, or work will slow to protect work already in place. When areas become too congested, with more than one trade working in the same area, productivity is greatly affected. None of the trades can accomplish their tasks as efficiently as they can in an area where they may be the only trade. When working around the client's employees, special precautions may have to be implemented, again slowing productivity.

In remodeling, addition, or renovation projects, the productivity rates in the estimate normally reflect the project conditions. As congested and occupied conditions can significantly impact productivity rates, careful planning and crew supervision is necessary. Even when work is required in occupied spaces, planning can isolate construction work from other functions in the space. Isolation of construction activities from occupant functions is important in achieving the necessary productivity. Thorough planning is necessary to maintain productivity rates under these conditions, such as having the appropriate amount of materials transported to the work location; having the proper construction equipment for conditions; isolation of the workforce from distraction; having the proper environmental conditions—light, heat, ventilation; control of dust and noise from occupied areas; and timely clean-up of congested work areas.

The occurrence of any of these factors that affect productivity often are beyond the control of the superintendent and foreman, but understanding their potential impact on productivity may help mitigate the problem. Also, in discussing these troublesome situations beforehand, any problems outside the scope of the contract can be discussed as possible adjustments to the contract. Awareness of these factors by jobsite management can help eliminate their negative impact.

Jobsite Labor Organization

An understanding of the typical hierarchy of labor and management on a construction project is necessary when building teams to construct the project. Each of the major work activities of the project are packaged into smaller tasks, with assignments of specific trades or a mixture of trades to accomplish these tasks. The superintendent manages jobsite activities. Some projects with distinct project elements, such as separate buildings or structures, will have more than one superintendent. On large projects, assistant superintendents, project engineers, or area superintendents may be used to control segments of the work. These assistants may even have some specific specialties themselves, such as mechanical or electrical skills. Field engineers assist other jobsite management with layout, shop drawings, and miscellaneous project activities. They usually are salaried employees, at the bottom tier of the management level.

Labor consists of four tiers: foreman, lead craftsperson, journeyman craftspeople, and apprentice. All of these positions are hourly, and when the crew is unionized, all belong to the same union. Foremen and lead craftspeople normally receive a premium, per hour, for their responsibilities, such as one or two dollars per hour above the craftsperson's wage level. The craftspeople are broken into skill and training levels, the highest being a journeyman, the next being an apprentice (first-year, second-year, etc.), the last, a helper. Where there is no formal training, the term **apprentice** is not applied. The definition of an apprentice is one who is in training under the supervision and mentoring of a person skilled in his or her trade. Most apprenticeship programs include regular training off of the jobsite. The term **helper** is used for an unskilled person who has no training and is not enrolled in a training program for a specific trade. Not all trades recognize the helper classification. Figure 7–1 indicates a typical jobsite labor hierarchy.

Labor Agreements

Commercial building construction is accomplished by both union and open shop (non-union) construction trades. Union affiliation is a decision made by the employees. Many areas of the country have both union contractors and non-union contractors competing for the same construction work. Different mixes of union and open shop contractors exist throughout the country, varying from all union to all non-union construction workers.

When a contractor employs union labor, a **labor agreement** exists between the contractor and the union. This agreement is a contract between the contractor and union, requiring certain responsibilities of both parties. The contractor and union, upon signing a labor agreement, are contractually bound to each other. The union agrees to provide trained craftspeople and the contractor agrees to provide specific wage rates, fringe benefits, and working conditions. The union maintains a trained hiring pool of journeymen and apprentices who are furnished to the contractor.

The agreement specifies procedures for hiring; work rules, such as hours and working conditions; and compensation to the employee, such as wages, fringe benefits, travel pay, and subsistence reimbursement. The agreement between the union and the contractor may be individually

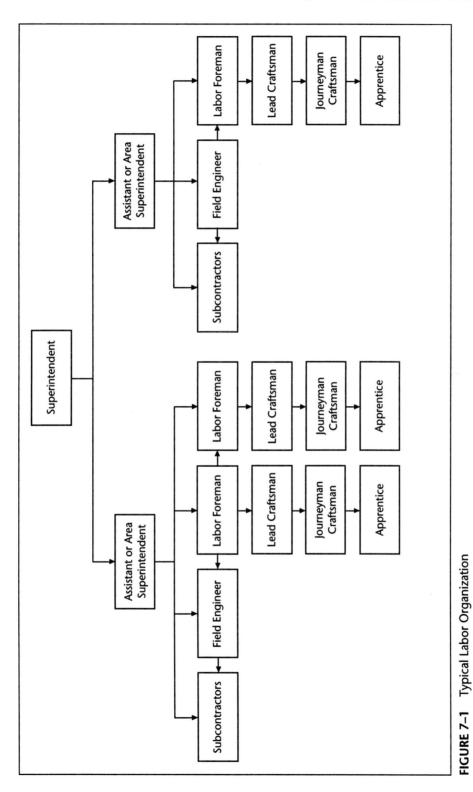

FIGURE 7–1 Typical Labor Organization

negotiated between separate contractors and the union, or between a management group and the union. A management organization, such as the Associated General Contractors of America, Inc., can represent its member contractors, who assign their bargaining rights to the management organization.

The labor union utilizes business agents who represent union members, ensuring that contractors live up to their agreement. Shop stewards are union members who are employed on the jobsite and who represent union employees on the jobsite. Union leadership can provide an adversarial role or a facilitator role. With increased competition by open shop contractors, union leadership can soften its role and actually help the contractor by furnishing highly trained and qualified labor.

The typical labor agreement covers work, conditions, and wage rates. The following articles are contained in the agreement between the Inland Northwest Chapter of the Associated General Contractors of America and the Washington State Council of Carpenters:

- a statement of purposes—promote cooperation between the contractor and local union organization
- the parties' and management's rights under the labor agreement
- rules governing strikes, lockouts, and slowdowns
- the agreement's area of jurisdiction
- the issue of using only subcontractors that also are signatory
- definition of pay day, holidays, work week, work day, shifts, when overtime is defined
- agreement on how grievances and jurisdictional disputes will be settled
- employee substance abuse, safety and accident prevention issues
- wage rates for different classifications of carpenters
- hiring hall procedure, work rules (tools, transportation, appointment of foreman)
- apprenticeship training and classification

A general contractor typically will have labor agreements only with a few unions, which will be hired directly. A general contractor might have labor agreements with the following unions: carpenters, laborers, cement masons, ironworkers, operating engineers, and teamsters. A contractor may only have agreements with carpenters and laborers, subcontracting out tasks that involve other trades. The labor agreement will contain a **subcontractor clause**, which will prohibit the contractor from subcontracting work to non-union firms for a particular labor classification. For instance, if the contractor has an agreement with the carpenters' union, the contractor is required to either accomplish tasks such as concrete forming with its own forces or with a subcontractor that employs union carpenters. This contractor can, however, hire a non-union excavator, as the contractor does not have an agreement with the operating engineers' union. Despite having agreements with one or more unions, the contractor can subcontract work to either union or non-union contractors who do not accomplish work listed under the work description in the union agreements.

Dual gates may be necessary on projects using both union and non-union workers. These are used to separate potential union and non-union workforces, pursuant to National Labor Relations Board procedures, to avoid secondary boycotts by one group over another nonparticipating firm. Dual gates apply to admission of the general contractor's craftspeople as well as subcontractors' employees and the delivery of materials to the project. Merit shop and non-union construction firms will plan into the job the use of dual gates, which includes security fencing, location of entrances, parking, deliveries, job trailer locations, and other pertinent site layout. One gate is set up for the entrance and exit of union firms and employees, the other for non-union firms and its employees. Laws and rules governing picketing of a project require that picketing cannot occur at the gate reserved for the neutral employer, the non-union firms not involved in the labor dispute. Picketing can only target the specific or primary firm engaged in the labor dispute. The use of dual gates should be discussed with all contractors and subcontractors at the beginning of the project, or as subcontractors move onto the site. Gates can be adjacent to each other, but by placing them apart potential picketing conflicts are eliminated. See Figure 7–2 for a sample gate sign.

Supervision and Control of Labor

The Superintendent

The superintendent is the contractor's jobsite representative and quite often the only person who can truly control costs, time, and quality. The superintendent is responsible for timely productivity within the company and for coordinating subcontractors on the job, which requires subcontractors' work to be completed on time and within the specifications and subcontracts.

A superintendent's typical day, if there is such a thing, starts with a walk on the jobsite but the superintendent is the last person on the site to leave. The superintendent's first concern is to lay out the day's schedule for his own forces, who usually begin work at 7:00 A.M. If the work is not ready and preplanned, worker productivity is at a standstill. Foremen must be informed and ready to direct craftspeople to the area where they will be working and to the particular tasks they will be expected to accomplish. Most work activities utilize a combination of trades and take multiple days to complete. The superintendent must possess both the understanding and visualization of what work needs to be done, the ability

FIGURE 7–2
Gate Sign

THIS GATE TO BE USED ONLY BY THE FOLLOWING EMPLOYEES, MATERIAL PEOPLE AND SUPPLIERS OF THE STATED FIRMS:

FGH Construction Company, Inc.

RANDOM Electric Company

JOY SHEETMETAL SUPPLY COMPANY

ALL FIRMS NOT LISTED ABOVE MUST USE OTHER DESIGNATED GATES.

Specifications:

Dimensions: 4' by 6'
2" Lettering
Bold typeface in red
Firms in black
Background, white

to plan the progression and completion of work activities, and the ability to tie each work activity to the next. All employees who are in supervisory roles must have an overall vision of the project, as well as an understanding of the thousands of individual tasks and work items that must be accomplished to complete a project. Work items for the general contractor and subcontractor must all be coordinated due to their interrelationship. For instance, the electrical conduit may pass through the concrete floor done by the general contractor into a metal stud wall done by another subcontractor or into a ceiling space occupied with HVAC ductwork, water lines, drain and vent line, sprinkler lines, conduit for cabling, lighting, and a number of other items. This coordination and completion of work is the main responsibility of the superintendent, jobsite management personnel, and foremen.

What follows is a description of a superintendent's ordinary day on the construction site.

On this particular day, the weekly toolbox safety talk is scheduled to start. The superintendent leads the discussion on the topic of confined spaces. He will then describe where and how this is applicable to the project the craftspeople are working on. It takes about fifteen minutes. The crafts then break up for direction on work activities of the day. The superintendent has prepared formwork detail drawings for one of his foremen and has given it to him. In passing, the superintendent mentions that he is expecting a delivery of miscellaneous metals later in the day and will need the rough terrain crane for unloading. Later on a job walk, a short discussion will occur between the subcontractor's foreman and the superintendent, as he observes the subcontractor's work, the number of craftspeople the subcontractor employs, the location of the project where they are working, and other coordination problems that may occur. On the way back to the trailer, the field engineer stops the superintendent with a layout question about a wall that the carpenters will be starting in two days and questions the arrival status of the inserts from the miscellaneous metals fabricator. Back at the trailer, the superintendent checks the scope of work and looks for an approved shop drawing from the fabricator. A call to the fabricator indicates that some inserts will arrive today but they are not sure if they are the correct ones. The project engineer asks for verification of shop drawing approval and checks the truck's delivery slips for the proper inserts. A call then comes over the radio from the carpenter foreman concerning a location question on mechanical sleeves that will be installed by the subcontractor in the concrete wall the next morning. Finally, the superintendent sits down for his first cup of coffee when low and behold, the owner, construction manager, and inspector all want to have a meeting. The day will continue at this hectic pace until the crafts leave at 3:30 P.M. Then, and only then, can the superintendent start his planning for the next day, week, and month. Prior to leaving for home, the superintendent will record his day's activities in reports and in his daily diary.

The superintendent is the principal manager of the work, oversees the subcontractor as the work is being accomplished, supervises the general contractor's own labor forces, and correlates all of the functions of the site construction staff. Along with the foreman on the project, the superintendent provides the leadership and motivation to accomplish assignments. Individuals who aspire to become superintendents will display the following characteristics:

- visualization and planning skills
- the ability to organize and motivate people
- understanding of technical and mechanical subjects
- ability to make decisions
- innovative and resourceful
- ability to adapt to changes
- along with leadership, the ability to instruct, teach, and train
- strong work ethic
- the capability to work with other team members as an equal partner

The Foreman

The foreman is the critical link to the craftsperson. The foreman is in a unique position—on one job, he is the foreman, while on the next job with a different company, he may be one of the crew members.

The construction industry and the companies within the industry hope to keep all of their trained craftspeople and key foremen working year round, and year to year, but, realistically, only some firms maintain a steady workload. The job of the foreman is to "push" the work. **Pushing the work** can be defined as evaluating the workman, making sure that his skills fit the work, and that he aggressively completes the work at hand. A good foreman will know how to efficiently use the labor, equipment, and materials to accomplish day-to-day tasks. The foreman is the individual who understands the technical part of the work. The foreman should develop loyalty and responsibility to the company. Companies have found that when the foreman is isolated or not considered part of the company, theft and poor workmanship from other trades increases, friction or polarization of trades starts to occur, and material waste increases. All of this points to the important role of the foreman. The foreman is the eyes, ears, and often the spokesperson for the superintendent. On smaller projects, the superintendent can be a "working" superintendent. The term **working** refers to the superintendent who is actually performing work with his tools. In many instances, the foreman is elevated to part-time superintendent. In that case, the foreman also can be called a working superintendent.

Overtime

One key decision made by the superintendent is the use or authorization of overtime. The reason for overtime is project-, task-, and time-specific. There is no set of rules to govern when to use overtime. The following list provides guidelines to help the superintendent make this decision.

- If the equipment used is very costly, compared to the labor expended, overtime may be appropriate.
- If bad weather will greatly affect the time to complete an activity and the cost to stop and start the activity is costly, overtime may be appropriate to complete the activity.
- Only truly critical activities should be viewed for overtime.
- Overtime should be used selectively to bonus a foreman or specific craftsperson. (Special care should be taken when used in this manner.)
- On some tasks, labor shortages can be anticipated, such as finish carpenters, thus overtime may be used.
- Some tasks cannot be done during poor weather or specific equipment will not function, thus overtime should be considered.

Employee Relations

The superintendent has the responsibility of hiring and firing employees. Generally it is the superintendent's direct responsibility to determine whether an employee has the skills to perform the required task, if the employee is working up to a normal productivity level, or if the employee is performing to the required level of quality workmanship. A major time when employees are laid off is when work is nearing completion. The superintendent will determine, with the help of the foreman, which workers are to be kept and which will be terminated for "lack of work." Termination for lack of work is not viewed as a negative discharge.

Employee Training

Formal craft training has typically been done in the union sector of the industry. Most states allow only one approved active apprentice training program. Most apprentice training will be done by the appropriate union with which the apprentice is affiliated. Today, this is being challenged by the merit shop organizations. Associated Builders and Contractors (ABC) is the leading organization to develop training programs outside of the unions. An example of the type of training that occurs is represented by the carpenter's union. An individual is an apprentice for four years prior to becoming a journeyman carpenter. The apprentice will take classes while he or she works in the trade, giving the individual the dual "real" world work and the "hands-on" lab and classroom training. Many firms prefer to have some apprentices on the job because they are generally eager to learn, work hard, and they have a sliding pay scale based on the number of years of apprentice training they have accomplished. This scale is negotiated in the master agreement, discussed earlier.

Historically, some vocational training has occurred at community colleges, and in some states, special vocational training centers and schools are active in craft training. Generally these programs do not have a mandatory work rule, so training only occurs in the lab and classroom. Many studies have noted that the construction industry critically needs skilled tradespeople.

Tools

The use of tools is another area of concern to the superintendent and foreman, who must ensure that the proper tools are available to the craftsperson. In many areas, trade practices or labor agreements establish what tools the craftsperson will provide and what tools the construction company is responsible for providing. In most commercial projects, all power tools will be provided by the company, along with cords and ladders. The craftsperson may be required to have a specific selection of hand tools, which varies from craft to craft and from residential to commercial work. Tools are almost never provided or even loaned to subcontractors and their craftspeople. A typical list of hand tools a carpenter would provide includes:

- hand-held hammers
- hand saws: ripping, finish, hacksaw, coping saw
- screwdrivers, chisels
- wrenches, pliers
- hand levels
- miscellaneous hand tools

  Smaller tools typically furnished by the contractor include:

- hand shovels, picks, sledge hammers
- circular saws, drill motors
- electrical cords
- pneumatic tools and compressors
- powder actuated tools

The foreman also is responsible for tool security. Two levels of security need to be established. One, to prevent loss of tools when the project is shut down after work hours and weekends, and two, to minimize minor theft of small items during the business day. Most large losses occur from theft after-hours and on weekends. Project sites are often isolated, dark, and lack security personnel. Tool security generally is accomplished with the use of lockable tool boxes, which must be left in secure places but left close enough to the work area to provide reasonable tool distribution in the morning and easy access to putting away tools at the end of the shift. If the tool box is near the area where crafts are leaving the jobsite through uncontrolled exits, theft may occur. All tools should be marked with company identification. A company should not allow workers to take tools home for personal use, as they may be forgotten and cannot be used the next day.

Labor Records

In conjunction with the processes of leading, directing, organizing, and planning work to be accomplished, all firms must implement a labor reporting system that will provide information to accounting for payroll, information for historical reporting of labor productivity, and information pertaining to cost and production control during the job. This system and the information generated must be simple to use and must provide for the level of data needed for future estimating. Because activities can be quickly completed and productivity influenced by so many factors, the system must also provide timely information for the control of the labor or cost items involved. Most direct labor on projects today are paid weekly from information generated by time cards. Additionally, the time card will show the labor distribution or project cost account that the labor is applied against. The report generated from the time card is called a weekly labor report, which shows weekly and cumulative labor costs versus budgeted hours and dollars. An additional report used in combination with the weekly labor report is a weekly quantity report, also based or collected on a project cost account basis, which allows a comparison to be made of man hours or the cost of labor to the quantities. The list of project cost codes is carefully assembled and is unique to each project. Similar work from project to project will have the same cost distribution code or project cost account code, making the historical comparisons somewhat easier. This comparison is a difficult task, as jobsite conditions vary from project to project. If one project is done in the rain while another is completed during a dry time of year, increased productivity would be expected from the craftsperson doing the work in the dry weather conditions. Labor cost accounts concern "self-work" or work that will be performed by the firm's own personnel. These craftspeople will then be on payroll and will be subject to direct control of the field personnel, as most work activity has additional costs associated with it. A separate code must be developed for differentiating between labor costs (-01), material costs (-02), equipment costs (-03), subcontractor costs (-04), and other costs (-05). The following list illustrates the cost codes for project elements on a particular project.

Typical Master List of Project Cost Accounts For Concrete Only

Code	Description	Code	Description
03.00	**CONCRETE WORK**	03.170	Concrete Stairs
03.010	Continuous Footing	03.180	Precast Concrete Stairs
03.020	Grade Beams	03.190	Apron Slabs
03.030	Column Footings	03.200	Sidewalks
03.040	Pilaster Footings	03.210	Concrete Driveways
03.050	Piers	03.220	Mechanical Bases/Pads
03.060	Pilecaps	03.230	Transformer (Electrical) Pads
03.070	Thickened Slab @ Bearing Partition	03.240	C. I. P. Curbs
		03.250	Pilasters
03.080	C. I. P. Walls	03.260	Tiltup Walls
03.090	C. I. P. Columns	03.270	Other Precast Work
03.100	C. I. P. Beams		
03.110	C. I. P. Spandrels		Open for Additional Codes
03.120	Interior S. O. G.	03.400	Spandeck
03.130	Floor Hardener/ Sealer @ S. O. G.	03.600	Inspection and Testing of Concrete
03.140	Structural Slabs	03.750	Reinforcing Steel
03.150	Lightweight Concrete Fills/Topping	03.760	Reinforcing Accessories
		03.770	Slab Dowels
03.160	Grout Column Bases	03.780	Wire Mesh

It should be noted that the type of work this contractor may be performing is defined by the codes he has chosen to track his costs. Also note the lack of codes in the area of form work, finishing, and rubbing and curing.

The originating document for labor information is the employee's time card. Time cards can be individual, grouped on a daily labor time card by trade, or all crafts may appear on one time card. If multiple trades are involved and each trade has its own foreman, grouping by trades works well. If the project size is very small, the other two forms work equally as well. It is important that the distribution of time, as applied to the various codes, be accurate. This means the person in charge of supervision who is closest to the work should fill out the distribution. Typically, this is done at the foreman level. Figure 7–3 is a typical sample daily labor time card that mixes carpenters and labors, with their hours worked, spread to multiple codes.

The time card is used in two ways by supervisors assigned to the project. First, management or office personnel use the time card to verify ac-

Project Creston Water Treatment Plant
Date 5/6/97

Weather Sunny
Prepared by T. Bono

No.	Name	Craft	Time Classif.	Hourly Rate	Cost Code 3.010	3.020	3.030	3.120	3.801	Total Hours	Gross Amount
24	Ramsey	CF	RT	$25.00	8					8	200.00
			OT								
13	Hojas	C	RT	$22.00	2	6				8	176.00
			OT								
55	Gin	C	RT	$22.00			5	3		8	176.00
			OT								
122	Drew	L	RT	$15.00		1		2	5	8	120.00
			OT								
145	Elliot	L	RT	$15.00		2	2		4	8	120.00
			OT								
			RT								
			OT								
			RT								
			OT								

FIGURE 7–3　Time Card

Project	Creston Water Treatment Plant					
Week Ending	5/8/97		Weekly Quantity Report			
Prepared By	K. Franklin					

Cost Code	Work Activity	Unit	Total Last Week	Total This Week	Total To Date	Total To Complete
3.010	Continous Footing	Cyd.	60	20	120	0
3.020	Grade Beams	Cyd.	85	10	95	0
3.030	Columns	Cyd.	0	0	0	125
3.120	S.O.G.	Cyd.	150	60	280	340
3.801	Finishing - Trowel	Sft.	2200	1250	4400	6550

FIGURE 7–4   Weekly Quantity Report

curate hours worked by employees. If time is being shown by employees who are absent from the project, theft of company money is occurring. This is a serious crime on the employee's part and may result in the company requesting law enforcement. Second, the information is used to set productivity standards for future estimates in the bidding process. These labor unit costs become one of the most important pieces of information the field can supply the company when matched with weekly quantities. The weekly quantity report, shown in Figure 7–4, compares work activities with cost codes and the amount of work accomplished last week, this week, to date, and amount to complete.

The weekly quantity report and time card are two pieces of information combined by the accounting office and returned to the field in the form of a weekly labor cost report, shown in Figure 7–5. This is one of the main tools used by the field staff to analyze the progress and project the final cost and schedule for the different work activities that are occurring.

Weekly Labor Cost Report

Project __Creston Water Treatment Plant__

Date__ 5/6/97 __

Week ending ____ 5/8/97 ___   Prepared by _K. Franklin_____

Cost Code	Description	Unit	Quantity			Labor Costs			To Date		Projected	
			Budget	This Week	Todate	Budget	This Week	Todate	+	( )	+	( )
3.010	Concrete Foot	Cyd	400	20	200	4000	200	2200	200		200	
3.020	Gr. Beams	Cyd	100	10	100	2000	176	1656		344		344
3.030	Columns	Cyd	57	0	0	1710	176	176		176	176	
3.120	SOG	Cyd	257	60	60	5140	1200	1200				
3.801	Finishing	Sft	7000	810	810	7000	700	700		200		200

FIGURE 7–5   Weekly Labor Cost Report

Figure 7–6 illustrates the paper flow for labor control, and the results—a payroll check to the employee and a labor cost report to the field and office staff.

Because of the potential for abuse, the project must have a system in place to prevent misuse, to solve problems of timekeeping, to check employees in and out each day, to verify and confirm an employee's time if mistakes are made on a payroll check, and to establish an audit procedure for accurate recording of quantities and cost codes. Toward this end, some companies use a system of brass tags for checking individuals in and out of jobsites. Tags sit on a numbered board; as the employee removes the numbered tags, an empty spot is left. The tags are individually assigned and often the number represents a craft as well. This craft numbering system makes it easier to determine the number of different craftspeople on the project, for daily job reports and for the foreman's use. The foreman will know instantly who is missing, along with the timekeeper, if the project warrants such a position.

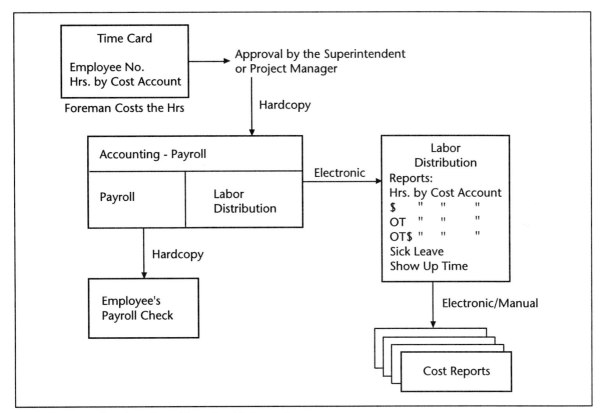

FIGURE 7–6 Flowchart From Time Card Information

Summary

The major issues concerning labor during the past decades have been maintaining an adequate number of skilled craftspeople on projects; creating a productive workforce; employing a trained workforce; and employing trained supervisors. All of these concerns still apply to today's workforce, which will be more diverse, not as mobile, with supervisors who may have less technical knowledge, and craftspersons who are not motivated by the same techniques used in the past. Projects have become more complex. Company management is beginning to rely on more information generated at the job level for competitiveness in the marketplace. Supervisors must be able to lead and motivate employees, as well as satisfy the owner and architect. Companies are becoming more client-based, requiring the project management team to become more responsive to the owner's needs, and at the same time remain as competitive and productive as possible. The craftsperson is increasingly becoming a more important member of the contractor's team.

CHAPTER 8

PERSONNEL AND SAFETY MANAGEMENT

CHAPTER OUTLINE

Corporate Safety Policy

Accident Prevention

Substance Abuse

Personal Protective Equipment

Hazardous Materials Communication

Safety Communications

Accident Reporting and Investigation

Summary

Personnel and safety management is a *must* for the construction industry—an industry with a notoriously high accident and fatality rate. Accidents that result in injuries, illnesses and fatalities are created in many ways:

- the inherent hazardous nature of construction work.
- the many methods and types of operations needed by construction companies to meet and complete assignments, resulting in confusion about safe methods to accomplish work activities.
- each construction project is unique, thus poses separate circumstances and safety hazards; work is not confined to a single workplace, but multiple jobsites.
- each construction activity has a relatively short duration, thus safety procedures are normally not well-established; the "learning curve" for safe and productive work often is not optimized because of this short duration.
- as construction is a very mobile industry, the workforce changes frequently, with varied levels of competence and expertise among employees; these employees may not be trained in safe work methods prior to arriving on the jobsite.

The construction employer is responsible for providing a safe working environment, as well as safety training for the work activities. Through the federal Occupational Safety and Health Act (OSHA), construction industry firms have become aware of the need for highly visible and proactive safety programs. These firms also have developed the attitude that all employees, from the president to the carpenter, are responsible for safety.

Studies show that fatal accidents do not discriminate against a worker based on age, experience, union versus non-union, day of the week, craft, or position in the company. Figure 8–1 illustrates the relationship of construction fatalities with the numbers of employees from different age groups.

The lack of safety guidelines and awareness affects all aspects of a construction company, from overall profits to employee and family morale. OSHA was passed in 1970—it was the first time the federal government began imposing national safety regulations on all industries and businesses. This act permits states to pass and implement their own state OSHA bills, using the federal act as a minimum standard. California and Washington are among the states that have opted for their own method of regulation. There are many issues that fall into the area of environmental concerns or hazardous materials use and handling and/or disposing of materials, regarding safety. Many of these regulations are rooted in the Williams-Steiger Occupational Safety and Health Act (1970) and the Environmental Protection Act (EPA). An example of the combined agencies working together is the asbestos issue. OSHA is the body that protects workers who are using asbestos, while EPA sets the compliance codes for the removal and disposal of the material.

OSHA and each state affiliate require construction firms to create and implement a safety and health program, following OSHA guidelines. To

FIGURE 8–1
Analysis of Fatalities
By Age Group and
Percentage of Work-
force (Courtesy of
OSHA)

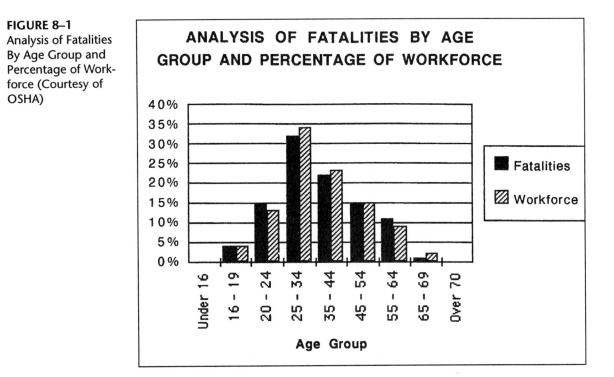

comply, this program must be a written plan containing the following key elements:

- management commitment
- hazard assessment and control
- safety planning, rules, and work procedures
- safety and health training

A suggested outline for a safety and health plan that incorporates the key elements and important subcategories has been developed for the contractor's use by the California Occupational and Health Administration (CAL/OSHA) Consultation Service. The following company safety policy outline is adapted from CAL/OSHA:

I. Corporate and Management Commitment
   A. Policy Statement
      May include safety and health goal
      Illustrates management involvement in workplace safety and health
   B. Objectives for the Safety and Health Program
      Based on the priorities of your workplace
      Should be measurable with time frames for completion
   C. Assignment of Responsibility for S and H
      Descriptions of duties
      Policy on accountability

II. Hazard Assessment and Control
  A. Hazard Assessment and Correction
     Initial survey by safety consultant or other professional
     Periodic surveys and samplings
     Employee reporting of hazards
     Tracking of identified hazards and their correction
  B. Accident Investigation
     Identification of causes and their correction
     Preventive actions
     Monitoring workplace injuries and illnesses
  C. Record Keeping
     Log and Summary of Occupational Injuries and Illnesses
     Material Safety Data Sheets
     Employee access to personal medical and exposure records
     Other required or appropriate records
  D. Equipment Monitoring and Maintenance Program
     Production equipment
     Personal protection equipment

III. Safety Planning, Rules, and Work Procedure
  A. Control of Potential Hazard
     Regarding equipment design, purchasing, engineering, maintenance and use
  B. Safety Rules
     General
     Specific to tasks, based on safe work procedures
     System for informing employees
  C. Work Procedures
     Analysis of tasks to develop safe work procedures
     Implementation
  D. Employee Involvement
     Reporting hazards
     Enforcement of rules
     Disciplinary procedures and reorientation
  E. Emergency Procedures
     First Aid
     Emergency Medical
     Fire, egress

IV. Safety & Health Training, Initial and Refresher
  A. Supervisor
     Safety and Health policy, rules and procedures
     Hazards of the workplace and how they are best controlled
     Accident Investigation
  B. Employees
     New employee safety orientation
     General and specific rules
     Use of personal protective equipment
     Preparation for emergencies

Training required by OSHA requirement
Safe work procedures

Specific language of many health and safety laws states that the employer will instruct employees in general safe work practices; provide specific instruction with regard to hazards unique to any job assignment; and schedule periodic inspections.

Construction accidents are expensive, thus the losses associated with these injuries must be controlled.

In addition to the previous key elements of the safety and health program required by OSHA and individual states, a company's in-house personnel and safety management plan should consist of the following additional subcategories:

- Corporate Policy on Safety
- Safe Practice and Operations Code
- First Aid
- Fire Protection
- Substance Abuse
- Personal Protective Equipment Requirements

- Protection of the General Public
- Hazardous Materials Communication
- Safety Communication
- Project Surveys
- Accident Reporting and Investigation
- OSHA Records and Regulations

Corporate Safety Policy

Safety is a company-wide issue. Creating a safety program and safety manual is only the beginning of developing safe jobsites to encourage employees to think "safety" as they start and complete project tasks. Corporate safety policy begins with a company declaration, stated objectives, and both accountability and authority for the implementation of safety regulations for a safe site.

The safety program manual generally provides written guidelines for a comprehensive safety program. A typical policy declaration from the president of the firm might include the following information:

1. The idea that safety is the highest priority in the workplace. Employees should not expect to work in an unsafe manner or environment. The company will not fire an employee for bringing to the attention of his supervisors unsafe conditions or actions.
2. The company should state who is accountable for providing a safe environment. This authority and accountability generally is directed to

the superintendents and foremen, as they are the ones in charge of day-to-day activities. Some firms also will designate an on-site safety engineer or safety manager who has specific duties.

3. The preface of all company manuals should include an open letter from the president to all employees addressing their responsibilities and their mandated use of the safety manual and information about the company's safety program. The following statements should be included as a way of encouraging employee participation to help provide a safe working environment.

- A safety policy and manual is provided to the employee so he will have a complete understanding of the companies' efforts to create a safe work place. It also is the employee's responsibility to himself, his family, the company, and his fellow workers to be aware of and practice safety procedures.
- Poor planning, lack of awareness, or failure to adhere to safety standards while an employee is performing his job or visiting the jobsite, or an overall indifference to safety procedures can cause jobsite accidents.
- Safe practices, use of safety equipment, following safety rules, pointing out unsafe acts by others or unsafe conditions on the jobsite, and being a full participating member of the company's safety team is necessary for all employees.

Safe Practice and Operations Code

The section on safe practice and operations details safety rules and guidelines concerning the individual jobsite. Accidents can be classified into two major categories: 1) Unsafe actions by people, and 2) Unsafe conditions left uncorrected. The following list includes some of the most common causes of accidents.

Unsafe actions by people:

- failure to correct and tell others about existing unsafe conditions
- use of tools in a manner not intended
- using tools that have not been maintained or serviced properly
- removal of safety guard or other safety equipment
- working at heights without proper safety equipment
- unsafe use of a ladder
- unaware of other working conditions in the area where working
- ignoring posted safety warnings or notices
- working with people under the influence, or being under the influence of a controlled substance
- leaving exposed live electrical connections

Unsafe conditions left uncorrected:

- jobsites without adequate fire protection equipment
- jobsites without readily available personal protection equipment
- inadequate or poor project housekeeping
- open or unprotected openings
- live and accessible electrical wiring
- not providing first aid equipment or training
- allowing employees who are substance abusers to be on the jobsite
- moving materials without the ability to control their movement or final placement

Accident Prevention

As part of the creation of a safe workplace, safety awareness has become an integral part of the construction jobsite. One of the ongoing safety training and awareness exercises is the "tailgate" or "toolbox" safety meetings, which are on-site meetings held with all workers and their supervisors in an attempt to reinforce to each individual the many jobsite hazards that may exist, and the consequences of unsafe actions for individual workers. In California, CAL/OSHA regulates and requires the construction industry to perform weekly tailgate safety meetings. (Sections 8406 and 1509 of Title 8 of the California Administrative Code.) These meetings usually run between ten to fifteen minutes long. Areas that should be discussed include work practices, tools and equipment, and their correct usage, attitudes, and/or other relevant activities. Topics should be current and relate to the tasks the workers now perform and will perform in the near future. The meetings also should provide individual workers the opportunity to discuss or relay needed safety corrections to all other workers. Only major issues should be discussed at these meetings, avoiding unnecessary exchanges and complaints.

The meeting should be led by supervisors. The toolbox meeting is intended to increase communication about safety issues so future accidents can be prevented. Each meeting should follow a set agenda. The following items can be used as an outline.

1. Points to remember regarding when and where to hold the meeting:
   - Limit to 10–15 minutes
   - Pre-schedule for both date and time
   - Hold meeting at the beginning of the shift
   - Meet in a place where everyone can hear
   - Take the time to show examples of problems being discussed

2. Topics to be discussed:
   - Topics should relate to the craft and the work being performed at the time of the meeting
   - Topics should be specific in nature and be the most critical in a list of topics
   - Topics need to be completely discussed before employees begin to interact

3. Preparation of subject matter:
   - Prepare to discuss the why, what, and how of the topic
   - Use terminology the employee or craftsperson will understand in his daily work

4. Record of meeting:
   - Record the people present
   - Record the topics discussed
   - Record the date and time
   - Use a standard form and format (see Figures 8–2 through 8–5)

Tailgate Safety Meeting

Attendance Sheet

Name of Employee	Meeting No.	Date	Topic Discussed

Shift _____
Supervisor_____
Date_____

FIGURE 8–2   Tailgate Safety Meeting Attendees Form

Safety Meeting Report Form

Date of Meeting:_____
Supervisor :   _____
Project Number:_____

Topic or Area of Discussion:

_____
_____
_____
_____
_____

Summary of Specific Items Discussed:

_____
_____
_____
_____
_____

Conclusion or changes to the jobsite to create a safer work place:

_____
_____
_____
_____
_____

Action Items needing immediate abatement:

_____
_____
_____
_____
_____

Remarks or comments:

_____
_____
_____
_____
_____

FIGURE 8–3   Safety Meeting Report Form

Fact sheets provided by the agency that has jurisdiction over the construction site should be made available to employees. CAL/OSHA provides fact sheets about specific requirements, for instance on compression gas cylinders, describing their proper handling and usage regarding hazard control. (See Figure 8–6 on page 202.)

<u>Employee Safety Record Card</u>
(Employee is to sign and return this card)

Name (Please Print) _____

Address _____

Home Phone _____

Social Security Number _____

Person To Be Contacted In Case Of Emergency _____

_____

Phone _____

I agree to report any injury received during the course of work to my supervisor. I have received a copy of the safety Manual for Maintenance and General Construction and agree to follow the rules.

        Employee's Signature_____
Employee Safety Courses
List The Safety Courses Received:

        First Aid _____

        OSHA 10 Hour Course _____ Yes _____No

        Other _____

Employee Safety Contacts
List the Tool Box Talks, Job Safety Analyses, etc. the employee has received and the date. On-the-job safety training is also to be listed.

<u>Date</u>   <u>Subject</u>   <u>Date</u>   <u>Subject</u>

_____

_____

_____

_____

_____

_____

_____

FIGURE 8–4   Employee Safety Record Card (Courtesy of Associated Builders and Contractors)

Employee Safety Discussion Attendance

Project Start Date_____

Project Manager _____

Project Number _____

Name of Employee	Safety Discussion Code Number								
	1	2	3	4	5	6	7	8	9

Each Employee Should Be Contacted At Least Once During the Month

Place Subject Code Numbers Listed Below In Proper Square Above as Each Employee is Contacted.

1. Housekeeping   2. Prompt reporting of accidents   3. Eye protection   4. Hearing protection

5. Horseplay   6. Bypassing safety devices   7. First aid   8. Safe lifting procedures

9. Past Accidents   10. Using unsafe equipment   11. New employee safety orientation

Note: Add more items here and more item numbers across

FIGURE 8–5   Employee Summary Safety Discussion Attendance

Pocket-sized safety rule books are available from state agencies or commercial publishers. Employees can be required to read and carry these pocket references on the jobsite. Any tool or reference that will help the employee follow safety regulations, whether on the job or in safety meetings, will help reduce jobsite accidents.

Compressed Gas Cylinders

Overfilling is a major cause of occupational injuries associated with the handling of compressed gas cylinders. If the specified filling density is exceeded, the cylinder becomes "liquid filled" and too little vapor space is left in the cylinder preventing expansion of the gas at higher temperatures. This hydrostatic pressure can increase to the point where the cylinder ruptures.

Overfilling is caused by failure to determine the capacity of the cylinder (expressed in cubic inches or water weight) or failure to properly determine the tare weight of the empty cylinder.

To avoid overfilling, know what the various markings stamped on the compressed gas cylinder mean, especially the specification number and the service pressure number. (See illustration A for the location of these marks on the cylinder.) Both marks provide information which is essential for the safe handling of compressed gas cylinders.

SPECIFICATION NUMBER

The specification number refers to the specific regulations under which the cylinder was manufactured. The regulations detail what inspections are required, whether the cylinder may be refilled or must be disposed of after a single use, how it is to be shipped, and for what gases it is authorized. The regulations also state the authorized service pressure of the cylinder.

Although cylinders are now manufactured according to specifications set by the Department of Transportation (DOT), in the past specifications have been set by both the Interstate Commerce Commission (ICC) and by the Bureau of Explosives (BE).

The specification number is preceded by the letters of the agency which established the specification. On Canadian cylinders the specification number is preceded by either CRC (Canadian Regulatory Commission) or BTC (Board of Transport Commissioners).

SERVICE PRESSURE

When a cylinder is authorized for use at only one service pressure, no pressure is marked. In order to determine the authorized service pressure, consult the appropriate specification. Some specifications may authorize cylinder use at various pressures. In this case the design service pressure is marked on the cylinder immediately following the specification number. For example, in illustration A the number DOT 3E1800 indicates Department of Transportation specification 3E with a service pressure of 1800 psig. Certain cylinders (3A and 3AA) may be filled to 110% of marked service pressure if they qualify by retest and are provided with frangible disc safety devices without fusible metal and are charged with non-liquid, non-flammable gas. These cylinders are marked with a plus sign following the hydrostatic retest date.

Charging the cylinder in excess of its service pressure rating is unsafe.

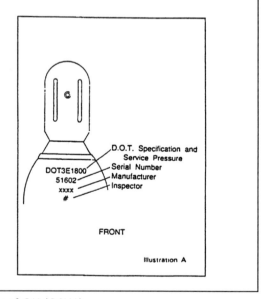

D.O.T. Specification and Service Pressure
Serial Number
Manufacturer
Inspector

DOT3E1800
51602
xxxx
#

FRONT

Illustration A

FIGURE 8–6   CAL/OSHA Safety Fact Sheet No. 15 (Courtesy of CAL/OSHA)

Medical and First Aid Facilities and Services

As projects become larger and more complex, an emergency services plan must be completed. This plan must address how an injury will be treated and, if transportation of an injured individual is required, how that will occur. OSHA requires the contractor to meet certain requirements for applying first aid, for the availability and amount of first aid kits, and for the amount of supplies in each kit. It is important that specific injuries be treated only by physicians, such as eye injuries or removal of foreign objects from the eye. OSHA also provides first aid training and certification of trained workers in first aid. First aid courses are available in most communities, either through a contractor's association or through the Red Cross. Most jurisdictions require one person on each crew to have full, updated first aid training, and most contractors require all field management personnel to have current first aid cards.

Protection of the General Public

Most construction activities occur where the general public has some form of access to them. The nature of construction work attracts people of all ages to jobsites. Often people will comment, "I wonder what they are building there?" It is the contractor's responsibility to make sure the general public is warned of all hazards on the construction site as well as to isolate the site from casual passersby. At all times, signs, barricades, and public protection devices must be visible and in place. Most urban areas have regulations about sidewalk protection walkways and vehicle access in and out of the site.

Some basic safeguards to protect the general public include:

* barriers in the form of fences and gates
* overhead protection from falling objects
* signage—both hazard and directional
* traffic controls
* security and surveillance
* walkway protection
* solid walls for protection from flying objects

Fire Protection

As a major loss prevention activity, fire prevention and protection rank among one of the top areas in safety control. The project superintendent must assess the project for potential fire hazards and scrutinize areas and activities that can cause fires to reduce property damage and possible burn injuries to workers. The key to fire prevention is early detection of potentially hazardous areas and conditions, such as plumbing, welding, smoking, and so on. When developing a loss prevention plan for fire protection, the first step involves posting emergency numbers and knowing the response time from local fire departments. Good housekeeping provides the basis for sound fire prevention. Storage of combustibles in an area away from and out of the building is important.

Fire extinguishers must be placed in the same approximate place throughout the project area, making sure these places are near areas or activities that may be prone to fires. Fire extinguishers are rated by the type of fires for which they are effective, and this information must be posted as well as understood by all craftspeople. These fire extinguishers are dated for current charge and must be maintained at all times. Random inspection by local fire officials often is made to construction sites. Charged fire extinguishers and other means of fire suppression normally are reviewed during these inspections.

Designated smoking and no smoking areas should be part of a fire prevention plan. Hazardous areas with flammable chemicals should be marked as "no smoking" areas.

Electricity is a major source of fires, as is combustible liquid, thus both need to be properly handled. The correct storage and handling of combustible liquid should be reviewed on a weekly basis. Flammable liquids must be carefully stored on the jobsite, usually away from other materials. A complete inventory of flammable and hazardous materials must be kept on the jobsite.

Flame-cutting and welding areas are the major causes of site fires, so control of work areas for these activities is necessary. Flame-cutting and welding should be isolated from combustible materials such as lumber. Many owners require the general contractor or subcontractor to obtain special mandatory cutting and burning permits before beginning any renovation or new construction projects.

Subcontractor activities must comply at all times with the general contractor's fire prevention plan and also must be policed by all of the general contractor's site management personnel.

During some months of the year, and with specific types of work that require drying, heaters will be used, which have the potential to become fire hazards so should be checked on a regular basis to ensure their safe operation. Some types of heaters are safer than others in construction areas. A variety of heaters that use fuel oil, propane, natural gas, or electricity is available, but some may be inappropriate for interior use. Ventilation of fumes also is necessary to provide a safe environment for temporary heaters.

Substance Abuse

Substance abuse usually is divided into three categories, including:

- alcoholic beverages
- legal drugs
- illegal drugs

Substance abuse can be a significant safety problem for the contractor. Quite often, habitual abuse and being "under the influence" of illegal drugs and/or alcohol is a sickness that requires outside treatment. The use of prescription drugs also can pose a hazard for the worker, causing side effects such as drowsiness, hyperactivity, or possible dependency. Supervisors on the job must always be aware of the physical and mental condition of employees. Too often a "problem" employee is ignored, resulting in a serious accident where an employee might injure himself or others in the work area.

Substance abuse programs should contain three elements. The first element should address a **standard of conduct** that defines and creates a zero tolerance level for the use of alcohol and drugs while on the job, as well as coming to work under the influence. When the results of off-duty excessive use of alcohol or drugs cause absenteeism, tardiness, or the inability to perform one's work, a zero tolerance level of acceptance also must be enforced. An employee must realize that his actions can result in an injury or fatality to themselves or others. Illicit activities, such as selling, distributing, or possessing illegal drugs should be grounds for dismissal and for referral to local law enforcement authorities. Many firms are developing drug testing requirements, which are used prior to employment.

The second element should create an **employee awareness** of drug and alcohol abuse and establish a prevention program. Employees should be required to attend formal sessions that discuss the dangerous effects of using drugs and alcohol on the jobsite. As part of a company's educational program, employees should be provided with information concerning the availability of drug and alcohol counseling, the type of treatment available, and rehabilitation facilities.

The last element of a formal substance abuse program is an **assistance program** for employees who are addicted to either drugs or alcohol. This program must be affiliated with a licensed professional in alcohol and drug treatment. The program also should ensure that the employees learn to take responsibility for their actions and it should clarify the written and stated company policy concerning second-time abuse of alcohol or drugs. This part of the program must be treated with strict confidentiality.

Random drug testing also is used in substance abuse programs to determine drug use. Drug testing is often used to determine suitability for employment. It also is used in random situations, and usually includes *all* employees. Some legal issues exist when employing a drug testing and substance abuse program, thus legal counsel is necessary when setting up this program. Only certified testing services should be used in drug testing programs. The applicable labor unions must be aware of these programs.

The conditions set forth in the standards of conduct and the need for mandatory employee awareness education must be a "condition of employment" by which all employees must abide. Many times key employees abuse one or more rules during a project, but the misuse of alcohol and drugs on the construction site must not be tolerated, as the consequences can be disastrous.

Personal Protective Equipment

As part of an overall safety plan, a supervisor must understand the importance of personal protective equipment. Individual protection equipment falls into eight categories, including:

- eye and face protection
- head protection
- hand protection
- foot protection
- respiratory protection
- protection from falls when working at heights (safety harnesses, lifelines and lanyards)
- hearing protection
- body protection

The contractor is obligated to furnish special protective gear for employees, such as respirators, hearing protection, safety harnesses, and so on. The employee is expected to wear appropriate clothing, such as steel-toed boots.

Proper use of protective equipment should be fully explained in the jobsite safety manual and reinforced in weekly toolbox safety meetings. Compliance with rules for proper use of protective equipment should be strictly enforced on the jobsite.

Hazardous Materials Communication

By law (29 CFR Sec. 1910.1200—The Federal Hazard Communication Standard), every project employee has the right to information concerning all chemicals being used on a project and any harmful effects they may cause. The construction process utilizes many different types of materials, all of which can cause potential harm to employees. An in-house system to document and post a Material Safety Data Sheet (MSDS) on all materials that will be used on the site during the construction process and the permanent materials that will be incorporated into the building should be developed and maintained on the project site.

The responsibility for such a program generally falls on the project engineer, office engineer, or a designated safety administrator. One requirement for the program is employee accessibility to the MSDS. A system must be devised to file this information, such as the Construction Specifications Institute (CSI) format for filing materials, which is the most appropriate system to use in the building industry.

The necessary steps in a right-to-know (MSDS) program are:

1. *Inventory all chemical products on the jobsite.* An up-to-date list of chemical products for each jobsite is necessary. Many construction chemicals fall within the hazardous categories. Some of the most common include acids and cleaning agents, adhesives, degreasing agents, detergents, gasoline, fuel oil, janitorial supplies, paints, shellacs, varnishes and lacquers, solvents, copy machine fluid, wood preservatives, and many others.

2. *Label hazardous chemical containers.* Each container must be labeled correctly. The warning must convey the specific hazard of the chemical, for example, if inhaled, this chemical will cause lung damage. In this case, lung damage is the hazard.

3. *Material Safety Data Sheet.* A detailed Material Safety Data Sheet (MSDS) is required for each chemical on the jobsite. These sheets, which are available from the chemical manufacturer, must be kept on file on the site. Information contained in the MSDS information includes the chemical makeup of the substance, fire and explosion hazard data, reactivity data, health hazard data, precautions for safe handling and use, and control measures.

4. *Inform all employees about the hazard communication program; then identify and train employees who may be exposed to hazardous chemicals.* The contractor is responsible for training employees in the proper use of chemicals. Management is responsible for the safe storage, use, and disposal of the chemicals, and also must furnish all protective clothing and gear to protect employees from the hazards of the product. Management also is responsible for the proper ventilation of chemicals, as well as for the environmental conditions in which chemicals are being used. Management should evaluate whether the product is appropriate for particular conditions of use.

5. *Develop and maintain at the jobsite a written program that explains how employees are informed and trained about hazardous chemicals in the workplace.* Sample programs are available from OSHA and state safety agencies.

Figure 8–7 provides a sample MSDS, used on a typical project.

Safety Communications

All states require that safety communications be conveyed to employees in a variety of ways. Previously, this was discussed in tailgate or toolbox meetings. The posting of federal- and state-required posters and forms on the jobsite concerning safety will be examined later in this section, however employers can also develop their own safety awareness posters and announcements, which can be visible in company magazines,

MATERIAL SAFETY DATA SHEET

("ESSENTIALLY SIMILAR" TO FORM OSHA 20)

OSHA 29 CFR 1910.1200

SECTION 1 - MATERIAL OR PRODUCT IDENTIFICATION AND USE

MANUFACTURER'S NAME	EMERGENCY TELEPHONE NO.

ADDRESS (Number, Street, City, State, and ZIP Code)

CHEMICAL NAME AND SYNONYMS	TRADE NAME AND SYNONYMS
CHEMICAL FAMILY	FORMULA

SECTION II - CHEMICAL DATA AND COMPOSITION (HAZARDOUS INGREDIENTS)

CHEMICAL FAMILY

FORMULA

CHEMICAL SUBSTANCES	CAS NO.	TLV	OSHA PEL
HAZARDOUS MIXTURES OF OTHER LIQUIDS, SOLIDS, OR GASES		TLV	

SECTION III - PHYSICAL DATA

BOILING POINT(F.)	SPECIFIC GRAVITY (H2O = 1)
VAPOR PRESSURE (mm Hg.)	PERCENT, VOLATILE BY VOLUME (%)
VAPOR DENSITY (AIR+1)	EVAPORATION RATE (_____ = 1)
SOLUBILITY IN WATER	APPEARANCE AND ODOR

SECTION IV - FIRE AND EXPLOSION HAZARD DATA

FLASH POINT (Method used)	FLAMMABLE LIMITS

EXTINGUISHING MEDIA

SPECIAL FIRE FIGHTING PROCEDURES

UNUSUAL FIRE AND EXPLOSION HAZARDS

Page (1)   (Continued on reverse side)

FIGURE 8–7   Material Safety Data Sheet (Courtesy of OSHA)

SECTION V - HEALTH HAZARD DATA

THRESHOLD LIMIT VALUE

EFFECTS OF OVEREXPOSURE

EMERGENCY AND FIRST AID PROCEDURES

SECTION VI - REACTIVITY DATA

STABILITY	UNSTABLE		CONDITIONS TO AVOID
	STABLE		

INCOMPARABILITY (Material to avoid)

HAZARDOUS DECOMPOSITION PRODUCTS

HAZARDOUS POLYMERIZATION	MAY OCCUR		CONDITIONS TO AVOID
	WILL NOT OCCUR		

SECTION VII - SPILL OR LEAK PROCEDURES

STEPS TO BE TAKEN IN CASE MATERIAL IS RELEASED OR SPILLED

WASTE DISPOSAL METHOD

SECTION VIII - SPECIAL PROTECTION INFORMATION

RESPIRATORY PROTECTION (Specify type)

VENTILATION	LOCAL EXHAUST	SPECIAL
	MECHANICAL (General)	OTHER

PROTECTIVE GLOVES	EYE PROTECTION

OTHER PROTECTIVE EQUIPMENT

SECTION IX - SPECIAL PRECAUTIONS

PRECAUTIONS TO BE TAKEN IN HANDLING AND STORING

OTHER PRECAUTIONS

PAGE (2)

FIGURE 8–7   continued

FIGURE 8–8
Examples of Safety
Signage

at company–sponsored events such as picnics, and at award banquets. Some specific requirements for posted safety communications, illustrated in Figure 8–8, include:

- safety and health protection on the job
- emergency telephone numbers

FIGURE 8–8
continued

- exiting and exit routes
- no smoking signs
- traffic direction signs
- industrial welfare commission orders/Davis Bacon requirements
- discrimination in employment
- notice to employees regarding unemployment and disability insurance

- payday notice
- summary of occupational injuries and illnesses posted during the month of February
- location of the MSDS
- general safety awareness
- crane and hand signals for controlling crane operations
- fire protection and prevention

Accident Reporting and Investigation

Accident reports are specific accounts describing particular incidents or accidents and are part of a company's safety program. The governing safety organization, whether it is OSHA or the state, requires detailed periodic reports and the use of prescribed forms. In addition to the required forms, the contractor should develop a form to document accidents and use the information to prevent future mishaps.

The construction company's accident form must be completed within twenty-four hours of the occurrence. The superintendent or field engineer should prepare the report, obtaining information from the accident victim, the foreman of the crew, and witnesses. Photographs of the conditions surrounding the accident should be taken and attached to the report. The report should be kept in the project safety file, then distributed to the company safety officer or safety committee, and the project manager.

The following information should be contained on an accident report form, as shown in Figure 8–9:

- Name of the individual: personal data, including address, social security number, employee number, and other related information
- Location of the accident: project and exact location in the project
- Description of the accident: description of the accident, first aid administered, and medical attention needed, if any
- Crew foreman: name and location of foreman at time of accident (include or attach foreman's statement to form)
- Witnesses: names of witnesses to accident (include or attach their statements to form)
- What caused the accident: It is essential in each accident investigation to determine what caused it. Although an accident occurs due to a combination of factors, there probably is an identifiable reason why it happened and how it can be avoided in the future. Accidents commonly happen because of incomplete knowledge about how to safely perform a particular task, faulty equipment, lack of safety equipment,

FGH CONSTRUCTION COMPANY, INC.
ACCIDENT REPORT

Project: New City Plaza, New City, CA   Proj.No.: 2-98   Date: 5/26/98

Injured Employee: Harold A. Johnston II, 5390 Dulzura , Atascadero CA   534-90-9078

Name   Address   Soc.Sec.No.

ACCIDENT LOCATION: Building A, forming tilt-up panels

ACCIDENT DESCRIPTION: Carpenter Johnston cut right hand with circular saw while cutting
lumber for tilt-up forms. Guard jammed on saw. Cut on right side of hand, just skin, no bone

FIRST AID APPLIED?_____ Wrapped immediately, pressure applied at wrist BY? J. Anderson

MEDICAL TREATMENT REQUIRED: Taken by J. Anderson, in company pickup, to City
Hospital Emergency room. Wound was cleaned, sewed, 6 stitches, bandaged. Johnston given tetanus shot.
Returned to jobsite. (Hospital report attached)

LOST TIME DUE TO ACCIDENT (ESTIMATE): None estimated, employee returned to work

Foreman: J. Anderson   WITNESSES: W. Mincks
          (statement attached)   (statements attached)

Cause Of Accident:
Guard jammed on circular saw. Carpenter continued to work.

ACTION TAKEN: Circular saw #110 taken out of service; sent to Repair shop. Toolbox talk
scheduled for 5/27/98 -- safety concerns when using saws and tools.

REPORT BY:_____
DATE:_____

FIGURE 8–9   Accident Report

negligence by the employee or another party, and failure of temporary or permanent construction.

- Action taken following the accident: After determining what caused the accident, certain steps need to be taken quickly and decisively. If a ladder fails and causes the accident, the ladder must be removed. If an employee appears negligent, further investigation may be necessary, followed by dismissal if the employee is found negligent.
- Time lost by the accident: Time lost is a determining factor regarding the severity of an accident to the contractor, as well as the amount of time an employee is off duty while recuperating from the accident. After minor medical attention, an employee may be able to return to work immediately, without any lost time; or, an employee may also be able to continue on the job, but perform a different work task until the injury has healed, which is not normally considered a "lost time" accident. If an employee cannot return to work for one week, the lost time would be five days, or forty hours.

OSHA Records and Regulatory Requirements

Construction firms must keep detailed records for OSHA as well as report specific types of illnesses or injuries. If a firm employs eleven or more employees at any one time in the previous calendar year, it must file OSHA Form 200, and 101. OSHA 200 is the Log and Summary of Occupational Injuries and Illnesses, and OSHA 101 is a supporting document that is used to record additional information pertaining to each injury and illness logged on OSHA Form 200. Both must be filed within twenty-four hours of the accident.

OSHA 200, shown in Figure 8–10, consists of three distinctive parts:

1. A descriptive section, describing the injury/illness and identifying the employee
2. An extent of injury section
3. The type and extent of illnesses section

The supplementary record form, OSHA 101, shown in Figure 8–11 on pages 217 and 218, consists of the following three parts:

1. Description of how the accident or illness occurred
2. List of objects or substances involved
3. Nature of and location where the injury/illness occurred

OSHA now permits other forms to be used in lieu of OSHA 101, for example, workers' compensation and private insurance forms. To qualify and be in compliance, these forms must contain all of the items stated on the OSHA 101 Form. An example of an alternate report is Form 5020 from the State of California, shown in Figure 8–12 on page 219.

Log and Summary of Occupational
Injuries and Illnesses

U.S. Department of Labor

Page ____ of ____

For Calendar Year 19 ____

Company Name ____

Establishment Name ____

Establishment Address ____

Form Approved
O.M.B. No. 1218-0176
See OMB Disclosure
Statement on reverse.

NOTE: This form is required by Public Law 91-596 and must be kept in the establishment for 5 years. Failure to maintain and post can result in the issuance of citations and assessment of penalties. (See posting requirements on the other side of form.)

RECORDABLE CASES: You are required to record information about every occupational death, every nonfatal occupational illness, and those nonfatal occupational injuries which involve one or more of the following: loss of consciousness, restriction of work or motion, transfer to another job, or medical treatment (other than first aid). (See definitions on the other side of form.)

(A)	(B)	(C)	(D)	(E)	(F)	Extent of, and Outcome of INJURY						Type, Extent of, and Outcome of ILLNESS								
Case or File Number	Date of Injury or Onset of Illness	Employee's Name	Occupation	Department	Description of Injury or Illness	Fatalities	Nonfatal Injuries					Type of Illness							Fatalities	Nonfatal Illnesses

ILLNESSES

INJURIES

PREVIOUS PAGE TOTALS

TOTALS (Instructions on other side of form)

Certification of Annual Summary Totals By ____   Title ____   Date ____

POST ONLY THIS PORTION OF THE LAST PAGE NO LATER THAN FEBRUARY 1.

OSHA No. 200

FIGURE 8–10   OSHA Form 200 (Courtesy of OSHA)

Public reporting burden for the collection of information is estimated to vary from 4 to 30 (time in minutes) per response with an average of 15 (time in minutes) per response, including the time for reviewing instructions, searching existing data sources, gathering and maintaining the data needed, and completing and reviewing the collection of information. If you have any comments regarding this estimate or any other aspect of this information collection, including suggestions for reducing this burden, please send them to the OSHA Office of Statistics and/or the Department of Labor, Office of IRM Policy, Room N-1301, 200 Constitution Avenue, N.W. Washington, D.C. 20210

Instructions for OSHA No. 200

I. Log and Summary of Occupational Injuries and Illnesses

Each employer who is subject to the recordkeeping requirements of the Occupational Safety and Health Act of 1970 must maintain for each establishment a log of all recordable occupational injuries and illnesses. This form (OSHA No. 200) may be used for that purpose. A substitute for the OSHA No. 200 is acceptable if it is as detailed, easily readable, and understandable as the OSHA No. 200.

Enter each recordable case on the log within six (6) workdays after learning of its occurrence. Although other records must be maintained at the establishment to which they refer, it is possible to prepare and maintain the log at another location, using data processing equipment if desired. If the log is prepared elsewhere, a copy updated to within 45 calendar days must be present at all times in the establishment.

Logs must be maintained and retained for five (5) years following the end of the calendar year to which they relate. Logs must be available (normally at the establishment) for inspection and copying by representatives of the Department of Labor, or the Department of Health and Human Services, or States accorded jurisdiction under the Act. Access to the log is also provided to employees, former employees and their representatives.

II. Changes in Extent of Outcome of Injury or Illness

If, during the 5-year period the log must be retained, there is a change in an extent and outcome of an injury or illness which affects entries in columns 1, 2, 6, 8, 9, or 13, the first entry should be lined out and a new entry made. For example, if an injured employee at first required only medical treatment but later lost workdays away from work, the check in column 6 should be lined out, and checks entered in columns 2 and 3 and the number of lost workdays entered in column 4.

In another example, if an employee with an occupational illness lost workdays, returned to work, and then died of the illness, any entries in columns 9 through 12 should be lined out and the date of death entered in column 8.

The entire entry for an injury or illness should be lined out if later found to be nonrecordable. For example, an injury which is later determined not to be work related, or which was initially thought to involve medical treatment but later was determined to have involved only first aid.

III. Posting Requirements

A copy of the totals and information following the fold line of the last page for the year must be posted at each establishment in the place or places where notices to employees are customarily posted. This copy must be posted no later than *February 1 and must remain in place until March 1.*

Even though there were no injuries or illnesses during the year, zeros must be entered on the totals line, and the form posted.

The person responsible for the *annual summary totals* shall certify that the totals are true and complete by signing at the bottom of the form.

IV. Instructions for Completing Log and Summary of Occupational Injuries and Illnesses

Column A – CASE OR FILE NUMBER. Self-explanatory.

Column B – DATE OF INJURY OR ONSET OF ILLNESS.
For occupational injuries, enter the date of the work accident which resulted in injury. For occupational illnesses, enter the date of initial diagnosis of illness, or, if absence from work occurred before diagnosis, enter the first day of the absence attributable to the illness which was later diagnosed or recognized.

Columns C through F – Self-explanatory.

Columns 1 and 8 – INJURY OR ILLNESS RELATED DEATHS. Self-explanatory.

Columns 2 and 9 – INJURIES OR ILLNESSES WITH LOST WORKDAYS. Self-explanatory.
Any injury which involves days away from work, or days of restricted work activity, or both must be recorded since it always involves one or more of the criteria for recordability.

Columns 3 and 10 – INJURIES OR ILLNESSES INVOLVING DAYS AWAY FROM WORK. Self-explanatory.

Columns 4 and 11 – LOST WORKDAYS—DAYS AWAY FROM WORK.
Enter the number of workdays (consecutive or not) on which the employee would have worked but could not because of occupational injury or illness. The number of lost workdays should not include the day of injury or onset of illness or any days on which the employee would not have worked even though able to work.
NOTE: For employees not having a regularly scheduled shift, such as certain truck drivers, construction workers, farm labor, casual labor, part-time employees, etc., it may be necessary to estimate the number of lost workdays. Estimates of lost workdays shall be based on prior work history of the employee AND days worked by employees, not ill or injured, working in the department and/or occupation of the ill or injured employee.

Columns 5 and 12 – LOST WORKDAYS—DAYS OF RESTRICTED WORK ACTIVITY.
Enter the number of workdays (consecutive or not) on which because of injury or illness:
(1) the employee was assigned to another job on a temporary basis, or
(2) the employee worked at a permanent job less than full time, or
(3) the employee worked at a permanently assigned job but could not perform all duties normally connected with it.
The number of lost workdays should not include the day of injury or onset of illness or any days on which the employee would not have worked even though able to work.

Columns 6 and 13 – INJURIES OR ILLNESSES WITHOUT LOST WORKDAYS. Self-explanatory.

Columns 7a through 7g – TYPE OF ILLNESS.
Enter a check in only *one* column for each illness.

TERMINATION OR PERMANENT TRANSFER–Place an asterisk to the right of the entry in columns 7a through 7g (type of illness) which represented a termination of employment or permanent transfer.

V. Totals

Add number of entries in columns 1 and 8.
Add number of checks in columns 2, 3, 6, 7, 9, 10, and 13.
Add number of days in columns 4, 5, 11, and 12.
Yearly totals for each column (1-13) are required for posting. Running or page totals may be generated at the discretion of the employer.
If an employee's loss of workdays is continuing at the time the totals are summarized, estimate the number of future workdays the employee will lose and add that estimate to the workdays already lost and include this figure in the annual totals. No further entries are to be made with respect to such cases in the next year's log.

VI. Definitions

OCCUPATIONAL INJURY is any injury such as a cut, fracture, sprain, amputation, etc., which results from a work accident or from an exposure involving a single incident in the work environment.
NOTE: Conditions resulting from animal bites, such as insect or snake bites or from one-time exposure to chemicals, are considered to be injuries.

OCCUPATIONAL ILLNESS of an employee is any abnormal condition or disorder, other than one resulting from an occupational injury, caused by exposure to environmental factors associated with employment. It includes acute and chronic illnesses or diseases which may be caused by inhalation, absorption, ingestion, or direct contact.

The following listing gives the categories of occupational illnesses and disorders that will be utilized for the purpose of classifying recordable illnesses. For purposes of information, examples of each category are given. These are typical examples, however, and are not to be considered the complete listing of the types of illnesses and disorders that are to be counted under each category.

7a. **Occupational Skin Diseases or Disorders**
Examples: Contact dermatitis, eczema, or rash caused by primary irritants and sensitizers or poisonous plants; oil acne; chrome ulcers; chemical burns or inflammations; etc.

7b. **Dust Diseases of the Lungs (Pneumoconioses)**
Examples: Silicosis, asbestosis and other asbestos-related diseases, coal worker's pneumoconiosis, byssinosis, siderosis, and other pneumoconioses.

7c. **Respiratory Conditions Due to Toxic Agents**
Examples: Pneumonitis, pharyngitis, rhinitis or acute congestion due to chemicals, dusts, gases, or fumes; farmer's lung; etc.

7d. **Poisoning (Systemic Effect of Toxic Materials)**
Examples: Poisoning by lead, mercury, cadmium, arsenic, or other metals; poisoning by carbon monoxide, hydrogen sulfide, or other gases; poisoning by benzol, carbon tetrachloride, or other organic solvents; poisoning by insecticide sprays such as parathion, lead arsenate; poisoning by other chemicals such as formaldehyde, plastics, and resins; etc.

7e. **Disorders Due to Physical Agents (Other than Toxic Materials)**
Examples: Heatstroke, sunstroke, heat exhaustion, and other effects of environmental heat; freezing, frostbite, and effects of exposure to low temperatures; caisson disease; effects of ionizing radiation (isotopes, X-rays, radium); effects of nonionizing radiation (welding flash, ultraviolet rays, microwaves, sunburn); etc.

7f. **Disorders Associated With Repeated Trauma**
Examples: Noise-induced hearing loss; synovitis, tenosynovitis, and bursitis; Raynaud's phenomena; and other conditions due to repeated motion, vibration, or pressure.

7g. **All Other Occupational Illnesses**
Examples: Anthrax, brucellosis, infectious hepatitis, malignant and benign tumors, food poisoning, histoplasmosis, coccidioidomycosis, etc.

MEDICAL TREATMENT includes treatment (other than first aid) administered by a physician or by registered professional personnel under the standing orders of a physician. Medical treatment does NOT include first-aid treatment (one-time treatment and subsequent observation of minor scratches, cuts, burns, splinters, and so forth, which do not ordinarily require medical care) even though provided by a physician or registered professional personnel.

ESTABLISHMENT: A single physical location where business is conducted or where services or industrial operations are performed (for example, a factory, mill, store, hotel, restaurant, movie theater, farm, ranch, bank, sales office, warehouse, or central administrative office). Where distinctly separate activities are performed at a single physical location, such as construction activities operated from the same physical location as a lumber yard, each activity shall be treated as a separate establishment.

For firms engaged in activities which may be physically dispersed, such as agriculture, construction, transportation, communications, and electric, gas, and sanitary services, records may be maintained at a place to which employees report each day.

Records for personnel who do not primarily report or work at a single establishment, such as traveling salesmen, technicians, engineers, etc., shall be maintained at the location from which they are paid or the base from which personnel operate to carry out their activities.

WORK ENVIRONMENT is comprised of the physical location, equipment, materials processed or used, and the kinds of operations performed in the course of an employee's work, whether on or off the employer's premises.

FIGURE 8–10 continued

Bureau of Labor Statistics
Supplementary Record of
Occupational Injuries and Illnesses

U.S. Department of Labor

This form is required by Public Law 91-596 and must be kept in the establishment for *5 years.* Failure to maintain can result in the issuance of citations and assessment of penalties.	Case or File No.	Form Approved O.M.B. No. 1220-0029

See OMB Disclosure Statement on reverse.

Employer

1. Name

2. Mail address *(No. and street, city or town, State, and zip code)*

3. Location, if different from mail address

Injured or Ill Employee

4. Name *(First, middle, and last)*   Social Security No.

5. Home address *(No. and street, city or town, State, and zip code)*

6. Age   7. Sex: *(Check one)*   Male ☐   Female ☐

8. Occupation *(Enter regular job title, not the specific activity he was performing at time of injury.)*

9. Department *(Enter name of department or division in which the injured person is regularly employed, even though he may have been temporarily working in another department at the time of injury.)*

The Accident or Exposure to Occupational Illness

If accident or exposure occurred on employer's premises, give address of plant or establishment in which it occurred. Do not indicate department or division within the plant or establishment. If accident occurred outside employer's premises at an identifiable address, give that address. If it occurred on a public highway or at any other place which cannot be identified by number and street, please provide place references locating the place of injury as accurately as possible.

10. Place of accident or exposure *(No. and street, city or town, State, and zip code)*

11. Was place of accident or exposure on employer's premises?   Yes ☐   No ☐

12. What was the employee doing when injured? *(Be specific. If he was using tools or equipment or handling material, name them and tell what he was doing with them.)*

13. How did the accident occur? *(Describe fully the events which resulted in the injury or occupational illness. Tell what happened and how it happened. Name any objects or substances involved and tell how they were involved. Give full details on all factors which led or contributed to the accident. Use separate sheet for additional space.)*

Occupational Injury or Occupational Illness

14. Describe the injury or illness in detail and indicate the part of body affected. *(E.g., amputation of right index finger at second joint; fracture of ribs; lead poisoning; dermatitis of left hand, etc.)*

15. Name the object or substance which directly injured the employee. *(For example, the machine or thing he struck against or which struck him; the vapor or poison he inhaled or swallowed; the chemical or radiation which irritated his skin; or in cases of strains, hernias, etc., the thing he was lifting, pulling, etc.)*

16. Date of injury or initial diagnosis of occupational illness   17. Did employee die? *(Check one)*   Yes ☐   No ☐

Other

18. Name and address of physician

19. If hospitalized, name and address of hospital

Date of report	Prepared by	Official position

OSHA No. 101 (Feb. 1981)

FIGURE 8–11   OSHA Form 101 (Courtesy of OSHA)

SUPPLEMENTARY RECORD OF OCCUPATIONAL INJURIES AND ILLNESSES

To supplement the Log and Summary of Occupational Injuries and Illnesses (OSHA No. 200), each establishment must maintain a record of each recordable occupational injury or illness. Worker's compensation, insurance, or other reports are acceptable as records if they contain all facts listed below or are supplemented to do so. If no suitable report is made for other purposes, this form (OSHA No. 101) may be used or the necessary facts can be listed on a separate plain sheet of paper. These records must also be available in the establishment without delay and at reasonable times for examination by representatives of the Department of Labor and the Department of Health and Human Services, and States accorded jurisdiction under the Act. The records must be maintained for a period of not less than five years following the end of the calendar year to which they relate.

Such records must contain at least the following facts:

1) *About the employer*—name, mail address, and location if different from mail address.

2) *About the injured or ill employee*—name, social security number, home address, age, sex, occupation, and department.

3) *About the accident or exposure to occupational illness*—place of accident or exposure, whether it was on employer's premises, what the employee was doing when injured, and how the accident occurred.

4) *About the occupational injury or illness*—description of the injury or illness, including part of body affected; name of the object or substance which directly injured the employee; and date of injury or diagnosis of illness.

5) *Other*—name and address of physician; if hospitalized, name and address of hospital; date of report; and name and position of person preparing the report.

SEE *DEFINITIONS* ON THE BACK OF OSHA FORM 200.

OMB DISCLOSURE STATEMENT

We estimate that it will take an average of 20 minutes to complete this form including time for reviewing instructions; searching, gathering and maintaining the data needed; and completing and reviewing the form. If you have any comments regarding this estimate or any other aspect of this recordkeeping system, send them to the Bureau of Labor Statistics, Division of Management Systems (1220-0029), Washington, D.C. 20212 and to the Office of Management and Budget, Paperwork Reduction Project (1220-0029), Washington, D.C. 20503.

*U.S. Government Printing Office: 1990-262-256/15418

FIGURE 8–11   continued

State of California
EMPLOYER'S REPORT
OF OCCUPATIONAL
INJURY OR ILLNESS

Please complete in triplicate (type, if possible). Mail two copies to:

OSHA
Case No.

☐ Fatality

Any person who makes or causes to be made a knowingly false or fraudulent material statement or material representation for the purpose of obtaining or denying workers' compensation benefits or payment is guilty of a felony.

NOTICE: California law requires employers to report within five days of knowledge every occupational injury illness which results in lost time beyond the date of the incident or requires medical treatment beyond first aid. If an employee subsequently dies as a result of a previously reported injury or illness., the employer must file within five days of knowledge an amended report indicating death. In addition, every serious injury/illness, or death must be reported immediately by telephone or telegraph to the nearest office of the California Division of Occupational Safety and Health.

		DO NOT USE THIS COL.
1. FIRM NAME	1A. POLICY NUMBER	Case No.
2. MAILING ADDRESS (Number and Street, City, ZIP)	2A. PHONE NUMBER	Case No.
3. LOCATION, IF DIFFERENT FROM MAILING ADDRESS (Number and Street, City, ZIP)	3A. LOCATION CODE	Ownership
4. NATURE OF BUSINESS, e.g., painting contractor, general contractor, etc. 5. STATE UNEMPLOYMENT INSURANCE ACCOUNT NUMBER		Industry
6. TYPE OF EMPLOYER ☐ PRIVATE ☐ STATE ☐ CITY ☐ COUNTY ☐ SCHOOL DIST. ☐ OTHER GOVERNMENT - SPECIFY _____		Occupation
7. EMPLOYEE NAME	9. DATE OF BIRTH (mm/dd/yy)	Sex
10. HOME ADDRESS (Number and Street, City, ZIP)	10A. PHONE NUMBER	Age
11. SEX ☐ MALE ☐ FEMALE	12. OCCUPATION (Regular job title - NO initials, abbr.. or no.)	13. DATE OF HIRE (mm/dd/yy)	Daily hours
14. EMPLOYEE USUALLY WORKS ____ hours per day ____ days per week ____ total weekly hours	14A. EMPLOYMENT STATUS (check applicable status at time of injury) regular full-time ☐ temporary ☐ part-time ☐ seasonal ☐	14B. Under what class code of you policy were wages assigned?	Days per week
15. GROSS WAGES SALARY $ ____ per ____	16. OTHER PAYMENTS NOT REPORTED AS WAGES/SALARY (e.g., meals, overtime, bonuses, etc.)? Yes ☐ $ ____ per ____ No ☐	Weekly hours
17. DATE OF INJURY OR ONSET OF ILLNESS (mm/dd/yy)	18. TIME INJURY/ILLNESS OCCURRED ____ A.M. ____ P.M.	19. THE EMPLOYEE BEGAN WORK ____ A.M. ____ P.M.	20. IF EMPLOYEE DIED, DATE OF DEATH (mm/dd/yy)	Weekly wage
21. UNABLE TO WORK FOR AT LEAST ON F DAY AFTER DATE OF INJURY Yes ☐ No ☐	22. DATE LAST WORKED (mm/dd/yy)	23. DATE RETURNED TO W (mm/dd/yy)	24. IF STILL OFF WORK CHECK THIS BOX ☐	County
25. PAID FULL WAGES FOR DAY OF INJURY OR LAST DAY WORKED? Yes ☐ No ☐	26. SALARY BEING CONTINUED? Yes ☐ No ☐	27. DATE OF EMPLOYER'S KNOWLEDGE /NOTICE OF INJURY ILLNESS	28. DATE EMPLOYEE WAS PROVIDED EMPLOYEE CLAIM FORM (mm/dd/yy)	Nature of Injury
29. SPECIFIC INJURY/ILLNESS AND PART OF BODY AFFECTED, MEDICAL DIAGNOSIS, if available, e.g. second degree burns on right arm.		Part of Body
30. LOCATION WHERE EVENT OR EXPOSURE OCCURRED (Number, Street, City, County, ZIP)	30B. ON EMPLOYER'S PREMISES? Yes ☐ No ☐	Source
31. DEPARTMENT WHERE EVENT OR EXPOSURE OCCURRED, e.g., shipping department, shop.	32. OTHER WORKERS INJURED/ILL IN THIS EVENT? Yes ☐ No ☐	Event
33. EQUIPMENT, MATERIALS AND CHEMICALS THE EMPLOYEE WAS USING WHEN EVENT OR EXPOSURE OCCURRED, e.g., acetylene, torch.		Sec. Source
34. SPECIFIC ACTIVITY THE EMPLOYEE WAS PERFORMING WHEN EVENT OR EXPOSURE OCCURRED, e.g., welding seams of metal forms.		Extent of injury
35. HOW INJURY/ILLNESS OCCURRED. DESCRIBE SEQUENCE OF EVENTS. SPECIFY OBJECT OR EXPOSURE WHICH DIRECTLY PRODUCED THE INJURY/ILLNESS, e.g., worker stepped back to inspect work and slipped on scrap material. As he ell, he brushed against fresh weld, and burned right hand. USE SEPARATE SHEET IF NECESSARY.		
36. NAME AND ADDRESS OF PHYSICIAN (Number & Street, City, ZIP)	36A. PHONE NUMBER	
37. IF HOSPITALIZED AS AN INPATIENT, NAME & ADDRESS OF HOSPITAL (No. & Street, City, ZIP)	37A. PHONE NUMBER	
Completed by (type or print)	Title	Date	Signature	

FILING OF THIS REPORT IS NOT AN ADMISSION OF LIABILITY

FIGURE 8–12   Employer's Report of Occupational Injury or Illness (Courtesy of CAL/OSHA)

These two forms must be kept at every physical location when the projects are designated fixed establishments. Records for employees of non-fixed establishments, such as electricians performing service work out of a van, can be kept at the field office or mobile base of operations or at the established central location. The address and telephone number recorded on the form is that of the non-fixed site. These records are to be kept for a period of five years.

Recording Injuries and Illnesses

Not all injuries or illnesses must be recorded. When an illness or injury occurs on the jobsite, many factors must be considered before the case can be classified. The following questions must be analyzed to ensure the proper procedure in completing OSHA Form 101 and OSHA Form 200 for compliance with state and federal regulations:

1. Who employs the injured?
2. Was there a death, an illness, or an injury?
3. Was the case work-related?
4. Was the case an injury or illness?
5. Was the injury recordable, based on medical treatment, loss of consciousness, restriction of work or motion, or transfer to another job?

These questions, which have been placed in chart formation in Figure 8–13, will help determine recordable and non-recordable cases.

To aid in establishing a work relationship to an injury or illness, OSHA has developed the guidelines, shown in Figure 8–14.

The employer has certain obligations to report injuries and illnesses, but not all must be reported to federal or state authorities. Two categories exist in which recorded cases must be reported: 1) audits of recorded injuries or illnesses for a certain time frame, requested by federal or state health and safety organizations, and 2) all accidents resulting in one or more fatalities or hospitalization of five or more employees in a given year. Some states have more stringent reporting requirements. This report must be provided within forty-eight hours after the occurrence of an accident resulting in an injury or illness, or within forty-eight hours after an accident causing a fatality. Again, some states have more stringent requirements, for example, CAL/OSHA requires immediate notification of a death or serious injury or illness.

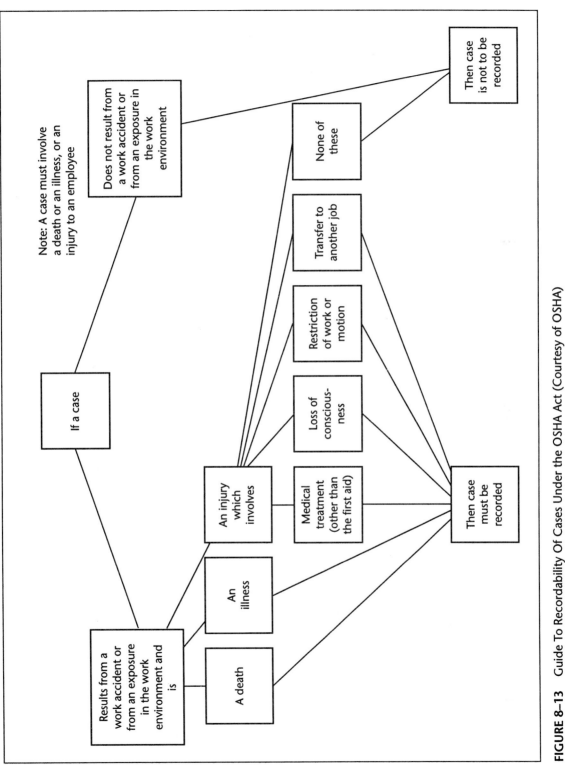

FIGURE 8–13 Guide To Recordability Of Cases Under the OSHA Act (Courtesy of OSHA)

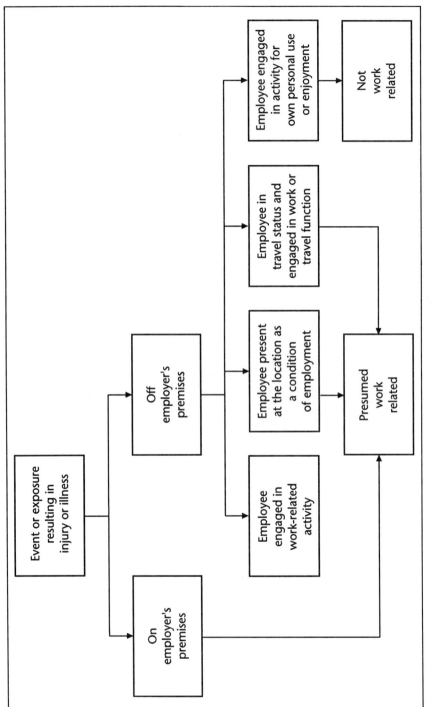

FIGURE 8–14   Guidelines For Establishing Work Relationship (Courtesy of OSHA)

Summary

In today's construction environment, the contractor can no longer passively promote safe work practices. With the passage of the Occupational Safety and Health Act of 1970, employers were mandated to follow and adhere to federal and state safety regulations. In addition, construction companies also realized the need to develop and implement their own in-house safety regulations. Whether safety requirements are federal-, state-, or self-imposed, employers must demonstrate a commitment to safety, must have in place hazard assessment and control programs, must develop and implement safety planning, must have rules and work procedures that are pro-active and pertain to the job site, and must provide safety and health training that is thorough and applicable to the work performed by its employees. It is not enough to make one's project engineer the safety manager at the site and expect to be in compliance with safety regulations and prevent accidents. One of the most aggressive and proactive safety programs developed for the construction industry's use is Dupont's STOP safety program. The acronym stands for Safety, Training, Observation, and Prevention. What makes this program so effective is its added key ingredients to the already developed and implemented safety programs. The key ingredients include positive intervention into the unsafe act that caused the accident, resulting in the illness or injury, education of the employee regarding why the act causing the accident is unsafe, and a method of tracking the occurrence and developing ways of ensuring positive improvement. Employers are realizing that the consistent and regulated monitoring of safety practices does pay. Construction firms that have poor safety records are becoming less competitive in the marketplace due to higher costs of insurance, lower morale of employees, increased costs of fines at the jobsite, and overall loss of job profits.

CHAPTER 9

SUBCONTRACTING AND PURCHASING

In the past, general contractors used subcontractors only to perform the "specialized" tasks of the construction project, and accomplished most of the tasks with their own labor. The current general contractor subcontracts much of the project work, providing primarily supervision for the construction project. Subcontractors are used on the construction project for the following reasons:

1. *Specialized labor for particular construction tasks*
   Skilled craftspeople trained in the specific assembly perform the task correctly, enhancing the quality of the installation. They also complete the task efficiently, with the minimal amount of labor for the appropriate task.
2. *Lower cost for subcontract work*
   As the labor for the subcontractor is specialized and only a narrow range of work is done, subcontractor costs are generally less. When utilizing subcontractors, the contractor has reduced overhead costs, as well as reduced labor and payroll costs.
3. *Reduced risk for the contractor*
   Several areas of risk are reduced by subcontracting work rather than accomplishing it with the contractor's own forces. The risk of labor productivity, where actual labor costs need to be less than or equal to the estimated labor cost, is eliminated for the contractor, the subcontractor now being held responsible. Other types of risk general contractors attempt to avoid by using subcontractors include:
   - productivity and/or cost control, from the estimate to the project's completion
   - lack of expertise in specialized areas
   - control of potential liquidated damages
   - cash flow
   - quality from lack of expertise or the right craftspeople
   - clean-up, warranties, and other general condition areas

Subcontract management is extremely valuable in today's legal climate. Because much of the risk has been shifted to the subcontractor, contracts between contractors and subcontractors have increased in importance. Informal agreements and understandings generally have been replaced by contractual agreements. The subcontract agreement is a serious legal contract, with binding provisions for both parties. Fulfillment of the obligations of the subcontract agreement is necessary for completion of the construction project. Subcontractors can affect the project's profitability by delay, poor workmanship, and default of contract. Subcontractor management relates directly to the profitability of the project.

Purchasing control also is necessary for a profitable project. **Purchasing** is the activity of procuring material or equipment for the project. The primary difference between purchases and subcontracts is that purchases do not contain jobsite labor, while subcontracts do. Purchasing procedures are part of a controlled and well-managed project. They can include:

- pricing and purchasing material and equipment, both from the central office and from the jobsite
- tracking and expediting the purchase to coincide with the construction schedule
- control of the material at the jobsite
- purchase or rental of construction equipment for the project

This chapter discusses the project management function of writing subcontracts, purchase orders, and material contracts, the administration of subcontracts, and the area of purchasing and expediting.

Subcontract Management

The Subcontractor

The contractor's field personnel should have a complete understanding of subcontractors and the subcontracting business to effectively manage subcontractors. Field personnel should understand the following factors about each subcontractor:

- the nature of the subcontractor's business
- the scheduling demands on the subcontractor
- the risks the subcontractor is facing on the project
- the equipment and safety concerns of the subcontractor

The development of a working relationship between the general contractor's staff and the subcontractor's staff can often make accomplishing the goals of both companies easier. The goal for both companies is to make a profit, expand their businesses, move into a more profitable market, and hopefully enjoy the process.

Subcontractors vary from very small firms to large international firms. They specialize in a certain area of the construction project, such as floor covering or mechanical or electrical work. Due to the smaller portion of projects, most subcontractors are concerned with various concurrent projects. The subcontractor's focus is to make a profit from all business activity, while servicing customers with a finite amount of resources.

Subcontractors have a wide variety of attributes, which help explain the nature of each subcontractor's business:

- Subcontractors may engage in a number of different types of business in their field: retail, subcontracting, and contracting as a prime contractor on some projects. They might perform residential, commercial building, industrial, or heavy construction.

- Most subcontractors have higher indirect overhead than general contractors. They may have inventory, fabrication facilities, shop facilities, or retail facilities, resulting in higher overhead for physical facilities and personnel.
- Capital requirements for a subcontracting business can be fairly high. Liquid assets, such as inventory and accounts receivable, are higher than that of the general contractor. Some subcontracting businesses, such as excavation and masonry subcontractors, have a fairly large long-term debt for equipment.
- Subcontractors are paid for their work considerably after the completion of that work. "Contingent Payment Clauses" in subcontract agreements state that the subcontractor will not receive payment until the contractor receives payment. Subcontractors may have retainage held for long periods of time, particularly when completing an early portion of a project, such as excavation.
- Many subcontractors have additional subcontractors working for them. For example, a mechanical contractor may complete the plumbing and piping work with its own forces, while subcontracting HVAC work, pipe insulation, fire sprinkler, excavation, temperature controls, and other portions of the work.
- Scheduling crews for the subcontractor often relates to keeping crews busy and fulfilling company-wide obligations, rather than meeting particular project requirements.

As a separate business entity, the subcontractor has many pressures that are not apparent in the project relationship alone. This "hidden agenda" requires that the contractor be clear, concise, and forthright in contractual and jobsite relationships with all subcontractors.

The Subcontract Agreement

The subcontract agreement is an agreement or contract between the contractor and subcontractor. It is based on the construction documents prepared by the architect and engineer and is an agreement for a specifically described portion of work.

Subcontract Agreement Amount

The amount of the subcontract agreement is normally based on a lump sum bid by the subcontractor to the contractor. This bid usually is submitted in a competitive environment, with the contractor receiving several subcontract bids for each subcontract area. The contractor requires competitive bids to maintain a low, competitive bid for the owner. Most contractors will choose the low subcontract bid. Negotiation after the receipt of bids is normal, to determine any missing or overlapping areas between subcontracts.

Selecting Subcontractor Bids

Fair and ethical bid processing is necessary for the contractor and subcontractors to remain competitive in the marketplace. **"Bid shopping,"** the practice where contractors supply bid amounts from subcontractors to other subcontractors, is not ethical in most markets. This practice tends to decrease subcontract bids to cost or below cost, resulting in jobsite problems. Other unethical practices, such as contractors demanding cuts in bids, happen occasionally, but limit the ability of the contractor to receive competitive bids for future projects. Fair treatment of subcontractors when receiving bids should establish a relationship of trust between the contractor and subcontractor.

One of the consequences of bid shopping is the increasing use of listing laws by states and other governmental agencies. The listing requirement is a part of the bid and is later incorporated into the contract itself. California requires a list of all subcontractors and material and equipment bids in excess of .5 percent of the total bid. This percentage is so small it requires the general contractor to list most of the subcontracts and purchases on the project. In other states, only the major subcontractors are listed, that is, mechanical, electrical, and so on.

Many subcontractors submit bids late in the bid period, resulting in an incomplete analysis of subcontract bids. These bids are normally made by telephone or fax. But both methods can create confusion that will need to be clarified, usually after the bid period. Comparison of scope and items covered and excluded is necessary prior to writing subcontracts. Negotiations for adding or deleting portions of the work usually are required prior to finalizing subcontract agreements.

A typical telephone subcontractor quote sheet is shown in Figure 9–1. The telephone quote usually does not contain the detail and contract modifications contained in the faxed bid. The trend to fax quotes in the past decade has complicated scope definition at bid time and has caused the general contractor potential problems with subcontract negotiations. Figures 9–2 and 9–3A and B illustrate a typical subcontract bid quote transmitted by fax. The process of scope clarification after the bid becomes very complicated and often is confused with "bid-shopping."

Subcontractors describe the scope of their work in relation to construction documents, usually by the section number in the specification. For instance, a masonry subcontractor might describe their scope of work as follows:

> All brick and CMU masonry work, as per Section 04000 of the Specification. Place reinforcing in masonry only. No weather protection.

A short description of the technical specifications is needed for writing scope of work clauses in the subcontract. The general contractor, at

Telephone Subcontractor Quote

Project:_____   Bid Date:_____
Company Name:_____   Estimator:_____
Address:_____   Phone:_____
City:_____   License:_____
State/Zip:_____   Called in By:_____

Division/Scope of Work:_____
_____

Inclusions:_____
_____
_____
_____
_____
_____
_____
_____
_____
_____
_____

                    Total Base Bid:_____

Exclusions:_____
_____
_____
_____
_____
_____
_____
_____
_____

Alternates
or Bid Items:_____
_____
_____
_____
_____

_____**Total Amt. of Alternate**_____

Bid includes products that meet or exceed specifications?   _____yes   _____no
Is bid per plans and specifications?   _____yes   _____no
Is your firm DBE / WBE / DVBE? (circle one)
Union labor   _____yes   _____no
_____Addenda?   _____yes   _____no   Installed?   _____yes   _____no
Tax?   _____yes   _____no   Erected?   _____yes   _____no
Bond?   _____yes   _____no   Furnish only? _____yes   _____no
                                              F.O.B. _____jobsite _____Trucks
                                              _____nearest R. R. crossing
Received by_____   _____Other_____
                                              Delivery_____
Date_____Time_____   Weight_____

FIGURE 9–1   Telephone Subcontractor Quote

Borland Mechanical Contractors
2839 South 1st Ave.
Atascadero, California 93422
(805) 462-1520
FAX (805) 462-2000
Contractor's License No. #195231

FAX COVER SHEET

This is a Subcontract Bid!

FAX (408) 584-9250

DATE: February 9, 1997   TIME: 1:45 PM

TO (Company): FGH Construction Company, Inc.

ATTENTION: John Anderson

FROM: Bill Franks - Estimator

REFERENCE: Naval Shipyard - Point Magu, Mechanical Repairs to Bldg. 210

The attached is Borland Mechanical Contractors scope letter for the above referenced project. Please review in depth. Any questions should be directed to Bill Franks prior to bid opening. Thank you for this opportunity to bid this project with you.

NUMBER OF PAGES INCLUDING COVER PAGE 3

If you do not receive all the pages, please call: (805) 462-1520

FIGURE 9–2   Fax Cover Sheet

Borland Mechanical Contractors
2839 South 1st Ave.
Atascadero, California 93422
(805) 462-1520
FAX (805) 462-2000
Contractor's License No. #195231

FGH Construction Company, Inc.
893 Higuria Street
Monterey, CA 93940

Attention: John Anderson

Reference: Naval Shipyard - Point Magu, Mechanical Repairs to Bldg. 210

Gentlemen:

We are pleased to quote you the Mechanical Proposal for the referenced project based upon the following scope of work listed below:

Division 15 - Mechanical Complete with the exception noted under work included, work not included and standard clarification of our proposal as outlined in attachment "A".

Work Included:

1. All work shown on the drawing as it related to the systems we are furnishing and installing.
2. Louvers as shown on the Mechanical Drawing.

Work excluded:

1. Formed Concrete
2. Painting
3. Temporary plumbing, electrical, and lighting for our construction use.
4. Sewage usage Fees
5. Premium Time
6. Parking Fees

Our lump sum bid is: **$ 345,500.00**
Should you need clarification or have question regarding this fax bid proposal, please call.

Sincerely,

Bill Franks
Estimator

FIGURE 9–3A   Fax Subcontractor Quote

Attachment "A"

Standard Clarification

The following conditions:

1. Cleanup - Borland Mechanical will remove all rubbish and debris generated by our operations to your trash bins onsite and leave our work area in a clean condition (not broom cleaned). Our bid does not include and we are not to be charged any prorated, general clean-up cost or haul off charges.

2. Back charges - FGH Construction will not perform any portion of Borland Mechanical's work or incur any costs for our account, except upon written order and unless Borland Mechanical is in default and have been given written notice under our subcontract.

3. Progress Payments - Progress payments shall be paid to Borland Mechanical in the percentages approved by the Owner for Borland Mechanical' work within five working days following FGH Construction's receipt. Any retention reduction given to FGH Construction will be passed on to Borland Mechanical.

4. Insurance, Liability, and Bond - This bid does not include the cost of Builder's Risk Insurance or the cost of a bond. Borland Mechanical can furnish a bond at your request and costs. Borland Mechanical shall be only liable for acts of our own employees.

5. ....

6. ....

7. ....

FIGURE 9–3B   Fax Subcontractor Quote

bid time, wants to quickly determine the individual subcontractor's proposed scope of work, then make comparisons to other subcontractors who are bidding the same work. If an electrical subcontractor bids on Section 16 complete, without any other inclusions or exclusions, and another electrical subcontractor does the same, the question of which subcontractor to use becomes much easier to answer. If both prices are within 10 percent of each other and both subcontractors are bondable and of good reputation, then the lowest bid of the two would be the logical and correct choice. Scope comparisons and discussions with subcontractors often are necessary to ensure full coverage of the subcontracted area.

Subcontract Agreement Contract Form

Many firms use a standardized contract form for the subcontract agreement. These forms are the products of construction organizations such as the Associated General Contractors of America, Inc. and the

American Subcontractors Association. An example of a subcontract agreement is shown in the Appendix.

The advantage of using standard contracts is that they are "tried-and-true," both with parties in the industry and in the courts. These contract forms are familiar to contractors and subcontractors, containing standard language and requirements. Most contracts have some bias, depending on who writes them. Standard forms co-written by contractor organizations and subcontractor organizations reduce this bias, however. Both parties have the right to modify sections of the subcontract agreement, subject to specific approval by the other party. Discussion about contract language and modifications is part of the negotiation process. When a discussion cannot be resolved the contractor must decide on the feasibility of awarding the subcontract to the second bidder. Both parties need to carefully consider their position in these discussions, determining the value of the contract language. The contractor may incur additional cost by subcontracting with another team, and the subcontractor may lose the contract. Many contractors will use an in-house subcontract, written by their attorneys. These contracts usually are biased heavily in favor of the contractor. Most contractors will not allow modifications to this custom agreement. Some subcontractors will present their own subcontract agreements, particularly in residential construction. These contracts are usually brief, intended to protect the subcontractors' right to payment. In commercial building construction, some large specialty subcontractors, such as elevator subcontractors, will prefer to use their own subcontract agreement. The bargaining power of subcontractors is enhanced when they are in an exclusive area and/or are a large firm.

The subcontract agreement contains provisions from the contract between the contractor and owner and provisions for the relationship between the contractor and subcontractor. The subcontract should contain the following articles or areas:

- preamble and date/parties to the agreement
- reference to the construction documents, and applicable sections of the specification
- scope of work
- subcontract price
- payment provisions
- changes, claims, and delays (including damages caused by)
- insurance provisions
- bonding provisions, if required by the contractor
- materials and workmanship
- time and schedule
- obligations of the contractor and subcontractor
- labor provisions
- contractual requirements, indemnification and recourse by the contractor
- remedy for solution of disputes
- termination of the agreement

Scope Definition in the Subcontract Agreement

Proper scope definition in the subcontract agreement is one of the keys to successful subcontract management. Each subcontract needs to be fully defined, then compared with the other subcontracts to ensure coverage of all items. All subcontracts should be adequately described, compared, and checked prior to writing and issuing any individual subcontract. Well-defined subcontract scope statements will avoid unnecessary arguments about who will handle certain items, reduces change orders, and gives the project management team tools to control the cost, time, and quality of the project.

The scope of work must be carefully described for each subcontract. Even when using the standard preprinted subcontract agreement, the exact scope of the subcontract work needs to be added to the agreement. The scope of work description should include:

- description of the work relating to the construction documents, including drawing and specification reference; reference should include dates of the documents, document numbers, and dates of the addenda
- additional work to be performed by the subcontractor, beyond the specification sections
- exclusions from work described in the construction documents, as per the subcontractor's bid
- any additional specific information relating to the project, such as schedule dates or delivery dates
- description of included alternates and negotiated additions or deletions to the agreement

Standard clauses almost always precede or follow the scope definition, stating that all work will be "performed in accordance with the owner and contractor contract, contract documents, and the plans and specifications." The terms "per plans and specifications" and "per contract documents" are used to connect the subcontract to all of the same provisions contained in the contractor's contract with the owner. Phrases such as "including but not limited to" should be avoided in the subcontract agreement, as they can lead to multiple interpretations. Figure 9–4 shows a sample scope of work statement in the subcontract agreement.

The **subcontract price** should reflect the quoted price or newly negotiated price from the subcontractor. Many subcontract agreements require some negotiation on scope, price, or contract provisions. One would expect to see the subcontract price reflect these clarifications and negotiations, but this is not always true. The system that requires the general contractor to set his price using many diverse subcontractor bids, competition leaving small margins and an all-inclusive scope of work defined by the contract documents, does not leave the general contractor with much room to maneuver on price. These negotiations normally take the form of clarification of scope, rather than a change in subcontract price.

The **"schedule of work"** clauses relate to the time frame required for completion of subcontract work. Terms similar to "time is of the essence"

"SECTION 2. <u>SCOPE</u>

Subcontractor agrees to furnish all labor, materials, equipment and other facilities required to perform the work to complete:

Project completion date per project master schedule.

Item of work to include but not limited to: Subsurface Investigation, Demolition, Site Preparation, Earthwork, Trenching, Backfilling, and Compacting.

for the project in accordance with the Contract Documents and as more particularly specified in:

Section (s) 02010, 02050, 02100, 02200, 02221, 02700, 02721 of the Contract Specifications.

See Exhibit "A" which is attached hereto and made a part hereof.

See Attachment No. 2 dated August 25, 1996, for Continuation of Scope of Work."

FIGURE 9–4   Scope of Work

are often used, emphasizing the importance of time to the subcontractor. The subcontractor usually is required to follow the contractor's schedule, but most contractors will prepare their schedules with input from subcontractors, including activity duration, sequence with other activities, and material and equipment delivery dates. As the contractor needs to compile a schedule from all subcontractors and suppliers, the project schedule may not conform with the subcontractors' ideal schedules.

Subcontract agreements include **indemnification clauses** that "hold the contractor harmless" from actions of the subcontractor. These clauses give the contractor protection from claims and lawsuits arising from the subcontractor's actions in pursuit of the work covered under the subcontract agreement. The subcontractor is responsible for providing appropriate liability insurance to protect the contractor and owner from lawsuits connected with the subcontractor's workforce or work on the project.

Control of subcontractors on the project is difficult for the contractor. Several standard clauses are included in subcontract agreements to provide some tools for control of the subcontractor by the superintendent. Compliance to the schedule, jobsite cooperation, and quality of the installation are common concerns to the superintendent. The subcontract agreement usually contains the following type of clauses to allow control of the subcontractor:

- schedule requirements related to termination clauses if the subcontractor does not meet schedule and manpower requirements
- quality control, inspections, and repair of defective work, also related to termination clauses
- clauses related to acceptability of subcontractor jobsite personnel and the right of the contractor to have that individual removed from the jobsite

Remedies for solving disputes usually are stipulated in the subcontract agreement. Arbitration is the most common remedy to disputes, however litigation is occasionally stipulated. Arbitration is normally considered more expedient than is litigation, although the arbitration process can be slow and expensive. The claim procedure for the subcontract is similar to that for the general contract, requiring notification of claim, continuation of work during the claim, continuation of payments, and a step-by-step procedure. The final settlement of the claim will be accomplished by the method stipulated, either a form of arbitration or litigation.

Termination of contract clauses are essential to any agreement. Specific clauses should be included in the subcontract agreement, stipulating:

- occurrences that will cause termination of the agreement, such as non-compliance with contractual terms by the subcontractor or nonpayment by the contractor
- notice necessary for termination, including form and timing
- determination of costs of termination
- remedies for termination or to oppose termination

Termination of the contract by the contractor usually relates to subcontractors' lack of ability to perform their contractual obligations. In these cases, the contractor needs to carefully follow the terms of the agreement, preferably with an attorney's guidance. Replacing a subcontractor usually will result in delays and extra costs. Termination proceedings normally are a last resort measure, as they often cause more problems than they solve. If the subcontractor is a solvent business entity, it is usually easier to solve performance problems than it is to terminate the contract. In cases where the subcontractor is insolvent financially, termination and replacement of the subcontractor may be the only available remedy.

The major clauses of each subcontract have to be administered and various subcontracts and subcontractors managed. A major concern for the contractor is the subcontractors' meeting schedule, providing an adequate jobsite workforce to meet the schedule. The project superintendent needs to work with subcontractors on a daily basis, monitoring their progress and jobsite workforce. If changes are necessary, immediate action is necessary to avoid project delay. The quality of the subcontractors' installations also needs to be monitored on a daily basis. Poor-quality work must be stopped and corrected before additional work is done incorrectly.

Coordination Meetings

Weekly coordination meetings are normally held, with all applicable subcontractors in attendance. The focus of the subcontractor meeting is the exchange of information on scheduling, interface between the various subcontractors, and the review of updated information, changes, and potential effects on all companies involved. Like any meeting, this should be scheduled at the same time each week, with set agendas; the superintendent should control the flow of the meeting. Generally, all of the major subcontractors should be at the weekly subcontractor meetings for the duration of the job. This is not a difficult task once subcontractors are on site, but many are hesitant to come to meetings if they feel their time will be wasted. Every contractor has his own method of encouraging subcontractors to attend these meetings, from fining them for missing meetings to providing incentives. When subcontractors are not able to attend meetings, minutes should be sent to them.

Scheduling Subcontractors

Jobsite personnel must understand subcontractors' business motivations in order to manage subcontractors on the jobsite. Subcontractors are involved in many projects simultaneously. Despite trying to achieve a balance of work for personnel and equipment, they inevitably have several jobs that need attention at the same time, due to project delays and schedules. Most subcontractors are interested in completing their work in a timely manner. Furnishing subcontractors with schedules of the work and updating the schedule will help meet project demands.

Subcontractor Submittals

Subcontractors are required to submit numerous documents to the contractor during the course of work. Prior to initial payment, the subcontractor must submit certificates of insurance, bonds, material samples/shop drawings/product data, and various certifications and affidavits required by the owner. During the project, the subcontractor may be required to submit certified payrolls, partial lien releases, and other miscellaneous documents.

Changes to the Subcontract Agreement

Control of changes with subcontractors is critical in subcontractor management. Official changes in the contract follow a specific procedure, requesting pricing information, followed by official notice to the contractor in the form of a change order. The contractor, then, needs to provide the subcontractor with written authority to proceed with additional work for the amount negotiated. This change to the subcontract agreement must be in writing. Technically, the work should not be done until formal notification is made, however, the work is often completed prior to completion of the paper work. Payment is not made, though, until all of the paper work is complete. Other changes occur during the work on an informal basis. It is often necessary to modify work procedures to accommodate all elements of the work. These changes are normally done at the moment, without regard for cost of the change. Careful communication needs to be accomplished between the contractor and subcontractors to avoid confusion about payment for work. As contracts become more complicated, the subcontract agreement may be explicit about the parties responsible for authorizing additional work.

Quality Control in Subcontract Work

When discussing subcontractor management, another area that must be controlled is quality and workmanship. The contractor's field personnel need to be involved in day-to-day quality control with subcontractors, as well as in quality review at the end of the subcontract work or at the completion of the project. The contractor's field personnel must be able to recognize correct work procedures and quality of work for each subcontract. Education of field personnel needs to be done periodically for updates in new and changing techniques. As the contractor is ultimately responsible for the quality of the work in the project, field personnel must be continually involved with quality review of the subcontractors' work as it progresses.

A common dispute between contractors and subcontractors regards the acceptance of the subcontractor's work. As most subcontract work is completed or partially completed prior to the completion of the project, the subcontractors' installations must be accepted by the contractor prior to acceptance of the project by the architect. Often, a subcontractor's work needs to be accepted prior to the start of another subcontractor's installation. For instance, the gypsum drywall subcontract must be completed and accepted prior to the start of painting. Many times a subcontractor's work is completed before other work in the same area, requiring protection and care of the craftspeople working in that area. For example, ceramic tile floor covering is installed before several operations in a toilet room, specifically cabinets, and installation of plumbing fixtures, toilet partitions, and toilet accessories. Any of the other trades could potentially damage the ceramic tile installation.

Most contractors are reluctant to accept responsibility for approving work prior to approval by the owner. It is usually difficult for the subcontractor or the contractor to protect their completed work from other craftspeople on the project. Disputes can easily arise in this area. Many contractors will use clauses in the subcontract agreement requiring the subcontractor to protect their work, deferring acceptance of subcontract work until accepted by the owner. Realistically, the contractor and subcontractors should cooperate to achieve a successful project. The contractor's superintendent and other field personnel must be involved with subcontract installations, avoiding unnecessary disputes.

Some guidelines for accepting subcontractor work include:

1. The contractor's superintendent should be aware of the subcontractor's progress and quality of work during the installation.
2. Inspection of the subcontractor's work should be made prior to the subcontractor's departure from the project. If possible, the architect or owner's inspector should be involved in reviewing the subcontract work for quality, as well as the contractor's field personnel. If further work or remedial work is required, it is best to complete it before the installation crew leaves the jobsite. It is also best to have the work repaired prior to commencing adjacent or attached work.
3. When a subcontractor's work is the substrate for another subcontractor's work, such as completion of drywall installations prior to painting, the contractor should involve the other subcontractor with acceptance of surfaces and adjoining installations, before the start of that subcontractor's installation. Careful inspection and agreement need to be made between the contractor and subcontractors about the acceptability of substrate work.
4. Agreement on the protection of the subcontractor's work must be reached prior to the subcontractor's departure from the project. Some subcontractors and contractors like to photograph or videotape the condition of the work when turned over to the contractor. Finishes often receive minor damage during the final stages of a project. The contractor's field personnel should be active in the final stages of the project, preventing damage and determining the cause of any damage.

Subcontractor Payment

Subcontractors are very concerned about receiving payment for their work at the jobsite. As most subcontract agreements have a "contingent payment clause," which provides that the subcontractor will be paid after the contractor receives payment from the project owner, payments for the work, less the amount held for retainage, are normally received over ninety days after the work is performed. Subcontractors are interested in requesting payment for their work to coincide with the payment period conclusion. Most subcontractors are obligated to pay for materials, as well as labor, considerably before receipt of payment from the contractor. The contractor must provide the necessary payment "cut-off" dates to each subcontractor, allowing the subcontractor to bill the appropriate amounts at the proper time. The payment process is fairly inflexible once established, not allowing for billings between cut-off dates. The contractor is responsible for the amount billed, requiring careful monitoring of subcontractor work completed and the amount paid to the subcontractor. The owner also will monitor the amount of work completed. To avoid reprocessing payment requests and the associated delay, the contractor needs to verify the amount of work completed by subcontractors prior to submitting the payment request. A subcontractor's billing may be listed in several items in the Schedule of Values used by the owner to determine payment, and the subcontractors should list their completion in appropriate sections to facilitate payment processing. As subcontractors often overbill for their work, due to cash flow concerns, it is often necessary to verify the amount with material and equipment invoices, certified payrolls, and verification of physical work completed.

Subcontractors' requests for payment usually will include an amount for material stored on the site. Most owners will pay for material delivered and stored on the site. This facilitates having adequate material on the jobsite for the work tasks and does not delay the progress of the project. Most owners are concerned, however, that the amount paid for material coincides to the material actually stored on the jobsite. Verification of the quantity of material stored at the site and the cost of the material, by invoice, often is necessary. The subcontractor usually is concerned about the storage and handling of the material on the jobsite and is responsible for this until it is incorporated into the work. Subcontractors will want to store their material close to the point of installation, if possible, to facilitate efficient installation.

Subcontract Back-charges

Many contractors will **"back-charge"** a deduction from the amount owed to the subcontractor, costs that have been incurred by the contractor that should have been the subcontractor's responsibility. The back-charge normally is deducted from the subcontractor's progress payment. These back-charges can include a wide variety of items, such as:

- charges for equipment use
- charges for material used by the subcontractor, furnished by the contractor
- charges for labor furnished to the subcontractor for their contractual responsibilities
- charges for clean-up of waste on the jobsite
- charges for delays caused by the subcontractor
- charges for repair of the subcontractor's work by the contractor or other subcontractors

The contractor should notify the subcontractor concerning the reason and amount of the back-charge before applying it to the payment. The subcontract agreement will normally allow for back-charges, within specified limits. An agreement on the back-charge should be reached between the contractor and the subcontractor, although this is not always possible. Unnecessary disputes arise in this area, which could be resolved with communication between the contractor and subcontractor. Some guidelines for back-charges include the following:

1. Notify the subcontractor of cost of equipment use prior to the use of the equipment. Agreement should be made in each instance and recorded, with each party keeping a copy of the record of hourly cost and hours used.

2. Record the use of materials, with both parties receiving copies of the record. *Example:* The plumbing subcontractor needs a cubic yard of concrete for thrust blocks on a water line. The contractor is pouring concrete at the time, and arrangement is made for the cubic yard to come off of one of the contractor's concrete trucks. A record is made by the field engineer that a cubic yard of concrete was furnished, at market value, indicating the time and date and containing signatures of the contractor's field person and the plumbing foreman. The amount is then deducted as a back-charge from the next payment made to the plumbing subcontractor.

3. A similar procedure to the one just described will be used when a subcontractor uses contractor personnel for work in the subcontractor's scope.

Example:

A temporary ramp is needed by the mechanical subcontractor for installation of a piece of equipment. The contractor furnishes two carpenters for eight hours for construction of this ramp. Record is made of the hours and the applicable rate, with the date noted and signed by the superintendent and mechanical foreman. Deduction of the amount is then made at the next payment.

4. Back-charge for clean-up might be agreed upon at the start of the project, or might be made as a result of the subcontractor not cleaning up waste. Agreement for the scope of work and cost should be made prior to commencing the clean-up activity.

5. Back-charging for delay is difficult, as it is normally disputed by the subcontractor. Discussion of the impact of a delay should occur, resulting in an agreement between both parties.

6. Repair of a subcontractor's work by another party usually is done when the subcontractor refuses to repair the work. This repair cost is normally part of a dispute between the contractor and subcontractor. Careful documentation must be made indicating notices sent to the subcontractor, date of the repairs, exact nature of the repairs, and detailed listing of labor hours and material used. Photographs and videotapes can assist in this documentation.

Withheld Payments

Instances may arise when the contractor will withhold payments from the subcontractor, as a reserve to cover payment of obligations incurred by the subcontractor. The contractor and subcontractor normally agree in the subcontract agreement to an amount of retainage to be held, but occasionally additional amounts must be withheld from the subcontractor's payment. Some reasons a contractor would withhold payment include:

- as a reserve to cover repair or replacement of work in place
- as a reserve to pay second-tier subcontractors, when they have furnished notice that they have not received payment

- as a reserve to pay labor, fringe benefits, and labor taxes when the subcontractor is not making appropriate payment
- as a reserve to pay equipment or material suppliers when they have not received payment from the subcontractor

These payments withheld are dispersed as necessary, under provisions contained in the subcontract agreement. Because these instances contain many legal issues, the contractor should consult an attorney prior to withholding payments from the subcontractor.

Subcontractor Coordination

If subcontractors are able to pursue work in a productive and profitable manner, they normally cooperate with the contractor and other subcontractors. This facilitates an overall positive jobsite attitude, optimizes the work schedule, and increases quality in the project. The contractor, then, should try to facilitate subcontractors' work by being ready for the installation and by providing working conditions favorable to the installation and completion of the subcontract. On the surface, a little extra coordination and cooperation for the subcontractors may appear to increase jobsite costs for the contractor. In reality, the extra cooperation reduces costs related to project efficiency, project duration, and completion of punchlists and quality reviews.

Scheduling subcontractors and the use of the subcontractors in the schedule can be crucial to the profitability of subcontractors on the project. When preparing the schedule, the contractor should be aware of the sequential relationship of the subcontractors' work, the duration of the work, and the delivery of materials. Fair and honest representation of the subcontract work in the schedule provides the subcontractor with realistic expectations to complete the work profitably. There are a number of scheduling philosophies, using the subcontractor as a tool to push the schedule along, usually at the expense of the subcontractor. These tactics have short-term benefits, at best, resulting in disputes, poor quality work, and subcontractor default on the project. The project schedule should be used as a tool to realistically indicate the work plan and help all process participants achieve or exceed the plan. There are certain subcontracts in every project that can provide the pace of the project. The contractor needs to identify these pacesetters and facilitate their work, using the following techniques:

- Discuss the fastest course possible with the subcontractor, realizing that optimization of the time frame is beneficial to all parties. Determine a reasonable duration for the installation.
- Determine what needs to be done prior to the start of the subcontract.
- Establish achievable start and finish dates for the activity, well in advance of the start of the work.
- Determine material delivery constraints, such as shop drawing reviews,

and minimize any possible delays in ordering and delivering the material.

- Determine use of contractor's lifting equipment to facilitate the installation.
- Determine appropriate jobsite storage of material and equipment.
- Be prepared for the subcontract work, rather than just expecting it to happen.
- Be flexible to changes and seek quick and efficient solutions to any changes.

Today's general contractor is offering more preconstruction services, which include scheduling, value engineering, conceptual estimating, constructability studies, and life cycle costing. To accomplish these tasks accurately and professionally, a general contractor will rely on long-term subcontractor relationships. Full project information in the early stages of a project, including cost and schedule of subcontract work, can provide subcontractors with a competitive edge in negotiating contracts. In many private projects, owners and architects rely on bidder designs in mechanical and electrical specialty areas. In this type of bid work, the general contractor assumes a great deal of risk if a highly sophisticated subcontractor who understands the design and engineering of the job as well as the traditional engineering firm does is not used. In this type of arrangement, the contractor assumes responsibility for the design of the applicable portions of the work, as well as for the normal construction responsibilities. Mistakes in the design of a bidder-designed contract can cause major problems in the successful completion of the project. Both mechanical and electrical specialty contractors, capable of design and construction of the systems, are available in most construction markets. These specialty contractors can provide budget information during the early stages of a project and final construction estimates when designs are complete. These can be very efficient arrangements, saving cost and time, if specialty contractors are competent in their work and can work closely with the general contractor. General contractors should establish close working relationships with subcontractors in most specialty areas, for work in negotiated and "cost-plus" contracts.

Subcontractor Safety and Waste Management

Additional project management and subcontract management relates to safety, waste management, and clean-up activities of the subcontractor. General contractors have found that these key elements are often ignored by the subcontractor, particularly by those that emphasize high productivity from their crews. Subcontractors do not always understand the relationship of safety to project profit. Many subcontractors expect the contractor to clean up and dispose of waste created by the subcontractor's crews, saving the subcontractor the cost of clean-up. The contractor needs to discuss the responsibility of safety programs and project clean-up with subcontractors prior to the start of each subcontractor's work. Continual reminders of these responsibilities need to be made by the subcontractor's field personnel throughout the project. Back-charging the subcontractors for clean-up costs is commonly done when subcontractors do not implement their responsibilities in cleaning up after their work. Occasionally, contractors will negotiate with the subcontractors to clean up the entire jobsite on a periodic basis.

Waste management is a relatively recent development, a result of higher landfill costs and increased environmental regulations. Many owners are requiring construction waste to be managed to efficient reduction, reuse, and recycling. The appropriate clauses must be added to the subcontract agreements, requiring subcontractors to plan, separate, and recycle their construction waste. Coordination of this waste management effort by the contractor is necessary. The contractor may wish to assume all of the waste management responsibilities on the project, at a negotiated cost to the subcontractors. This central waste management effort may be more effective and evident than relying on individual subcontractors.

The Subcontractor's Subcontracts

Many subcontractors subcontract out portions of the work. These "second-tier subcontracts" are responsible for providing specific work to the subcontractor. As there is no contractual relationship between the contractor and the second-tier subcontract, the contractor has less direct control of that portion of the work, relaying all directions through the subcontractor. The subcontractor though is responsible for this second-tier subcontracted work to the contractor. Responsible full-service subcontractors, such as the mechanical subcontractor responsible for all of the work specified in Division 15 of the Contract Documents, are often more desirable to the contractor, as the superintendent deals with one foreman rather than four or five different foremen for different subcontractors. These situations rely on the ability of the subcontractor to manage subcontractors, similar to the responsibilities of a general contractor.

Purchase of Materials

Subcontracting is only part of the purchasing process. When only material is being procured, short forms and different methods are used to timely obtain and properly deliver the needed material or equipment to the jobsite. Material and specific building equipment to be incorporated into the project, without jobsite labor included, are purchased under a purchase order or a "material contract."

Material Contracts

The material contract is not used by all general contractors, but when it is, it usually is used for major purchases of bulk materials, materials and equipment to be furnished to others, or equipment to be installed by the general contractor's own forces. It should be understood that the majority of general contractors in commercial construction do not furnish a great deal of labor on the jobsite, as 75 percent to 100 percent of the work is accomplished by subcontractors. In the area of heavy and industrial construction, some contractors accomplish the majority of the work with their own forces, including mechanical and electrical work. As an example, the installation of waste and water treatment equipment can be a competitive advantage if the general contractor has the expertise to complete this installation with its own crews. The discussion of material contracts has been placed after subcontract writing because some of the same elements can be found on the material contract. The description of the materials or equipment to be furnished is analyzed much like a subcontract scope of work, relating it to both the original quote from the vendor and the specific specification furnished by the owner.

Key elements of material contracts are the delivery articles and use of liquidated damages for late deliveries. These articles often will specify particular dates for delivery of material or equipment to the jobsite. Tied to these dates are liquidated damages for late delivery. These liquidated damages are not a penalty, but rather anticipated costs for problems caused by the late delivery of goods ordered. The contract will generally carry a clause that allows changes to be made during the submittal and shop drawing process without voiding the contract. A key element when shop drawings are involved is that the coordination between this equipment and other elements of the structure is needed. This type of contract also could be used for supply only of specialty fabricated items. This contract should include an article of compliance to all applicable laws and should define how payment will be made. Material contracts should be carefully compiled, with the advice of an attorney, and contain the correct applicable wording. While subcontracts are covered by contractual law, material contracts are covered by the Uniform Commercial Code. The subtle differences between these two areas of the law require special attention to the material contract when custom contracts are used.

Figure 9–5 is an example of a material contract used for purchase of material at a predetermined cost.

AGC CALIFORNIA

SHORT FORM PURCHASE ORDER

Order No. _____

Name of Project _____

SELLER: _____   BUYER: _____
_____   _____
_____   _____

DELIVER TO: _____   DATE: _____
_____

1.   MATERIAL TO BE PROVIDED:

Description/ Item No.	Estimated Quantity	Unit Price	Extension*	Delivery Date

*Approximate Amount of Material Contract (Based on Estimated Quantity)

Ship to _____

Via _____   Mark _____

Deliver F.O.B. _____   With Freight Allowed to _____

Mail Invoices in Triplicate to _____

_____   Terms _____

2.   PAYMENT. The price specified shall include all taxes and duties of any kind levied by federal, state, municipal, or other governmental authority, which either party is required to pay with respect to the materials covered by this Agreement. The obligation of Buyer to make any payment hereunder is subject to the condition precedent of payment by Owner to Buyer therefor, except payments withheld by the culpable acts or omissions of Buyer.

3.   LIABILITY AND INDEMNIFICATION. Seller assumes all risks in furnishing the material and work ordered hereunder. Seller shall be liable to Buyer and shall indemnify Buyer for any added costs, loss, damages, claims, expenses or royalties Buyer incurs as a result of, arising out of, or incurred in connection with the performance or nonperformance of this Agreement, including actual attorneys' and experts' or consultants' fees incurred in good faith, except when arising from Buyer's sole negligence.

4.   DELIVERY. Time is of the essence of this Agreement. Should delivery for any reason fail to be timely, Seller shall be liable for all damages suffered by Buyer as a result of such failure, including, without limitation, any liquidated damages under Buyer's Prime Contract. Failure to furnish materials within the scheduled time shall give Buyer the right to cancel any undelivered balance of this order without additional charge.

© Associated General Contractors of California, Inc. 1988

Form AGCC-7
3/95

-over-

FIGURE 9–5   Materials Contract (Courtesy of Associated General Contractors of California, Inc.)

5.   **COMPLIANCE.** Seller's performance shall in all ways strictly conform with all applicable laws, regulations, safety orders, labor agreements and working conditions to which it is subject, including, but not limited to, all state, federal and local non-discrimination in employment provisions, and all applicable provisions required by the Prime Contract and by Buyer's own internal safety program, and all local regulations and building codes. Seller shall execute and deliver all documents as may be required to effect or evidence compliance.

6.   **COMPLIANCE WITH LICENSE LAW.** CONTRACTORS ARE REQUIRED BY LAW TO BE LICENSED AND REGULATED BY THE CONTRACTORS STATE LICENSE BOARD. ANY QUESTIONS CONCERNING A CONTRACTOR MAY BE REFERRED TO THE REGISTRAR OF THE BOARD WHOSE ADDRESS IS:

CONTRACTORS STATE LICENSE BOARD
P.O. BOX 26000
SACRAMENTO, CALIFORNIA 95826

7.   **BONDS OR INSURANCE.** Seller shall furnish one of the following to Buyer as required:

Supply Bond in an amount specified by Buyer (choose one)
___ The premium for the bond will be paid by Buyer
___ The premium for the bond is included in the price of this Agreement

Insurance in amount and type specified by Buyer; the specifications and policies are hereby incorporated by reference (choose one)
___ The premium for the insurance will be paid by Buyer
___ The cost of insurance is included in the price of this Agreement

8.   **TERMINATION.** Buyer may terminate or suspend at its convenience all or any portion of this Agreement not shipped as of the date of termination or suspension of this Agreement. Seller shall receive payment for work actually performed. Seller shall not be entitled to any recovery on account of profit or unabsorbed overhead with respect to work not actually performed or on account of future work, as of the date of termination or suspension. No termination or suspension shall relieve Buyer or Seller of any of their obligations as to any material shipped prior to Seller's receipt of the termination or suspension order.

If Seller fails to perform any obligation under this Agreement, Buyer may terminate this order for default. In the event of a termination for default, Buyer may, in addition to all other rights and remedies, purchase substitute items or services elsewhere and hold Seller liable for any and all excess costs incurred, including attorneys' fees and experts' and consultants' fees actually incurred.

9.   **ATTORNEYS' FEES.** In the event either party becomes involved in litigation or arbitration in which the services of an attorney or other expert are reasonably required arising out of this Agreement or its performance, or nonperformance, the prevailing party shall be fully compensated for the cost of its participation in such proceedings, including the cost incurred for attorneys' fees and experts' fees. Unless judgment goes by default, the attorneys' fee award shall not be computed in accordance with any court schedule but shall be such as to fully reimburse all attorneys' fees actually incurred in good faith. California law shall apply to all disputes arising out of this Agreement or its performance or nonperformance.

We acknowledge receipt of, and accept Purchaser's order: This order is hereby approved:

_____	_____
SELLER	BUYER
By _____	By _____
Title _____	Title _____
Date _____	Date _____
Contractor's License Number: _____	Contractor's License Number: _____
(If Required)	(If Required)

NOTE: Before execution, users should insure that this form meets their specific needs. Some construction material procurement agreements may require the use of specialized provisions not included in this form. This document has important legal consequences; users are encouraged to consult with an attorney with respect to its use or modification.

AGC
CALIFORNIA   -2-

FIGURE 9–5   continued

Purchase Orders

The purchase order is a control document used to define items to be ordered, the payment terms, the quoted price, how it may be shipped or delivered, the delivery time, and the cost account for the purchase. The purchase order system generally has the following components:

- the actual purchase order number
- a link to the cost control system to allow committed costs to be entered
- indication of who made the purchase and for which project
- coordination of delivery slips, purchase orders, and invoices

Most contractors will use a book of purchase orders that are serialized or numbered, as well as duplicates so loose purchase orders are not floating around. The purchase order number also is a way to assure payment to the vendor. The purchase order establishes accounting information for both the buyer and seller. For the vendor (seller), it may indicate a credit purchase. It establishes delivery and tax information for the seller. Purchase orders also can be used to establish a fixed unit price when the order quantity is not yet determined or when multiple deliveries are needed. Lumber for concrete formwork often is ordered this way. As businesses become more computerized, linking specific purchases to potential claims or changes can be accomplished without much effort, allowing the estimator to review costs in more detail. Figure 9–6 is an example of a purchase order for a construction project.

Expediting and Tracking Material and Equipment

The use of material contracts and purchase orders helps the general contractor expedite critical equipment and materials. Expediting can be defined as the tracking, the confirming, and/or the accelerating of delivery of an item that has been purchased. When purchasing is done from a central or home office, expediting is part of the purchasing agent's job description. When purchasing is site-based, the project engineer may track and expedite the items. If a company uses a central purchasing method, it never completely moves away from incidentals being purchased by the jobsite. Some firms believe central purchasing provides the company with greater control and a broader number of vendors' quotes from which to choose, thus giving the contractor opportunity for a better price. Figure 9–7 shows an equipment tracking or expediting form used for guaranteeing delivery of equipment on time.

Purchase Order

FGH CONSTRUCTION COMPANY, INC.

Purchase Order No. _____
All Invoices, Correspondence,
Packages, Etc. must carry P.O. No.

VENDOR:

_____
_____
_____
_____

SHIP TO:

_____
_____
_____
_____

To Be Delivered   _____   Picked Up _____
(Shipping Terms)
FOB Above Address   _____
Freight Collect   _____

Delivery Required By: P.O. Date   Terms   This P.O. is: _____
                                                                                Original
                                                                                Confirmation Only

Quantity Ordered	Unit	Description (note: Subject to terms & conditions on Reverse Side)	Cost Code	Unit Price	Total Amount

Total Cost   _____

Remarks:

_____
_____

Note: Send all invoices to: FGH CONSTRUCTION COMPANY, INC.
    Shipping Address   _____   By: _____
    Home Office Address   _____   _____
    Address Shown Below:

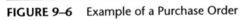

FIGURE 9–6   Example of a Purchase Order

Creston, California
WATER TREATMENT PLANT

EQUIPMENT AND MATERIAL EXPEDITING FORM

Qty.	Equip./ Material Desig.	Description	Mfg.	Spec. Section	Supplier	Contact	Phone Number	Lead Time (Wks.)	CPM Scheduled Date	Revised Date	Submittal Date	Received Date

FIGURE 9–7   Equipment and Material Expediting Form

Summary

Subcontracts, material contracts, and purchase orders are all used to control the cost of the project and work flow as defined by the schedule. All parts of project management are tied to one another. The way the subcontract is written, its scope for the subcontractor, and its general conditions all relate to the way the project manager and superintendent will accomplish the work on time, within the budget, and at the required quality. The use of a subcontract gives all parties a better understanding of what is expected, helping the general contractor manage the project more effectively.

Subcontractors are used extensively in construction projects primarily to:

- enhance project quality
- reduce cost through specialization
- reduce risk for the contractor

The subcontract agreement, the agreement between the contractor and the subcontractor, contains the following attributes:

- based on the bid and subsequent negotiations with the subcontractor
- standard forms are often used for the contract form
- careful scope definition is necessary
- remedies for problems are contained in the agreement

Subcontractor management by the contractor is necessary in the management of a successful project. Some areas of concern for subcontractor management include:

- coordination meetings
- scheduling work activities
- submittal control
- administering changes in the work
- quality control
- payment control, including back-charges and withheld payments
- coordination of work activities among subcontractors
- safety and waste management

Control also is necessary in the purchase of material and equipment. Purchase orders and material contracts are used to:

- control costs
- report costs to the appropriate project activities
- control delivery of material

Additionally, the tracking of material and equipment is necessary in controlling purchases and rentals for the construction project.

CHAPTER 10

PROJECT QUALITY MANAGEMENT

The construction industry often defines quality as conformance to standards and specifications. A short-term goal for the contractor is to meet contract requirements. However, a contractor's quality performance has long-term implications, with a contractor's reputation being partly based on the quality of the firm's previous projects. Productivity is also directly related to the quality of the contractor's work. The quality of a completed project is often used to measure the quality of a contractor.

The terms quality control and quality assurance are often specified as methods to attain quality in the construction project. The formal construction quality programs were adapted from the manufacturing industry and implemented in most segments of construction projects by government agencies such as the Army Corps of Engineers. Under these provisions, the contractor is required to control the quality of the installation and provide written documentation related to testing and inspection of the installation. The owner will often use a check mechanism, quality assurance, to monitor the contractor's quality control. In some contracts, the contractor is required to provide a quality control program and an independent quality assurance program from a testing agency. In other cases, quality assurance is contracted directly to a testing agency by the owner, or implemented by the owner's own forces.

The construction industry is realizing that quality must adopt a more important role and become a part of the construction company's culture. New terms, such as total quality management (TQM), continuous improvement, and partnering are being used to describe contractors' efforts to improve project quality and attitudes. Contractors are taking a proactive approach to quality, rather than a reactive one to contract requirements.

The contractor must meet or exceed the standards specified and comply to numerous regulations and standards, such as environmental and safety regulations and building codes. To meet these requirements, the contractor's team should be knowledgeable in materials, systems, methods, techniques, test procedures, industry practices, and applicable codes and regulations. Quality is achieved using current state-of-the-art knowledge regarding installations in the construction process.

Defining Quality

Quality is the characteristic element of an item that can be evaluated as meeting a standard. If the item meets or exceeds the standard, it is deemed to be of good quality, or high quality. If the item does not meet the standard, it is deemed to be of poor quality. Each installation in construction has several standards of quality: appearance, structural ability, composition, durability, suitability, and quality of workmanship. The acceptability of the installation, and of the project, relies on meeting or exceeding the standards specified.

Quality also can relate to the suitability of an installation for its intended use—meeting the owner's needs. Quality, as defined here, may not be the contractor's responsibility, as the specified material and system, selected by the designer, may not be appropriate to meet the owner's needs. The owner, however, may hold the contractor responsible for specific installations. The contractor must evaluate the quality of the material or installation prior to beginning a job.

Quality is an important aspect of the construction project. The diagram in Figure 10–1 illustrates this importance.

Time and cost are important elements of the performance of the construction project, equally as important as quality. A change in the construction cost or budget can have a marked impact on the quality of the project; the duration of the project also can influence the quality of the project.

Acceptable or exceptional quality needs to be achieved on the construction project with each installation, avoiding necessary rework to attain that quality level. Rework of an installation costs more and can also cause delays in the project.

A quality management program within the contractor's organization is important in reaching goals of quality with the construction project. Some elements of the quality management program include:

- establish company awareness of the intended quality standard
- develop and implement the continuous training of superintendents and project managers
- develop and train skilled craftspeople
- staff projects with an adequate number of workers so all management functions can be performed
- establish systems that will document and control the construction process, from purchasing through installation
- require superintendents to perform preliminary punch lists throughout the project
- use subcontractors that have reputations for high-quality work
- hold discussions about quality with subcontractors
- inspect materials for damage as they arrive on the site and ensure that they meet specifications
- work with inspection agencies so they can perform their job efficiently and accurately

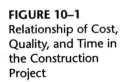

FIGURE 10–1
Relationship of Cost, Quality, and Time in the Construction Project

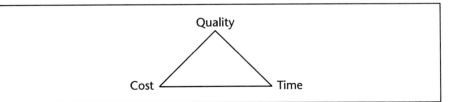

Total Quality Management

Total quality management is a process of bringing quality to the construction project. The quality of a construction project is the product of innovation, quality improvement, and productivity improvement, implemented by an involved and responsible workforce. The construction industry has always been involved in this, however, the industry has depended on decisions from strong leadership, rather than involving employees at all levels to participate in improving the quality of the product and process. The contractor can achieve a strong total quality management program that involves employees in improving the product. The measure of success of a quality management program is customer satisfaction. The continued success of the contractor is directly related to this.

Total quality management should create opportunities for improvement in construction methods, processes, and the delivery of the product to the customer. It should also provide a process for continuing development of these opportunities in the construction project. The workforce, field management, and company management need to be committed to providing a high-quality product to the customer.

The concept of total quality management comes from the work and leadership of the late Dr. W. Edwards Deming. Deming's formula for a company's business success is based on the relationship between improved quality and improved productivity. A successful project, then, would have the following elements:

- installations executed properly, one time only
- teamwork among project participants, including field management, skilled craftspeople, and subcontractors
- commitment of all who are involved to accomplish the task as efficiently as possible
- recognition of the quality of the project by the customer and potential customers, ultimately resulting in increased sales

Another leading scholar, Philip Crosby, author of *Quality If Free,* recognized the rewards and costs of adding quality to the process. He defined quality as something that was measurable in terms of costs.

Another major step in the recognition of the importance of quality to American industry was the establishment of the Malcolm Baldridge National Quality Award. One of the recipients of this prestigious award is a construction-related company, Granite Rock, Inc. of Watsonville, California, which won for excellence in quality production and management. This award illustrates that construction firms and construction-related firms can attain success in implementing quality control and enhancement programs.

The advantages of total quality management and continuous improvement of construction may be difficult to see in the short term—it may seem like adding costs without benefit. Construction owners, however, recognize the quality of the construction product and firms that are committed to satisfying their customers. These owners select construction firms that make their needs and the quality of the project top priorities.

The training and skill of the workforce is an important element of total quality management. The construction employer should train its employees in the quality installation of materials and systems, if the employee is not already so trained. Trade unions offer employees apprenticeship programs where they can acquire the necessary skills. Continuing training is necessary for employees, to keep them aware of changes in products and techniques. It also aids in achieving increased quality in the construction project.

The process of continuous improvement or the TQM process generally is defined by six broad steps.

1. Make suggestions for areas of improvement
2. Break down into parts and provide measurement tools
3. Create solutions to the suggestions
4. Implement and observe the solutions at work
5. Acknowledge the individuals and award their ingenuity
6. Incorporate the solutions on a broad scale

To implement the continuous improvement process, management, staff, and individual employees must understand the building blocks that constitute quality. These blocks include:

- construction work activities made up of smaller components
- quality defined by client satisfaction
- elimination of rework and mistakes achieved by prevention
- quality measured by attaching costs
- quality improvement pertaining to productivity
- quality improvement through individual commitment and teamwork

Dr. Deming lists the essential points a company must integrate into its system to become "total quality management" successful. They are:

- create consistency of purpose to improve product and service (plan to stay in business)
- adopt the new philosophy (stop tolerating poor quality)
- cease dependence on inspection to achieve quality (improve the process)
- end the practice of awarding business on the basis of price tag alone (seek longer-term supplier relationships; reduce the number of suppliers)
- constantly devise and implement improvements for every process in the systems of planning, production, and service
- institute modern training (for everybody)
- institute modern methods of supervision (the responsibility of foremen must be changed from sheer numbers to *quality*)
- drive out fear (encourage employees to speak up)
- break down barriers between departments
- eliminate slogans, exhortations, and targets for the workforce

- remove barriers to pride in workmanship (poor supervisors, poor materials, inadequate equipment, lack of training, etc.)
- institute a vigorous program of education and self-improvement for everyone
- require commitment by all employees to accomplish the transformation, and create a structure of top management that will stress the use of the aforementioned points

Some of Dr. Deming's essential points do not pertain to methods used by the construction industry in its business, but the majority of them can be applied when put into the context of an overall quality strategy, which begins with the management of the organization committing to a process of high-quality work attitudes and business practices. Training, the creation of vision and mission statements, and adherence to a high-quality principle are all elements to the implementation of a quality strategy. Surveys should be employed among customers and employees concerning their reactions and thoughts pertaining to quality. Teamwork and commitment is first provided by the establishment of continuous improvement groups. Quality improvement suggestions must be implemented and outside subcontractors and suppliers must be brought into the process. If the company creates a continuous improvement plan toward quality, the process will continue to expand and work.

Implementing a Quality Plan

A quality construction company can be defined by its customer and employee satisfaction levels. Accident-free jobsites, jobsite training sessions, retention of employees, jobsite teamwork with labor, management, architects, and owners, and jobsite efficiency are all indicators of the success of a quality improvement program.

A specific continuous improvement program toward increased quality starts with a vision and mission statement such as, "Our company strives to be a leader in providing the highest level of construction services to our clients. Our mission in providing these services is to satisfy our customers' needs and demands at all levels of their organization."

Using this as a starting point, specific areas of improvement or goals can be defined. Goals do not need to be specific in detail, but they do need to be measurable and achievable. Suggestions may be received concerning areas that have not been measured or have not been accurately measured previously. For example, if one of the goals is to make project punch lists shorter, review of past performance in the areas of punch lists must be done. As areas for improvement are developed, the company needs to develop standards for measurement, methods of implementing improvements, and methods of evaluating performance. A quality or continuous improvement committee should be formed to meet on a regular basis to brainstorm, discuss, make recommendations for, follow up on ideas implemented, and provide incentives for employees who make the suggestions.

The planning process of developing areas for improvement becomes as important as the idea itself. Without the use of brainstorming and creating flowcharts to analyze the process and cost benefits, the area cannot be developed to the implementation stage. It should be noted that when creating these areas or ideas, problems and situations that are a one-time occurrence must be separated from ongoing processes. The areas for improvement should relate to the goals of customer satisfaction, improved productivity, and the other key elements discussed earlier. The committee must create a systematic methodology for verifying and defining different areas of improvement. One important step in the process is the formal presentation to management of the committee's suggestions, which is no different from other types of presentations where one group sells to another. Presentations must have a clear definition of the area of improvement, the possible outcomes, cost of implementation, and a measurable cost savings or changes that will occur from the improvement. A documented record of the meeting should be kept. If implemented, the committee is responsible for tracking and measuring the results. Results should be published, and individuals who originated the ideas should be rewarded.

The Jobsite Quality Control Team

Prior to the start of the project, quality control responsibilities need to be assigned. Some projects, such as federal projects, may require that a full-time Quality Control (QC) engineer be assigned to the project. This position needs to recognized, along with a list of responsibilities for that position that is relevant to specification and jobsite requirements.

Quality control responsibilities often are allocated to superintendents, assistant superintendents, engineers, and field engineers, in addition to their other responsibilities. Their role in quality control should be explicitly expressed and related to their assigned activities. A checklist of specific project quality concerns can be used as part of this assignment. Figure 10–2 illustrates an example quality control assignment and checklist for concrete work.

The primary objectives of quality control assignments are to:

- avoid duplication of effort
- assure that every quality aspect is covered
- provide a clear delineation of responsibilities
- provide effective guidance on the project to achieve quality work
- provide documentation of materials, installation, and tests

Organization of the quality control effort before the project starts is essential for comprehensive quality control throughout the project.

Quality Control Checklist
Concrete

Project: City Hall Renovations, Springfield, CA

QC Assignment: Fred Smith, Field Engineer

Submittals:
- Request, Submit, Approve Mix Designs:
    - 3,000 psi
    - 4,000 psi
        - Standard
        - Exposed Aggregate
        - Colored (Red)
    - 5,000 psi
    - 6,000 psi

- Request, Submit, Approve Rebar Packages:
    - Footings, Foundations
    - Tilt-up Panels
    - Elevated Slabs
- Miscellaneous Product Submittals:
    - Sealer
    - Expansion Joint material
    - Water stop

General:
- Receive Concrete tickets from drivers, file
- Receive, organize, chart test results for cylinder breaks
- Organize rebar -- keep cut sheets and drawings for each area
- Integrate material handling with safety talks
- MSDS on jobsite for:
    - Concrete
    - Form oil
    - Sealers
- Provide training in proper concrete procedures as necessary.

Footings:
- Assure compaction in trenches
- Dewater footing trenches -- Do not pour in standing water
- Assure no frost in footing subgrade
- Forms to level and square
- Assure proper placement of reinforcing
- Assure ease of concrete placement
- Pour within temperature parameters: 40 deg. F to 80 deg. F
- Order trucks with sufficient time between trucks
- Assure imbeds and dowels are placed in correct locations

FIGURE 10–2   Partial Concrete QC Checklist

Partnering

This book has emphasized the team aspects of a project, the importance of the team's working relationship to satisfactory completion of the project, and the contractual relationship among different team members. Partnering is an attempt to improve quality through the improvement of the working relationship. **Partnering** has been defined by the Construction Industry Institute as:

> Partnering is a long-term commitment between two or more organizations for the purpose of achieving specific business objectives by maximizing the effectiveness of each participant's resources. This requires changing traditional relationships to a shared culture without regard to organizational boundaries. The relationship is based upon trust, dedication to common goals, and an understanding of each other's individual expectations and values. Expected benefits include improved efficiency and cost effectiveness, increased opportunity for innovation, and the continuous improvement of quality products and services.

The working relationship between contractor and owner differs markedly between the private and public sectors. Contractors are usually selected in the public sector on the sole basis of low responsive bid. The working relationship in many public projects is strained, with continual disagreements producing change orders and claims. The private sector, however, often selects contractors based on a variety of factors, including price, past performance, quality, and the ability to work with the owner and their architect. These projects are usually conducted in a cooperative atmosphere between all parties that are working toward accomplishing the project's goals, rather than individual goals. This atmosphere of cooperation can be termed "partnering." Public sector owners are realizing the disadvantages of the adversarial relationship in construction contracts that result in poor quality construction and longer and more costly projects. Partnering is being used on public and private projects to eliminate these adversarial relationships and to return the focus to the project goals.

Partnering is effective only when all parties accept the spirit of cooperation on the project and maintain that cooperation through the entire project. A partnering relationship is based on the common goals of trust and on an understanding of all of the parties' roles, responsibilities, and worth. Expected benefits are improved efficiency and cost-effectiveness, with the increased opportunity for innovative ideas that could increase a company's profit potential on a project.

Partnering is the commitment of organizations to change traditional relationships due to an increasing adversarial and litigious environment. Incentives in the public and private sector may be somewhat different, but the final goals are very much the same.

The key elements of partnering include:

- top-to-bottom commitment to the process
- respect for each person's work and expertise

- individual trust in the team as a whole
- setting mutual project goals and objectives
- implementation of the agreed-upon process
- timely information transfers
- taking responsibility
- evaluating the process on a continuous basis

As with many other quality improvement processes, the company's top management must be supportive and committed to the process for its success. The construction project is a complicated operation and its players must learn to respect one another for what they individually and collectively bring to it. Often, individuals are quick to blame others for problems that arise. During the partnering experience, both respect and trust are essential ingredients in the process of problem-solving, versus problem documentation and blame. By creating common goals that are measurable, and by committing to these goals, it is easier to implement openness in the information flow. Individuals who take responsibility for the common goals set forth in the formal partnering process and who agree to communicate in a timely manner and adapt problem solving skills should continue with the project. Problems between individuals can occur, but in evaluating the process one should see a collective movement from an adversarial attitude to a more cooperative one. Joint problem-solving generally produces a quicker, cheaper, and higher-quality solution.

Testing and Inspection

Both testing and inspection are required by most contracts to protect the owner, public, architect, and contractors. The inspection process should be done by independent agencies outside of the contract process to avoid any vested interest. Inspection and testing, if they are done correctly, will bring an unbiased review of the installation and product requirements. Inspections and tests must occur at many steps during the contractual process. Materials that are mass-manufactured may not have specified testing or quality control by field personnel, but in all cases, some form of in-house or outside testing and quality control has been performed at the manufacturing plant. Many products require labels from institutes or other agencies, such as UL labels on electrical devices to show conformance to product standards and federal regulations.

Most projects are subject to inspection by the local code authority or building inspector. Most municipalities or local jurisdictions have plan review personnel, as well as field inspectors, in their offices. The building inspector's main task is the enforcement of local building codes. Each municipality or local code authority has its own building code, which is normally based on a standard code. These building codes have been established, written, and continuously updated by national or regional code organizations. The three primary building codes currently used in the United States include:

1. The Uniform Building Code, published by the International Conference of Building Officials (ICBO), Whittier, California. This code is used mostly in the western United States.
2. The Basic/National Building Code, published by Building Officials and Code Administrators International (BOCA), Homewood, Illinois. This code is used primarily in the Northeast and in midwestern portions of the United States.
3. The Standard Building Code, published by the Southern Building Code Congress International (SBCCI), Birmingham, Alabama. This code is used primarily in the southern United States.

Other codes apply in mechanical, electrical, and life safety areas. The three primary national codes in these areas include:

1. National Plumbing Code, published by the American Public Health Association and the American Society of Mechanical Engineers.
2. The National Electrical Code, published by the American Insurance Association.
3. The Life Safety Code, published by the National Fire Protection Association International.

Additional codes are incorporated into local building codes, such as energy codes, access requirements (American Disability Act/ADA standards), zoning ordinances, health codes, and other regulations that protect the environment, safety, and health of the general public.

Building codes are written to protect the environment, health, life, and safety of the public. Other types of codes to protect the public include hospital, fire and fire suppression, restaurant health, and other specialized inspection areas. The building inspector is concerned with the general design, structural stability, and construction of a building. The first step in assuring compliance to building codes is submitting plans for a plan check. If the building is a mixed-use facility with many complicated or specialized functions, the designer will work with the city's or county's planning and zoning department, as well as the building department, throughout the initial design. The plan check fee and permit fee can be paid by the owner prior to beginning the work, or they can be included in the cost responsibilities of the contractor. Figure 10–3 is an example of typical building permit fees, based on the 1994 edition of the *Uniform Building Code.* Fees for the plan check normally are in addition to the basic permit fee. Many cities have cited higher costs in inspection and plan review and have enacted higher building permit fees. But these are not the only costs incurred in obtaining the permit; many jurisdictions today impose other fees to pay for the costs of increasing city infrastructure, such as school fees, floodplain fees, and road and bridge fees.

TABLE 1-A—BUILDING PERMIT FEES

TOTAL VALUATION	FEE
$1.00 to $500.00	$21.00
$501.00 to $2,000.00	$21.00 for the first $500.00 plus $2.75 for each additional $100.00, or fraction thereof, to and including $2,000.00
$2,001.00 to $25,000.00	$62.25 for the first $2,000.00 plus $12.50 for each additional $1,000.00, or fraction thereof, to and including $25,000.00
$25,001.00 to $50,000.00	$349.75 for the first $25,000.00 plus $9.00 for each additional $1,000.00, or fraction thereof, to and including $50,000.00
$50,001.00 to $100,000.00	$574.75 for the first $50,000.00 plus $6.25 for each additional $1,000.00, or fraction thereof, to and including $100,000.00
$100,001.00 to $500,000.00	$887.25 for the first $100,000.00 plus $5.00 for each additional $1,000.00, or fraction thereof, to and including $500,000.00
$500,001.00 to $1,000,000.00	$2,887.25 for the first $500,000.00 plus $4.25 for each additional $1,000.00, or fraction thereof, to and including $1,000,000.00
$1,000,001.00 and up	$5,012.25 for the first $1,000,000.00 plus $2.75 for each additional $1,000.00, or fraction thereof

Other Inspections and Fees:
1. Inspections outside of normal business hours .... $42.00 per hour*
   (minimum charge—two hours)
2. Reinspection fees assessed under provisions of
   Section 108.8 .... $42.00 per hour*
3. Inspections for which no fee is specifically indicated .... $42.00 per hour*
   (minimum charge—one-half hour)
4. Additional plan review required by changes, additions
   or revisions to plans .... $42.00 per hour*
   (minimum charge—one-half hour)
5. For use of outside consultants for plan checking and
   inspections, or both .... Actual costs**

*Or the total hourly cost to the jurisdiction, whichever is the greatest. This cost shall include supervision, overhead, equipment, hourly wages and fringe benefits of the employees involved.

**Actual costs include administrative and overhead costs.

FIGURE 10–3 Typical Building Permit Fees (Reproduced from the 1994 edition of the *Uniform Building Code,*™ copyright© 1994, with permission of the publisher, the International Conference of Building Officials.)

Often code compliance changes are required by the plan check. If this is completed prior to signing a contract with the general contractor, it will allow the owner time to have these items priced and incorporated into the contract before the main contract is signed, which eliminates the need for a change order the first day of the project. As one of the first requirements of the project, an official stamped set of plans must be kept on the jobsite for use by the inspector, while the inspection is made. Figure 10–4 is a typical flowchart for the building permit process.

Figure 10–5 on page 266 illustrates an example of a building permit that has been issued after the application and review has been completed. Figure 10–6 on page 267 shows a sample jobsite inspection card.

FIGURE 10–4
Flowchart of the
Building Permit
Process

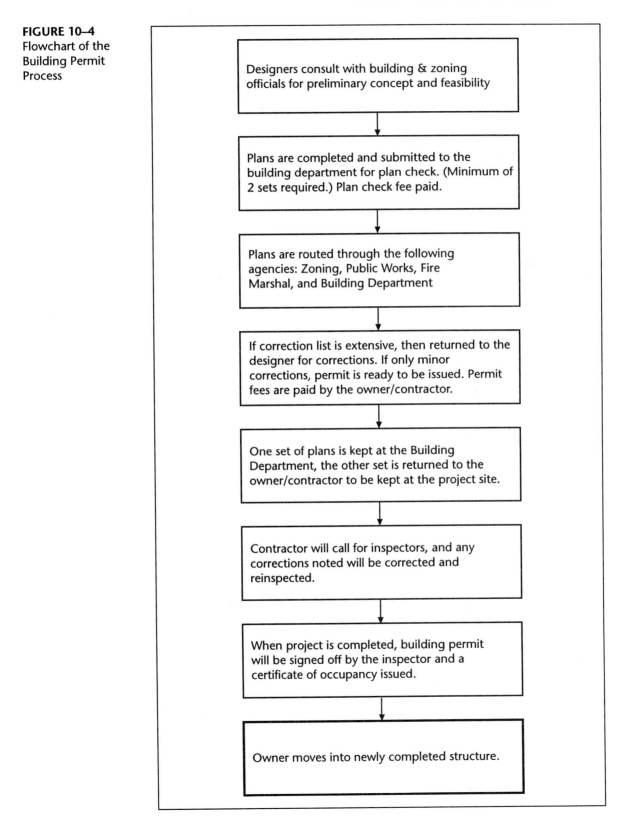

Designers consult with building & zoning officials for preliminary concept and feasibility

Plans are completed and submitted to the building department for plan check. (Minimum of 2 sets required.) Plan check fee paid.

Plans are routed through the following agencies: Zoning, Public Works, Fire Marshal, and Building Department

If correction list is extensive, then returned to the designer for corrections. If only minor corrections, permit is ready to be issued. Permit fees are paid by the owner/contractor.

One set of plans is kept at the Building Department, the other set is returned to the owner/contractor to be kept at the project site.

Contractor will call for inspectors, and any corrections noted will be corrected and reinspected.

When project is completed, building permit will be signed off by the inspector and a certificate of occupancy issued.

Owner moves into newly completed structure.

City of Pullman, CA

Construction Permit
Permit Number: _____

Project Address _____   **Date Received:** _____
Legal Description _____   **Date Issued:** _____
Assessor's Parcel No. _____   **Zone:** _____ **Lot Area:** _____

Contractor/Builder/Owner _____   Arch./Engr./Designer _____
Address(Mailing) _____   Address(Mailing) _____
(City/State/Zip) _____   (City/State/Zip) _____
License No. _____   License No. _____
Phone _____   Phone _____

Project Description: _____
_____
_____

Valuation: _____

Fees: **Requirements/Conditions:**

Building Permit _____   _____
Plumbing Permit _____   _____
HVAC Permit _____   _____
Electrical Permit _____   _____
Grading Permit _____   _____
Demolition Permit _____   _____
Sign Permit _____   _____
    Subtotal: _____   _____
Building Plan Check _____   _____
Fire Saftey Plan Check _____   _____
    Subtotal: _____   _____
Water Impact _____   _____
Water Meter Installation _____   _____
Wastewater Impact _____   _____
School Assessment _____   _____
Flood Assessment _____   _____
    Subtotal: _____   _____
Total Fee: _____   _____
Paid To Date: _____   _____
Balance Due: _____   _____

I have read and signed the "Owner Builder", "Workers Compensation Declaration" and the "Certificate Of Exemption From Workers Compensation Insurance".

Notice To Applicant:
If, after making any of the foregoing declarations, you become subject to any Labor Code or License Law provision, you must comply with such provisions or this permit shall be deemed revoked. I certify that I have read this application and state that the above information is correct. I agree to comply with all city ordinances and state laws relating to building construction and hereby authorize representatives of this city to enter upon the above-mentioned property for inspection purposes. Unless noted under "Requirements/Conditions", this permit will become null and void if work or construction authorized is not started within 180 days, or if construction or work is suspended or abandoned for a period of 180 days any time after work is commenced. A new permit will be then required to proceed with the work.

Signature _____   Date _____

Printed Name _____   Reciept No. _____

FIGURE 10–5 Example Building Permit

FIGURE 10–6
Jobsite Inspection
Card

City of Pullman, CA

Building Division:	533-8956	Permit No. _____
Inspection Requests:	533-8900	Date Issued _____

Notice: Inspectors Available: 8 - 8:30 a.m.   533 - 8958

Address of Project: _____

_____

Type of Project:

Building	New ( )	Add. ( )	Atleration ( )	Repair ( )
Electrical	( )			
HVAC	( )			
Plumbing	( )			
Change of Occupancy ( )				
Other ( )	_____			

Contractor/Building	Phone Number	License Number

Arch./Engr./Designer	Phone Number	License Number

Project Description: _____

_____

_____

Legal Description	Assesor's No.	Zone	Lot Area

Front Side of Card

SITE		ROUGH INSPECTIONS		SPECIAL REQUIRMENTS		
Grading		Roof Nailing		Description	Yes	Received
U'Grd. Utilites		Framing		Soils/compaction Report		
Septic/Sewer		Wiring		Rough Grading Verified		
Setbacks		Drain/Waste/Vent		Height & Setback Verified		
FOUNDATIONS		Water		Engineered Septic		
Footings		Gas		Special Inspection		
Piers		HVAC		Truss Plan		
Stem/Pad		Shear Nailing		Encroachment		
Hold Downs		Framing Hdwr.		Final Grading Verified		
Garage /Slab		Fireplace		Temporary Power		
RETAINING WALLS		INSULATION		Utility Agreement		
Footings		Underfloor				
Steel		Walls				
Drains		Ceiling				
Final		Other				
UNDER SLAB		SURFACE COVERING		OCCUPANCY APPROVAL		
Drain/Waste/Vent		Lath		Building (Temporary)		
Water		Drywall		Planning		
Gas				Public Works		
HVAC				Fire		
MISC.		POOLS & SPAS		Health		
Hood		Pre-Gunite		Police		
Suspended Ceilings		Pre-Deck		Electrical (REL)		
Shower Pan		Pre-Plaster		Gas (REL)		
Gas Test		Fence				
Sewer Lateral						
Sprinklers						
Misc.						

Back Side of Card

The building inspector can use two additional inspection forms. One is the Correction Notice, Figure 10–7, and the other is the Stop Work Notice, Figure 10–8. In some instances the on-site inspection record will provide space for correction comments, but the larger correction notice provides a more detailed comment area for noting necessary corrections. In most cases, a correction notice is issued, unless the correction is life-threatening or the correction requires immediate action. The stop work notice is used when public safety is threatened. If the work being performed is outside the permit, the stop work notice also is used.

The following list summarizes the most important aspects of the building inspector's responsibilities.

Inspectors are trained personnel in the specific area of inspection.

Inspectors are independent and do not enjoy any type of gain from their decisions.

City of San Pullman, California

Building Department

CORRECTION NOTICE

Project Address: ___5962 J street___

I have this day inspected this structure and these premises and have found the following violations of city and/or state laws governing same:

Platform for Hot water tank needs to be raised to the required height of 18" and seismic strap   *installed*
around tank.

When corrections have been made, please call for re-inspection

Date ___6/13/96___   Inspector ___Bill Jones___

DO NOT REMOVE THIS TAG

FIGURE 10–7 Example Correction Notice

City of San Pullman, California

Building Department

<u>STOP WORK</u>
Notice

Address: _____5962 J street_____

This building has been inspected and

_____ Your *building appears to encroach into the setback shown on your site plan*_____

_____

_____

is found to be in non-conformance and is not acceptable. All work is to stop and corrections immediately made. Before proceeding call _____634-9222_____ This a serious issue and must be addressed directly with our office.

Date _____6/13/96_____ Inspector _____Bill *Jones*_____

DO NOT REMOVE THIS NOTICE

FIGURE 10–8 Example Stop Work Notice

Inspectors do not direct the contractor.

Inspectors must make decisions regarding conformance to specifications or codes.

Inspectors can accept or reject work.

Other areas of compliance used by the outside inspecting agency, owner's representative, or project architect are codes and applicable laws, standards, and manufacturers' literature and recommendations. The building code is only one of many codes that may be part of the specifications. Some areas have adopted seismic standards that are not included in the building codes. Federal programs and laws such as the American Disability Act, Occupational Safety and Health Act, and Regulation By Environmental Protection Agency are among a few additional requirements followed. Most specifications reference explicit standards to be used as guidelines for inspections, comparisons to the norm, testing methods, and other foundations needed for inspecting. The list of reference standards and testing agencies is long, the most common being The American Society for Testing and Materials (ASTM), American Concrete Institute (ACI), and

Underwriters Labs (UL) label. Additionally, some specifications reference other specification standards, such as the Federal Specifications, Military Specifications (MIL specs.), and American Association of State Highway and Transportation Officials (AASHTO). Figure 10–9 is an example of specification language used to reference associations and their standard methods of installation and other quality standards.

Other field testing that may be required by the building department or specifications falls within the category of special inspections, which requires employment of an independent testing company by either the owner or contractor. Typical specification language could read as shown in Figure 10–10.

Areas that normally require outside laboratory and inspection services include:

- soil testing: compliance to compaction density requirements (An example of a typical specification section showing soil testing requirements is illustrated in Figure 10–11 on page 272.)
- concrete testing: slump, cylinder compressive strength
- reinforcing steel placement: confirmation of placement locations
- welding testing/inspection: visual, X-ray testing
- bolt torque: tested torque on example bolts
- aggregate testing: hardness, material composition
- asphalt testing: mix design, strength
- layout and vertical alignment (An example of a specification section for layout requirements is shown in Figure 10–12 on page 272.)

SECTION 00800
STANDARDS AND ABBREVIATIONS

Part 1 - General

1.01 Description

    The list below represents standards and abbreviations used in the Drawing and Project   Manual. These standards to be incorporated into these documents by name are   given for the contractor's use in providing installation and quality standards for the   contractor's work. When in conflict with the project manual, the project manual will govern.

Part 2 - Manufacturer, Associations, Institutes, and Standard - Abbreviations

Aluminum Association, Inc.	**AA**
American Arbitration Association	**AAA**
Brick Institute of America	**BIA**
Metal Building Manufacturers Association	**MBMA**
....	

End of Section 0800

FIGURE 10–9   Standards and Abbreviations, Example Specification Section

An independent testing lab could provide the following tests on cast-in-place concrete, using its own field inspector.

- placement and condition of all reinforcement
- verifying the mix used, the time from batching to placement, and the amount of water added, on all concrete trip tickets
- verify admixtures used
- check concrete slump
- cast cylinders for strength tests
- properly store and break cylinders

SECTION 01410
TESTING LABORATORY SERVICES

Part 1 - General

1.01 Work Included
    A. Work includes but is not limited to the following:
        1. Inspection and testing required by laws, ordinances, rules, regulations, orders, or approvals of public authorities: See conditions of the Contract.
    B. On site monitoring and Laboratory and Field tests required and standards for testing:
        1. See requirements specified this section and other Specification Sections throughout the Project Manual.

1.02 Related Work
    A. Coordinated related work ....
    B. Owner will employ and pay for an Independent Testing Laboratory or Laboratories to perform inspection and testing services specified herein unless otherwise note.
        1. Contractor's cooperation with the Laboratory is required to facilitate their required services.
        2. Employment of a Laboratory does not relieve Contractor's obligations to perform any contract work and any other testing as required deemed necessary   to   meet proper performance standards.
        3. In general, test requirement for individual section is found in Part 3 - Execution of specification section.

1.03 Quality Assurance

1.04 References

1.05 Duties of Laboratory:

1.06 Contractor's Responsibilities

    A. Cooperate with engineering and laboratory personnel; provide access to work, and to manufacturer's operations.

    B. Secure and deliver to laboratory adequate quantities of representational samples of   proposed materials.

    C. Provide to Laboratory:
        etc.

1.07 Tests, Inspections, and Methods required

1.08 Identification of Asbestos Materials

End of Section 01410

FIGURE 10–10   Testing Laboratory Service, Example Specification Section

"....... 5. Soils Consultant and Soil Testing

A. The Contractor ...

B. It is essential that subgrade preparation and compaction of fill be performed carefully and in accordance with the plans and specifications; failure by the Contractor to perform site preparation work properly could result in structural damage to the building and other construction feature which damage shall be rectified by the Contractor

C. A qualified soils consultant who is a registered professional engineer in California shall inspect stripping of topsoil and site preparation in the building and paving areas. A minimum notice.................

D. A qualified and bonded soil test firm shall:

1. Certify in writing that the imported fill material used by the contractor meets gradation and sand equivalent requirements of the specification hereinafter.

2. Do the necessary sampling and laboratory work to develop moisture/density curves for the fill material.

3. Make field density tests as specified in Sections 02210,   02221, and 02241, and submit all test results to the Architect   promptly.   Test ....

4. Furnish the Architect a letter upon completion of grading, signed and sealed by a registered Professional Engineer ....

5. Submit a final written report to the Architect covering ......

6. The cost of testing shall be paid for by the Owner, except as provided in Section 01410.

FIGURE 10–11   Soil Testing Requirements, Example Specification Section

"........ 3.   Construction layout and Staking

All construction grading and utility staking shall be done under the supervision of a Civil Engineer or Land Surveyor registered in the State of California. The use of lasers are encouraged for providing straightness and level work. The use of straight edges and carpenter level to set flow lines are not acceptable."

FIGURE 10–12   Layout Requirements, Example Specification Section

All contractor field team members have responsibilities for ensuring quality, with the construction superintendent assuming the primary responsibility for its implementation in the construction product. The superintendent should perform quality checks on the contractor's own work and the work of the subcontractors. Inspections should be similar for technical conformance to specifications and aesthetics. The material or equipment furnished, the installation of the material or equipment, and the final finishes are equally important. The quality of an installation relates to its durability as well as to its initial appearance. Instances of poor workmanship should be noted by the contractor, the architect, and/or the owner's representative. Immediate repair or rework must be done to correct work not meeting the proper standards. As mentioned in the previous discussion about TQM, rework is expensive, thus doing it right the first time is the best way to achieve a quality product. Generally, patching and other forms of corrective measures do not provide a satisfactory final architectural finish that pleases everyone, including the contractor.

Checklists can be developed and used to ensure that the superintendent or other quality control personnel include all aspects of an installation when preparing and implementing punch lists.

A preliminary punch list for an acoustical ceiling subcontractor involves the following:

1. Remove tools, equipment, and excess materials
2. Restore any damaged finishes from the installation
3. Properly store replacement tiles, required by specifications
4. Color quality should be from same manufactured lot, matching colors
5. Texture, finish should match specified and approved material
6. Is the ceiling flat, square, and true?
7. Do the cut tiles fill the spaces completely?
8. Are the joints at the ends of walls angle butt?

Figure 10–13 shows the specification on acoustical ceilings.

All technical specifications have quality issues detailed throughout each section. The specification notes allowable tolerances and defines shop drawing requirements and samples needed for review along with manufacturer's test data. The specifier also comments on job conditions needed for proper installation, and later walks through the installation required by the subcontractor while referencing the Acoustical and Insulating Materials Association performance data.

Short checklists can be taken from the specifications, manufacturer's literature, and other standards. With today's digital cameras and word processing programs, examples of acceptable workmanship and what specific standards should look like are much easier to produce. When completing a project, the review of what worked and problems that occurred should be documented and discussed as part of the process to eliminate any future quality issues.

Section 09510 - Acoustical Ceilings

Part 1 - General

1.01 Related work specified elsewhere

1.02 Description

Provide acoustic ceiling treatment where scheduled or indicated on drawings.

1.03 Allowable Tolerances

Unless otherwise noted, level within 1/8 inch in 12 feet.

1.04 Submittals

Before acoustical ceiling materials are delivered to the job site, the following shall
be submitted to the Architect.

A. Shop Drawings: ....

B. Samples: ....

C. Show proposed methods of seismic bracing to conform with requirements of
UBC Table 23-J. Submit complete calculations defining compliance with seismic
bracing requirements of UBC.

1.05 Job Conditions

Maintain temperature and humidity conditions approximating those in completed
building before, during, and after installation of acoustic ceiling materials. Delay
installation until building is enclosed and all "wet" finish work has dried.

Part 2 - Products

2.01 Suspended Tile Acoustical

Part 3 - Execution

3.01 Installation

In accordance with acoustical and insulating materials association performance data
and as follows:

Etc.

End of Section 09510

FIGURE 10–13   Acoustical Ceilings, Example Specification Section

Quality is specified in different places and in different ways. Most specifications have a clause that states that "all work will be performed in a workmanlike manner." This same clause further specifies that all work performed will "be of good quality and free from defects and defaults." If the quality has not been specified precisely, generally the implied warranty is not held.

Summary

This chapter has attempted to create a broad picture and understanding of quality and quality issues. Quality is not a simple goal to achieve. The work environment for construction is not the controlled setting of manufacturing plants. The product is always new in some form, and the details are somewhat different from the last project. Documents differ in detail and accuracy from job to job, the workmen are of varying skills, and the multitude of constructors utilize different information about specific construction techniques. The chapter has discussed the process of inspection and of implementing quality improvement. Additionally, the relationship of total quality management and the phrase "continuous improvement" were related to the construction industry, emphasizing the importance of these areas to the modern-day contractor. Quality, customer satisfaction, improved productivity, and future work are all interrelated in current and future construction practices.

CHAPTER 11

TIME AND COST CONTROL

The contractor must control both the duration of the project and the cost of the project, the major parameters of the project. The contractor commits to completing the project in a specific time frame and at a finite cost. Many contractors feel that both time and cost are naturally controlled by the construction process and field personnel. Although that premise is basically true, the contractor needs to use management techniques to plan and accurately monitor the actual progress and cost to that which is planned and estimated. The contractor is then able to modify operations to meet the time frame and estimated cost.

Traditionally, the contractor's field personnel did their best to complete the project as quickly as possible, with minimal cost. The results of this effort during the project were unclear. It would appear as though the project would be completed on time, and the project cost seemed to be within the budget. The effectiveness of control of time and cost on the jobsite was evident at the end of the project: if the job was completed on time, and if the project made a profit or loss. In recent years, however, the use of construction scheduling and computerized cost control accounting provide a means of monitoring the effectiveness of the controls during a project, allowing field personnel to adjust productivity factors, crew size, and subcontractor activity to bring the project to acceptable progress.

Small as well as large construction projects can benefit from time and cost control. Large projects use specific individuals dedicated to scheduling and cost control. This should not be thought of however as a luxury available only to larger projects. A construction schedule and cost control information can help any contractor achieve optimum results during the construction process.

The purpose of this chapter is to introduce control techniques and relate them to activities during the construction period. This chapter does not detail the use of these systems as there are numerous texts available on scheduling, project management, and cost control techniques that deal with these subjects.

Project Duration Control

The construction project owner will always need a project completed within a specific time period, whether explicitly stated in the contract documents or not. The owner's date for project completion establishes a parameter and a maximum duration for the project. The contractor must estimate the necessary means of achieving the owner's completion date. If there is not enough time to complete the project using normal construction shifts and techniques, the contractor is expected to use multiple shifts or other methods to complete the project by the specified date. There usually are liquidated damages, a daily monetary amount charged to the contractor for compensation of costs incurred by the owner, when the facility cannot be used.

The owner's completion date, however, rarely has a direct relationship to the actual amount of time needed to construct the project. This date is often based on the owner's need for the facility, rather than the expected construction duration. A school project will be required to be completed in July or August, allowing time for equipment move-in so the facility can be in operation in early September at the start of the school year. A retailer will want their facility completed in late summer, to be able to stock goods and train personnel prior to the holiday retail season. These completion dates establish a maximum construction duration, but the contractor may be able to complete the facility prior to the completion date.

Completion of construction in the minimum time will reduce the contractor's jobsite overhead and enable the contractor to pursue other work. To optimize profits in both the specific project and the aggregate, the contractor needs to complete the projects in the shortest amount of time, while not increasing cost with lost productivity or overtime costs. The contractor needs to plan the project carefully, for without planning and diligence in maintaining the plan, the construction project can extend for a long duration, costing the contractor considerable resources. A project planned according to construction sequence and realistic projected duration of activities, rather than one based on the prescribed completion date, can actually help the contractor complete the project earlier than scheduled.

The plan for the construction of the project includes careful consideration of the elements and relationships in the project. No two projects are ever alike. Each has its own set of materials, systems, subcontractors, weather conditions, soil conditions, delivery dates, sequences, and durations. Previous projects help one become aware of the duration of certain specific activities. However, two projects of similar scope might have completely different time frames for completion, due to the time of year started, weather delays encountered, sequencing, and delivery of materials. If two exact projects were started on the same day, the two superintendents would sequence the projects differently, encounter different problems during construction, and complete the projects at different times.

The project plan is the plan of construction conceived by the project superintendent, along with the purchasing restraints already in place (subcontracts and purchase orders) to most efficiently achieve the objectives of the project. It is important to remember that the project plan is generated by the personnel who will manage it. The project superintendent must be intimately involved in planning the project, in addition to other personnel. Too often the plan and subsequent schedule are developed by personnel who are not involved with the day-to-day management of the project, resulting in a wide difference between the schedule and the actual project sequence and duration.

The project plan involves more than just a schedule of activities. It must consider all project needs and how the solutions to those needs affect the duration of activities and the project itself. Some planning considerations include:

- activities on the project
- duration of the activities

- sequencing of the activities
- the interrelationship of subcontract activities
- equipment requirements for the project
- utilization of subcontractors
- field management personnel
- crew size and project staffing of craftspeople
- anticipated material delivery dates
- weather planning: anticipated lost days, weather protection
- anticipation of delays: other contracts, labor strikes, owner-based changes
- alternative solutions to sequencing

Careful analysis of the entire project is needed to tie activities together to form a plan and subsequent schedule. When a scheduling consultant is used, the superintendent and other field personnel must formulate the plan for the project and convey that information to the scheduling consultant.

Prior to scheduling, the activities must be identified, the sequences must be established, and the durations of the activities must be estimated. This information could be identified jointly by the scheduler and the superintendent, but the definition of the information should be done by the superintendent. Some considerations when compiling this data are discussed next.

Scope of Activities

Each activity in the construction schedule must be identifiable, with the following attributes:

- able to be described as a separate function, such as "formwork for footings"
- must have a labor crew, with attributable crew hours for the activity
- must have a specific duration attached definable for the work, such as "ten working days"
- must be able to be monitored by comparing planned duration with actual duration during the construction process

A construction activity has an identifiable block of time that must be included in the construction sequence. It can contain several different steps, trades, and estimate line items, or there could be a time restraint before further work can be done. Two typical construction activities and their components could be:

EXAMPLE

Prep and Pour Concrete Slab:
Subgrade preparation
Spread and prepare gravel base
Install embedded items
Install underslab electrical, plumbing, and ductwork

EXAMPLE continued

Set forms
Install reinforcing steel
Set screeds, expansion joints
Pour concrete
Finish concrete
Duration: 3 days
Cure Concrete Slab:
Wet cure concrete slab
(Time restraint—no relating work to be done until curing complete)
Duration: 10 days

The appropriate scope of the activity is important when formulating the plan and schedule. The size of the project helps determine the scope of the activities. When preparing a short-term schedule, small units are appropriate. When preparing a large project, larger units are appropriate. Short-duration items, however, can create some problems with the schedule. Any fluctuation can affect these items, easily throwing the actual construction off-schedule. Short durations also can inflate the schedule by requiring too much time for individual parts.

Using the previous "Prep and Pour Concrete Slab" activity, the following example illustrates the short time needed and inflated time created by adding all of the subactivities together:

EXAMPLE

Prep and Pour Concrete Slab:	
Subgrade preparation	2 hours
Spread and prepare gravel base	4 hours
Install embedded items	2 hours
Install underslab electrical, plumbing, and ductwork	8 hours
Set forms	4 hours
Install reinforcing steel	2 hours
Set screeds, expansion joints	4 hours
Pour concrete	2 hours
Finish concrete	6 hours
Total Hours	34 hours

The estimated 34 hours are more than three days (24 hours) previously established for this activity. Of course, some of the activities could be done concurrently, saving the overall duration; some of the activities are ample estimates for the work, resulting in an inflated duration. It is easy to see that if one activity takes a little longer, caused by a long coffee break, an unexpected material delivery, an illness, or a number of other factors, the construction would be easily thrown off-schedule.

An overly long duration also is not appropriate for monitoring actual against projected time. Assume the first example, "Prep and Pour Concrete Slab," is for a 5,000 SF concrete slab. If the concrete slab was much larger, for instance 50,000 SF, the item of "Prep and Pour Concrete Slab" may be too large to monitor and may be separated into five sections, of

10,000 SF each, and a duration of three working days per section, using a larger crew. "Prep and Pour Concrete Slab, Section 1," and so on would be more accurate for monitoring the construction, as the actual slab pouring and finishing will be accomplished in five sections here.

In most construction schedules, activity duration should be no shorter than one day. Situations exist, such as a plant shutdown for a short-term repair or for additions, that must be scheduled in small increments, such as hours. A guideline can be established for activity duration, such as between three and twenty working days, but this should not totally eliminate necessary activities that are shorter or longer.

A descriptive list of the items included in each activity is a good organizing tool for checklists to determine that all activities are included in the plan. Descriptions of the activities are also good references during the project, to determine if the activity is complete. Some activities may resemble others in a brief description, so the full description is necessary to identify the activity and its scope.

Sequence of Activities

The plan of the project will determine the sequence of the activity. In any project, several different sequences could be used to accomplish the project, depending upon the strategy of the superintendent, considering the project conditions. The example shown in Figure 11–1 illustrates two different sequences of a portion of the project, either of which might be applicable regarding crane mobilization costs, weather conditions at the projected time of the activities, subcontractor availability, crew availability, and duration available for the activities within the scheduled completion.

Distinct reasons may exist for the superintendent choosing one sequence over another. For quality reasons, the superintendent may want to pour the slab without a pour back between the existing and new slabs. Sequence 1 would facilitate that condition. When selecting a particular sequence, it may be necessary to compare durations and costs associated with the sequence, similar to the comparison in Figure 11–2. In this example, Sequence 2 has a shorter duration of 1.5 days and a cost that is lower by $1,900. In most circumstances, the shorter duration and lower cost would be the reasons to choose Sequence 2. Mitigating reasons may exist, however, making Sequence 1 more desirable, such as quality concerns, construction details and connections, availability of the crane, availability of the concrete finishers, and a variety of other reasons.

In the examples shown in Figures 11–1 and 11–2, the sequence was assumed as "finish to start," that is, the activities were completed before the next activity started. This is not always the case, though, as one activity may start after a portion of the preceding activity is completed. Numerous concurrent activities occur on the jobsite. If all construction sequences were the finish-to-start type, the project would be estimated at a much longer duration. What follows are common relationships available with most commercial scheduling software.

FIGURE 11–1
Example Sequence
Comparison

Project Description: This portion of the project involves an addition to an existing gymnasium. The existing building consists of concrete tilt-up wall panels and metal roof joist, with a concrete slab-on-grade. The existing building has a hardwood gymnasium floor applied above the concrete slab. The concrete panels at the end of the existing building are to be removed, and used as the panels at the end of the addition. The addition attaches to the existing building, and is constructed of tilt-up concrete wall panels and metal joists, with a concrete slab on grade, similar to the original building.

Sequence 1:
1. Mobilize crane
2. Remove tilt-up walls to be relocated, store on site
3. Build temporary weather partition at end of existing building
4. Excavate for footings
5. Prep and pour footings
6. Prepare subgrade
7. Pour slab on grade
8. Slab cure
9. Prep and pour panels on slab
10. Mobilize crane
11. Lift panels into place
12. Set joists and deck
13. Backfill at "pour-backs"
14. Pour concrete slab at "pour-backs"

Sequence 2:
1. Excavate for footings
2. Prep and pour footings
3. Prepare subgrade
4. Pour slab on grade
5. Slab cure
6. Prep and pour panels on slab
7. Mobilize crane
8. Remove and relocate existing panels
9. Lift new panels into place
10. Set joists and deck
11. Backfill at "pour-backs"
12. Pour concrete slab at "pour-backs"

Finish to Start: The second activity does not start until the first one is completed.

EXAMPLE

Foundation wall needs to be completed with forms stripped prior to setting the stud wall on the foundation wall.

A period of time can exist between the finish and start, or a lag between the two activities.

FIGURE 11–2
Example Sequence
Cost Comparison

Comparison of Sequence 1 and Sequence 2
Addition to Gymnasium

Sequence 1:

Activity	Duration (Days)	Cost
1. Mobilize crane	0.5	$ 1,000
2. Remove tilt-up walls	0.5	$ 1,100
3. Temporary partition	2	$ 1,300
4. Excavate footings	2	$ 840
5. Prep & pour footings	5	$ 2,500
6. Prepare subgrade	5	$ 1,200
7. Pour slab on grade	2	$ 7,500
8. Slab cure	5	$ 200
9. Prep and pour panels	15	$ 8,500
10. Mobilize crane	0.5	$ 1,000
11. Lift panels into place	0.5	$ 1,400
12. Set joists and deck	1	$ 2,600
13. Backfill at "pour-backs"	2	$ 1,000
14. Pour concrete "pour-backs"	1	$ 700
Total	42	$30,840

Sequence 2:

Acitivity	Duration (Days)	Cost
1. Excavate footings	2	$ 840
2. Prep and pour footings	5	$ 2,500
3. Prepare subgrade	5	$ 1,200
4. Pour slab on grade	2	$ 7,500
5. Slab cure	5	$ 200
6. Prep and pour panels	15	$ 8,500
7. Mobilize crane	0.5	$ 1,000
8. Remove and relocate ex. panels	0.5	$ 1,200
9. Lift new panels into place	0.5	$ 1,400
10. Set joists and deck	1	$ 2,600
11. Backfill at "pour-backs"	3	$ 1,300
12. Pour concrete "pour-backs"	1	$ 700
Total	40.5	$28,940

EXAMPLE

The concrete slab must be completed prior to installation of the interior studs. As the slab needs to cure and harden before framing labor and equipment is on the slab, the contractor could insert a five-day lag, allowing the slab to cure for a week prior to installation of the framing.

Start to Start: Two activities may need to begin at the same time, such as installation of carpet and resilient floorcovering, with the same subcontractor installing both, but using separate crews. A lag might occur between the two starts as well.

EXAMPLE

Drywall installation may have a fairly long duration, such as twenty working days. Drywall, however, will begin in a certain area and progress away from that area. Some areas will be ready for painting five days after the start of the drywall operation. Time would be wasted if the painter waited until all of the drywall was completed. The relationship, then, would be a start-to-start with a five-day lag, with painting beginning five working days after the drywall. Obviously, coordination is necessary to ensure that the painter knows where to begin and that the area is clean and ready for the painter when scheduled.

Finish to Finish: This relationship is used when two activities should be completed at the same time, despite the fact that they might have different durations.

Logical sequencing, however, is most important in planning and scheduling. Care should be taken to ensure correct sequencing, with the necessary activities preceding each activity. Considerable disruption can be caused by careless sequencing. If subcontractors are scheduled to arrive and work on the project, they will require that the necessary adjoining and substrate work be completed, and they will often leave the jobsite if this preliminary work is not done. The schedule is a guide and should be changed if the logic is not correct. The logic, or proper sequencing, is the key to a successful project. All parties involved should review the logic of the project prior to completion of the schedule.

Duration of Activities

The duration of an activity must be accurately predicted. When determining the duration of a construction activity consider these factors:

- The duration of an activity needs to be relative to the amount of labor hours included in the estimate. Input from the estimator is often necessary to determine the intended crew size and amount of labor. Construction activities do not always coincide with the estimated items, so some interpretation of the estimated hours is necessary. This information should be combined with the superintendent's view on how the work will be accomplished.
- The duration of the activity should relate to the quantity of work accomplished. It is important to factor crew size against the quantity of work to be performed. The duration of concrete formwork activities will certainly vary from a situation where there are two carpenters and a laborer to eight carpenters and four laborers. When the crew is properly sized for the project, the productivity rate should not be affected by using a larger crew, but the duration of the activity should be shorter.
- Weather concerns also should be considered in the duration of the activity. The time of year in which the activity will be accomplished has a large influence on the duration of the activity. For example, a built-

up roofing activity might have a ten-working day duration during warm weather. If this activity was intended to be completed in December in the northern half of the United States, the duration should be longer to allow for weather conditions appropriate for installation, without freezing or precipitation. A contractor might allow twenty days during that period, hoping the work would be done within the twenty-day window. Another contractor might change the sequence, either completing the roofing prior to November or after March.

- Consideration of delivery of all items may affect the duration of the activities. Delivery is a factor in the sequencing of the activity, but it also can affect the duration.
- Subcontractor involvement in the activity may affect the duration. Duration control can be facilitated by the contractor by ensuring that the crews and the number of crews are available at the proper time. The contractor cannot as easily affect the performance of subcontractors. Some cushion should be allowed in the duration for subcontractors arriving on the project.

Duration, in most schedules, is expressed in working days, not calendar days. Five working days would indicate a week in most cases. For cases on accelerated project phases, the construction week consists of seven days. Use of "working days" for the duration works best, with the project calendar being adjusted for longer work weeks or holidays.

The Project Schedule

The project schedule is a manifestation of the project plan—a written or graphical depiction. The schedule is a communication tool to share the project plan among project participants. It has many forms: a simple, noncalculated bar chart; the Gantt chart, a calculated bar chart; a network chart showing interdependencies, some time-scaled; and tabular schedules, listing dates. It should always be remembered that the contractor is trying to communicate his plan to other participants: architects, owners, construction managers, subcontractors, and suppliers. The exact method used may vary, depending on the target for the information. A construction manager, for instance, would normally be quite knowledgeable about schedules, therefore probably would request network reports, tabular reports, and the Gantt chart. The owner, however, may not be interested in reading anything other than the Gantt chart. The contractor should ensure that the appropriate report is given to each party, possibly with additional clarification. Some contractors will send everyone a three- or four-inch thick stack of reports, expecting the recipient to discern applicable information. These reports usually are either filed or thrown away, without any exchange of information.

As most scheduling software produces many reports and graphs, there is a tendency to become overwhelmed with the information, losing track of what is important during construction. The following list contains some schedule information relevant to the parties involved in the construction process.

Owner:

Completion dates: project and milestones for partial completion

Current status of the project: problem areas and projected changes to dates

Dates and relationship of owner-furnished items and separate contracts

Current activities in progress

Schedule for owner-provided services, such as testing

Architect:

Completion dates: project and milestones for partial completion

Current status of the project: problem areas and projected changes to dates

Relationship of submittal review to construction schedule

Current activities in progress

Schedule for inspection/observation, particularly for subconsultants

Contractor:

Detailed relationship of activities

Current status of activities

Relationship of projected delivery dates of material and equipment to activities

Schedule of subcontract activities for coordination

Duration of activities to schedule crews

Interrelationship of activities

Opportunities to condense construction duration without increasing cost

Opportunities to level crew size

Current activities that are behind or that are expected to exceed their projected duration

Subcontractor:

Expected schedule for particular activities: date required on jobsite and duration allowed for work

Relationship of other activities that directly affect the subcontract activities

Current status of project, relating to particular subcontract activities

Relationship of subcontract material and equipment to activity

Suppliers:

Date material or equipment due at jobsite

Anticipated order date, after review of shop drawings and submittals

The majority of scheduling is currently done with scheduling or project management software, which organizes, calculates, levels, and provides numerous views of the schedule, including network, Gantt charts, and tabular listings. (Chapter 12 describes some popular scheduling and project management software.) As mentioned previously, a plan should be established prior to inputting data into scheduling software, with a list of activities, durations, and predecessors prepared. A partial list of activities for input into scheduling software is illustrated in Figure 11–3.

Full scheduling software normally calculates the **critical path** for the project. The critical path is the calculated path for the project that considers the dependencies of predecessors. In considering the predecessors for each activity, the longest predecessor duration establishes the start date of the activity. The completion date established by the critical path is the earliest possible date, considering the dependencies of the activities. Any change to an activity on the critical path will directly affect the completion date of the project. Activities not on the critical path have a **float,** or a period of time in addition to the activity's duration, prior to the start of the successor activity. The float allows some flexibility with noncritical items and flexibility in duration and the actual start of the item. For example, if an item has five days float and its scheduled early start date is Monday, August 7, the item can start any day from August 7 through August 11, assuming that its duration remains constant. If the activity actually starts on August 7, five additional days could be added to the duration, possibly using a smaller crew than originally anticipated. If the activity is started ten working days later than initially anticipated, either the duration of the activity would have to be shortened by the additional five days, possibly by enlarging the crew, or the activity will affect the critical path, and thus the completion date, by the five days. Careful monitoring of activities and their effect on the critical path of the project is necessary throughout the project.

Updating of the project is necessary on a periodic basis, either weekly, biweekly, or monthly. The primary purpose of this update is to determine which activities are behind schedule and what their effect will be on the completion date of the project. After determining the activities that are behind schedule, the project management team must analyze the effect and what steps, if any, must be taken to bring the project back to a schedule that will achieve the projected completion date. As mentioned earlier, the project plan should be flexible to allow for changes in activity sequences, crew sizes, and shifts worked. The initial schedule should always be maintained as a reference for the project, comparing updated progress to the original plan.

Activity Number	Activity	Duration	Predecessors
1	Mobilize	5 days	
2	Clear site	5 days	1
3	Mass excavation	10 days	2
4	Structural excavation	10 days	3
5	Plumbing excavation	4 days	2
6	Prep & pour footings	7 days	4 + 5 days
7	Underslab plumbing Rough-In	10 days	5
8	Foundation wall	20 days	6 + 3 days
9	Prep. slab subgrade	4 days	7, 8

FIGURE 11–3
Schedule Data

A variety of methods are used to update the schedule and communicate the updates to those involved in the project. Most scheduling software programs have updating features. When using these features, always be sure the original schedule is used as a comparison. The revised completion dates from the computerized schedule updates should be viewed as data to compare to the initial completion date, allowing the contractor to manipulate activities to meet the initial schedule date. The target date can change, however, with project scope changes and project events that cannot be avoided. Change orders can change the duration of the project, by granting a time extension in a specific number of days. If a new completion date is established, justification for the new date should be made. Completion date changes should be infrequent, however, as the owner depends on the project being completed by the established date.

Because the schedule is a communication tool, the schedule updates should be done in a manner that effectively communicates the variance in schedule, if any, and the steps that are necessary to remediate any problems. Some schedulers will use the network diagram or bar chart to indicate the current status of activities by color-coding the charts. For instance, red would indicate behind schedule, green would indicate on schedule, blue would indicate ahead of schedule, and yellow would indicate that the activity was late, but had not yet affected the schedule. This color coding allows the project management team to quickly identify problem areas. Tabular reports usually are available to indicate the activities showing variance. As many of the construction activities are subcontract items, the contractor needs to work with the subcontractors closely, indicating the problem and identifying the solution. The example that follows indicates a comparison of the original schedule with the updated schedule and shows the remedial schedule that is necessary, including dialogue with subcontractors.

EXAMPLE

Figure 11–4 illustrates a partial construction schedule, showing the finish stages of a project. The completion date for the project is August 26. The owner has used this date for moving into the facility and hiring additional personnel. There is a liquidated damages clause in this project, stipulating $1,000/day liquidated damages for late completion of the project.

Figure 11–5 shows the schedule as impacted by actual progress. Activity 15401, Plumbing Rough-In for the interior walls, was completed a week later than scheduled, resulting in delaying the start of the drywall activity. As these activities were on the critical path, the project completion would be delayed until September 3. The superintendent for the contractor needed to analyze the alternatives caused by this delay:

- Could the completion date be extended to September 3? In this case, the completion date could not be changed due to the owner's arrangements and insistence to impose liquidated damages if the facility could not be used at the scheduled completion date.
- Can durations of activities be compressed to save the five working days from the schedule? The superintendent examined each item, starting with critical path items, and discussed activity durations with the applicable subcontractors. The superintendent came to the conclusion that scheduled durations could not be com-

Act ID	Activity Description	Orig Dur	Rem Dur	Early Start	Early Finish	Total Float	%
NCB1 - New City Office Building							
09251	Interior Metal Studs	10d	10d	10JUN96	21JUN96	0	0
15401	Plumbing Rough-In	10d	10d	10JUN96	21JUN96	0	0
16100	Electrical Rough-In	10d	10d	10JUN96	21JUN96	0	0
09255	Drywall	10d	10d	24JUN96	08JUL96	0	0
09900	Painting	15d	15d	09JUL96	29JUL96	0	0
09300	Ceramic Tile	5d	5d	09JUL96	15JUL96	5d	0
09500	Acoustical Ceilings	10d	10d	16JUL96	29JUL96	0	0
16500	Light Fixtures	10d	10d	23JUL96	05AUG96	0	0
15500	Diffusers, Grilles	3d	3d	23JUL96	25JUL96	7d	0
09650	Carpet	5d	5d	30JUL96	05AUG96	0	0
15420	Plumbing Fixtures	3d	3d	30JUL96	01AUG96	3d	0
10800	Toilet Partitions, Accessories	4d	4d	02AUG96	07AUG96	3d	0
15800	Testing, Balancing	5d	5d	06AUG96	12AUG96	0	0
16800	Test Fire Alarm	5d	5d	06AUG96	12AUG96	0	0
20000	Punchlist	10d	10d	13AUG96	26AUG96	0	0
		55d	55d	10JUN96	26AUG96	0	0

FIGURE 11–4   Original Construction Schedule

EXAMPLE
continued

pressed sufficiently to return the project to the original completion date.

- Can different sequencing be used to return the project to the scheduled completion date? The superintendent realized that by coordinating the area of work of the drywall contractor, the painting contractor could start work a week earlier. This earlier start was acceptable to the painting subcontractor. Ceramic tile would also start a week earlier than originally scheduled, which was also acceptable, as the ceramic contractor had received delivery of the ceramic tile. Figure 11–6 on page 292 illustrates the different sequence by starting the painting a week early. This remedial schedule results in the project's maintaining its original completion date of August 26.

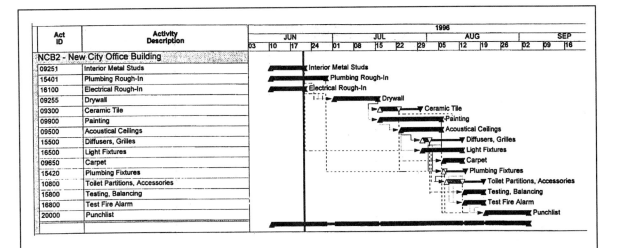

Act ID	Activity Description	Orig Dur	Rem Dur	Early Start	Early Finish	Total Float	%
NCB2 - New City Office Building							
15401	Plumbing Rough-In	10d	6d	10JUN96 A	28JUN96	0	41
09251	Interior Metal Studs	10d	0	10JUN96 A	21JUN96 A		100
16100	Electrical Rough-In	10d	0	10JUN96 A	21JUN96 A		100
09255	Drywall	10d	10d	01JUL96	15JUL96	0	0
09900	Painting	15d	15d	16JUL96	05AUG96	0	0
09300	Ceramic Tile	5d	5d	16JUL96	22JUL96	5d	0
09500	Acoustical Ceilings	10d	10d	23JUL96	05AUG96	0	0
16500	Light Fixtures	10d	10d	30JUL96	12AUG96	0	0
15500	Diffusers, Grilles	3d	3d	30JUL96	01AUG96	7d	0
09650	Carpet	5d	5d	06AUG96	12AUG96	0	0
15420	Plumbing Fixtures	1d	1d	06AUG96	06AUG96	5d	0
10800	Toilet Partitions, Accessories	4d	4d	07AUG96	12AUG96	5d	0
15800	Testing, Balancing	5d	5d	13AUG96	19AUG96	0	0
16800	Test Fire Alarm	5d	5d	13AUG96	19AUG96	0	0
20000	Punchlist	10d	10d	20AUG96	03SEP96	0	0
		60d	51d	10JUN96 A	03SEP96	0	21

FIGURE 11–5   Updated Schedule

A variety of methods can be applied to get the project back on schedule, including:

- Altering the sequencing of the activities, bringing some activities up in time, changing the critical path.
- Altering the duration of the activity or activities showing a problem, which can be done by increasing crew size, adding multiple shifts, or increasing productivity for the activity; an activity also can be split into other activities, separating the essential predecessor elements from those that could be done later.

Act ID	Activity Description			1996			
			JUN	JUL	AUG	SEP	
NCB3 - New City Office Building							
09251	Interior Metal Studs						
15401	Plumbing Rough-In						
16100	Electrical Rough-In						
09255	Drywall						
09900	Painting						
09300	Ceramic Tile						
09500	Acoustical Ceilings						
15500	Diffusers, Grilles						
16500	Light Fixtures						
09650	Carpet						
15420	Plumbing Fixtures						
10800	Toilet Partitions, Accessories						
15800	Testing, Balancing						
16800	Test Fire Alarm						
20000	Punchlist						

Act ID	Activity Description	Orig Dur	Rem Dur	Early Start	Early Finish	Total Float	%
NCB3 - New City Office Building							
15401	Plumbing Rough-In	15d	15d	10JUN96	28JUN96	0	0
09251	Interior Metal Studs	10d	10d	10JUN96	21JUN96	5d	0
16100	Electrical Rough-In	10d	10d	10JUN96	21JUN96	5d	0
09255	Drywall	10d	10d	01JUL96	15JUL96	0	0
09900	Painting	15d	15d	09JUL96	29JUL96	0	0
09300	Ceramic Tile	5d	5d	16JUL96	22JUL96	0	0
09500	Acoustical Ceilings	10d	10d	16JUL96	29JUL96	0	0
16500	Light Fixtures	10d	10d	23JUL96	05AUG96	0	0
15500	Diffusers, Grilles	3d	3d	23JUL96	25JUL96	7d	0
09650	Carpet	5d	5d	30JUL96	05AUG96	0	0
15420	Plumbing Fixtures	3d	3d	30JUL96	01AUG96	3d	0
10800	Toilet Partitions, Accessories	4d	4d	02AUG96	07AUG96	3d	0
15800	Testing, Balancing	5d	5d	06AUG96	12AUG96	0	0
16800	Test Fire Alarm	5d	5d	06AUG96	12AUG96	0	0
20000	Punchlist	10d	10d	13AUG96	26AUG96	0	0
		55d	55d	10JUN96	26AUG96	0	0

FIGURE 11–6   Remedial Schedule

- Reevaluating the duration of the successor activities, to see if time can be gained later in the project. (Note: this can be dangerous territory—rationalizing that things will get better. Careful analysis is necessary before making this decision.)

The schedule and progress updating of the schedule during the project provides the contractor with a tool to help control the project's duration. The schedule itself merely gives an indication of the project's status. The creativity, experience, and management skill of the project team allow for the effective management of project activities and control of the completion date.

Several occasions arise during the construction project when it is necessary to prepare short-term or special-use schedules. These usually are separate from the primary schedule and can be in a different scale, such as hours. Some occasions for using short-term schedules include:

- detailed clarification of a larger item in the construction schedule, which may be used to schedule crews, or coordinate several different crews
- schedule of a power shutdown or other utility interruption, which might be done in hours
- installation of a piece of equipment, coordinating the crafts involved
- schedule of the completion of a specific area of the project for early owner occupancy
- schedule of the move-in activities of the owner
- schedule for remediation of work behind schedule
- specific schedule for a trade or section of the work

These short-term schedules often are prepared in the field by project personnel. Because their purposes are specific, their formats will be customized to the particular situations.

Project Cost Control

Comparison of the actual cost with the estimated cost during the project provides an accurate measure of the current success of the project. Cost comparison data during the construction activity provides the contractor with opportunities to adjust the factors involved with the activity. Active cost control enables the contractor to achieve a profit on construction activities.

A number of computer software programs are available for cost accounting and cost control. These programs compare the actual cost with the estimate, providing accurate cost information to the project team immediately after input. The software, however, just computes and collates the data. The entire project management team must understand the parameters of effective cost control and follow these parameters as they set up the estimate, the cost accounts, labor, material, and equipment reporting, productivity management, and project management. Conscious effort by the project manager, superintendent, field engineers, and foremen can make a cost control system work.

The section that follows is a discussion of some of the important factors in a cost control system.

Realistic Cost Control Activities

The primary objective in setting up a project cost control system involves the correlation of the cost control activities from the estimate with reportable construction activities. The cost control activity should be related to a crew assignment, so reporting labor hours can easily be tracked. The activity should not be so small that reporting hours becomes difficult for the foreman. The items should be narrow enough to identify a problem, facilitating a solution. Cost control activities will be used as historical data in estimating future projects and can be applied to several projects, not just one. Obviously, the formation of these activities is very subjective. The example shown in Figure 11–7 examines concrete foundation walls for a project. The controllable cost here will be the formwork labor, as it continues for a period of time longer than the weekly labor (payroll) reporting.

The table in Figure 11–7, comparing actual with estimated hours, provides an idea about the productivity of the crews for the concrete foundation wall. Comparing the actual with the estimated figures results in a variance, which can be projected to a total variance if the same rate of productivity continues. Item 03113, Strip Formwork, is currently over its estimated cost by $471.36. Because only 16 percent of that item has been completed, management has a chance to review conditions and discover why that item is a bit behind. It is possible that the learning curve is impacting that item, but other conditions, such as crew size, equipment, weather, or other factors, may have affected that item's productivity. Management can take steps to modify the condition, such as reducing the crew size from three to two laborers, which may be more efficient.

Each firm will have its own format for a cost control system. The major concerns in maintaining an effective cost control system include:

- Ensuring that the cost control accounts are of the proper scope. The cost control account should be understandable. Labor should be able to be easily assigned to the cost control account. Estimate line items should be able to act as cost control accounts.
- Relating the crew assignment directly to the cost control activity. The crew assignment should be clear and understandable.
- Cost account results should be able to be used as historical cost data.
- Field personnel should realize the purpose of cost control and report the costs accurately.
- Comparison of actual and estimated cost should be done immediately, providing data to field management personnel.
- A feedback system should be in place to receive information from the crew relating to the cause of variances between estimated and actual costs.

The estimate was based on a unit price cost per cubic yard of concrete:
Unit Prices: Material: $ 60.00/CY; Labor, $ 80.00/CY; Equipment, $ 60.00/CY
Estimate line item:

Section	Item	Quantity	Material	Labor	Equipment	Total
03310	Concrete Foundation Walls	1,000 CY	$ 60,000.00	$ 80,000.00	$ 60,000.00	$200,000.00

Is this item adequate for cost control? Can we definitively track the costs, identify a problem, and be able to take action to solve the problem? Possibly, but definitive information is not really available. We need to look at the situation with definable crew tasks, which are not too small. The following is a breakdown of the concrete walls in definable categories associated with crew makeup and cost.

Section	Activity	Crew	Labor Cost	Crew Cost	Hrs	Total Cost
03112	Set Formwork	6 carpenters 2 laborers	$ 28.45 $ 25.30	$ 221.30	240	$ 53,112.00
03113	Strip Formwork	3 laborers	$ 25.30	$ 75.90	60	$ 4,554.00
03202	Install Rebar	3 ironworkers	$ 32.15	$ 96.45	68	$ 6,558.60
03302	Pour Walls	6 laborers	$ 25.30	$ 151.80	64	$ 9,715.20
03303	Sack Walls	2 laborers	$ 25.30	$ 50.60	120	$ 6,072.00
	Total foundation wall labor					$ 80,011.80

With this breakdown, it will be easier to track the cost progress of labor for the concrete walls. Each crew is definable, with definable crew hours for each activity. The crews will be doing tasks that can be described in a simple set of instructions and drawings. The progress can be measured by the quantity of work completed.

Section	Activity	Quantity	Quant. to Date	% Complete	Crew Hours	Crew Hrs to Date	Cost to Date	%comp x Budget	Variance	Projected Variance
03112	Set Formwork	2500 LF	900 LF	36%	240	80	$17,704.00	$19,120.32	$1,416.32	$3,934.22
03113	Strip Formwork	2500 LF	400 LF	16%	60	20	$1,200.00	$728.64	($471.36)	($2,946.00)
03202	Install Rebar	2500 LF	800 LF	32%	68	16	$1,543.20	$2,098.75	$555.55	$1,736.10
03302	Pour Walls	1000 CY	320 CY	32%	64	16	$2,428.80	$3,108.88	$680.08	$2,125.20
03303	Sack Walls	40000 SF	3200 SF	8%	120	8	$404.80	$485.76	$80.96	$1,012.00

FIGURE 11–7   Cost Account Analysis

FIGURE 11–8
Productivity Problems and Solutions

Productivity Problem	Possible Solution
Poor instructions/layout	Field management: improve instructions
Incorrect installation techniques	Field management: provide training to crew
Inexperienced/Untrained crew	Field management: train crew; hire new crew
Poor or incorrect details for the installation	Architect: clarification; possible change order
Incorrect sequencing	Field management: rearrange schedule
Poor, insufficient, or incorrect shop drawings	Supplier: revise drawings, possible back-charge
Insufficient time allowed	Field management: reassess crew time; rearrange schedule; revise historical cost data
Wrong or not enough tools and equipment	Field management: Supply crew with proper tools
Insufficient, incorrect, or late delivery of materials	Field management: assess problem—supplier or field management? Arrange for material ASAP
Defective materials	Supplier: replace materials ASAP; possible back-charge
Poor material location requiring rehandling	Field management: relocate material; revise jobsite plan for proper material handling
Wrong crew size	Field management: adjust crew size—could be larger or smaller
Attitude of crew	Field management: find source of problem; solve problem; may have to change some personnel
Unknown (latent) conditions	Field management: notify architect; possible change order
Poor environmental conditions: heat, light, ventiliation, etc.	Field management: provide necessary equipment to adjust enviromental conditions
Interference or additional instructions by other parties, outside of contract channels	Field management: discuss with appropriate party; possible change order
Late, unwarranted rejection, or beyond scope inspection	Field management: discuss with appropriate party; possible change order
Weather conditions	Field management: determine if weather is beyond normal; possible change order

- The crews need to be thanked when they accomplish the task under the estimated cost.
- Crews need to be encouraged when costs are over the estimate. There usually are factors that cause labor to overrun the estimate—active project management finds the problem and solves it.

Inefficient productivity usually has a reason or reasons for not achieving the target. Several different reasons exist for low productivity and several solutions exist to solving the problems. The most common response to low productivity is encouraging the crew to work faster. There may be other solutions, however. Management should examine possibilities such as improving its role in activities, obtaining additional funding for activities, such as change orders, and managing subcontractors, with possible back-charges for impacted delay costs.

The chart in Figure 11–8 includes the most common reasons for low productivity and offers some possible solutions to those problems.

Field management is not always aware of the problems or source of problems, because sufficient dialogue does not occur between the crew and management. Conversations with crew foremen and crew members often indicate the problem. Easy reporting, such as a questionnaire of yes-and-no questions can provide enough input to field management to start an immediate investigation of the problems.

Summary

Control of time and cost are two of the most important tasks of the field management team. Controlling the project to minimal time and cost are the contractor's prime objectives. Time relates to cost, as shorter durations should reduce the cost of jobsite overhead on the project. Time and cost control, however, are normally separate tasks on the jobsite. This can be accomplished by individuals being assigned solely to the controls area, or by project field engineers.

The project plan is the superintendent's method for constructing the project. It will have its own activities, durations, sequences, interrelationship of activities, equipment usage, personnel, material delivery constraints, weather constraints, and predictable delays. It is flexible and responsive to changes; its activities must be monitored, updated, and revised to reach each goal. The project's sequence, controlled by the superintendent, can be analyzed for time impact and associated costs.

The schedule is the communications tool of the project plan. Various forms of schedule representation are available and should match the comprehension level of the parties involved. Scheduling software can be used, which will calculate the critical path and float of noncritical items.

Cost control compares the estimated and actual cost. Cost control accounts should match activities done in the field. Prompt comparison of costs can provide field management with opportunities to adjust methods, with the actual cost meeting or being less than the estimated cost.

CHAPTER 12

COMPUTERIZED PROJECT ADMINISTRATION

CHAPTER OUTLINE

Computer use in the construction industry has grown rapidly over the past decade. Computerized scheduling, estimating, and cost accounting are common applications in the construction industry. The construction industry also is seeing the introduction of computer systems for project, document, and contract control. The proper maintenance, organization, and speedy access to jobsite records are key elements to effective project administration. Many different transactions happen with each exchange of questions, or when problems arise. These pieces of information take the form of correspondence, letters, and other written documents. Many times these documents and their issues must be linked or used together by the project team. The project team must be able to find solutions the quickest, best, and most equitable way for all parties. All of the project elements, such as changes, payments, and subcontractor relations, should be equally addressed. Documenting construction project activities has always occurred, but computers integrate and track the documentation process and more quickly and efficiently. Some of the documentation included in computerized applications are document and record keeping, subcontractor control, managing subcontractor quotations, tracking correspondence (letters, transmittal, submittals, etc.), maintaining daily reports, creating reminder or "dunning" letters, recording meeting minutes for quick distribution, and maintaining cost records on payments from owners, including payments to subcontractors, changes, claims, and other financial matters relating to the project.

Many individual functions previously mentioned are done on computers today, such as correspondence, application for payment, and so on. An integrated approach to training, standardization of software, and in-house coordination of computer services are needed to make project administration more efficient. Most project managers write their own letters to owners, architects, material suppliers, and subcontractors, but the correspondence is filed in one location without the ability to link issues or subjects between different systems or files. Quick retrieval is necessary for letters, correspondence, meeting minutes, and other documents relating to a specific issue. A computerized system can avoid incomplete documentation caused by misfiled documents, incomplete documentation, or poor organization of project information.

Productivity improvement is a prime objective of most construction companies. Management often feels, though, that computerized systems will reduce project personnel productivity, rather than improve it. They feel that training and retraining employees to keep abreast of the latest software detracts from the employee's job. The assumption that each new generation or version of the software must be purchased, and that training needs to be added as each new version is adopted, is the basis for some firms experiencing unneeded high costs related to computers and productivity. Quite often specific software versions have a much longer application life than the software company's distribution of new versions.

The illustrations and examples concerning document management software in this chapter generally will be based upon the following software products: *Super Prolog Plus/Prolog Manager* from Meridian Software, *Expedition* from Primavera Systems, Inc., *Primavera Project Planner for Windows* from Primavera Systems, and *MS Project* from Microsoft. Word processing and spreadsheet examples are found in *MS Word* and *Excel,* but any of today's more sophisticated word processors and/or spreadsheets produce the same basic look and functions. Primavera System's *Expedition* is one of the leaders in the development of software that helps project managers and staff manage their contract with the owner, subcontractor contract management, document processing, and record keeping. This program also has a broad cost module built into it that allows all associated cost controls required by even the most demanding of projects. *Expedition* also has a link with the *Primavera Project Planner. Super Prolog Plus/Prolog Manager* is a Windows-based program with an icon-based menu and "switchboard" for navigation through parts of the program. *Super Prolog Plus/Prolog Manager* is written with the contractor's project management team in mind. *Super Prolog Plus/Prolog Manager* has many of the forms needed by the superintendent, as well as project engineering documentation tools. It has a built-in word processor for further ease of use in corresponding with all parties. *Super Prolog Plus/Prolog Manager* has an integrator for data from *MS Project. MS Project* is a good example of the moderately priced planning and scheduling software available on the market. It is integrated with other leading MicroSoft product lines and is a featured project management planning and scheduling application. *Primavera Project Planner ("P3")* from Primavera Systems is considered by many to be the leading software in high-end planning and scheduling software. The number and level of resources that can be applied in *P3* is by far the largest. P3 has had a long history in project management software, *P3* for Windows being its latest version.

Document Flow

Project organizational charts are an important tool when creating a contract management system. Communication lines, as well as contractual responsibilities, must be established. The paperwork flow, whether through the office or jobsite, should be established and entered into the project software. Figure 12–1 indicates typical paperwork flow lines in a construction project.

Much of the industry's paper work is repetitive in nature, in that the type of form or reason for using the document will be accomplished many times during the project. Only a few forms are used during the project. Many of the contractor's operations are duplicated month after month, using the same format, but different data. "Application and Certificate for Payment" is a process that happens month to month, using the same form,

FIGURE 12–1
Typical Paperwork
Flow

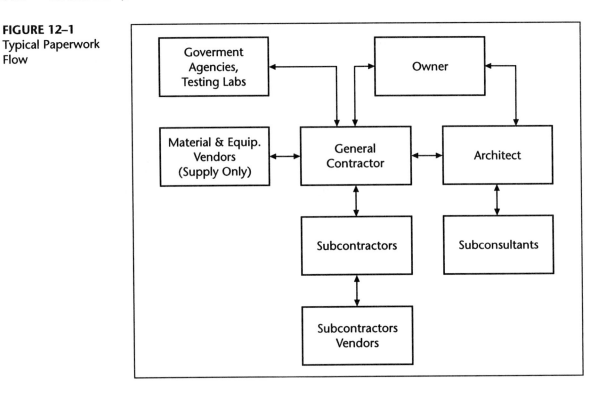

but changing data. Most of the information on this Certificate remains the same, with changes only in the work completed during the applicable period. Electronically, the transferring and recording of this information is done efficiently and accurately. Monthly billing becomes a much easier task when previous information is stored and available without copying and retyping. Contractors now regularly use computerized spreadsheets to produce and update monthly billings.

Integration of information could produce usable project management information in addition to billings, payments from the owner, and keeping track of monies owed to subcontractors. Electronic contract control software also can provide tracking for submittal, schedule, total cost, accounts, and document control.

Document Control

Document control can encompass all of the major individual categories, subjects, and types of paperwork pertinent to a company's daily business. The narrow definition of document control is used to describe the control of submittals; correspondence in the form of a letter; transmittals and their attachments; application and certificates for payments; owner and main contract development and control; subcontracts and scopes development and control; material contracts and purchase order documentation and control; material tracking, purchase, expediting, delivery, and installation; changes and change order documentation, cost development, tracking; problems documentation and issue linking, requests for information, both issuing and receiving of the information; meeting minutes taken and distributed; telephone logs, daily reports, and daily logs and diaries; and drawing controls. This narrow definition of document control separates other project control areas of schedule control or planning and scheduling activities, and the larger issue of cost control. Both document control and project control are major areas of project management systems and play key roles in the management and decision-making by the project team. Scheduling and cost control will only be briefly mentioned in this text, as others have thoroughly covered the subject.

Construction document classifications are fairly universal in the types of documents, the forms each take, and the reasons the documents are produced. Because documents can be detailed regarding classification and usage, a **database** of the different types of blank forms and a **document control** or storage system should be implemented into the contract management system. Typically, databases are used to organize, store, sort, manipulate, compare, and retrieve information groups. Many firms use databases to keep track of inventories of materials. An example of a database can be found at the local lumber yard. As an order is placed, the computer checks the inventory in the yard and confirms the availability of the material needed. When the order is placed, the computer immediately subtracts the items from its inventory (or database of items) and combines the prices from the accounting system, applies the total cost against the customer's account, if the order is charged, and summarizes the total order with the prices attached in the form of a receipt. The computer also prints a sheet (report) for the yard people so they can deliver material requested or refer the customer to the area where the material is located. Again, the location may be a part of this confirmation order form for the yard people. The location of the material is just another piece of information that can be attached to the item itself. These databases can be used to organize, standardize, store, sort, and retrieve documents used on a project. Figure 12–2 lists documents by type and the area or category in which they are used or tracked on many construction projects. In this document list, many documents are filed in different places, but issues that relate to more than one area are not filed together. This is where the concept of linking issues comes into play. The list also shows the static nature of paper files.

The many different management issues that can arise add a third dimension to the matrix, with many of the activities shown being repetitious. The letters and transmittals are sent over and over to the same addressees and addresses. The computer completes this mundane task of

FIGURE 12–2
Document Use List

> **Documents (By Type) To Manage**
>
> Letters
> (General, Informational, Dunning)
>   1. To Owner
>   2. From Owner
>   3. To Architect
>   4. To C.M.
>   5. To Subcontractors
>   a. Subcontractor #1
>   b. Subcontractor #2
>   6. To Vendor Submittals
>
> Shop DWG's
>   Samples
>
> Schedules
>
> Pay Requests
>   1. To Owner
>   2. From Subcontractors
>
> Transmittals
>
> Changes
>   1. Originator
>   a. Owner
>   b. Architect
>   c. Contractor
>   2. Subcontractor
>
> Contracts
>   1. Owner (Main)
>   2. Subcontracts
>
> Purchase Orders
>   1. Vendor
>   a. Materials
>   b. Equipment

typing the addresses each time by using an additional database of the parties involved in the project. This database can draw information from other databases to simplify the day-to-day work required in maintaining the document work flow. The database used should have the ability to create the document form and log, whether it is a transmittal, form letter, customized form letter, an original letter, or one of many reports and tracking forms, such as submittal registers.

The major problem facing the all-in-one document control system (database) today is the ability to store paper documents that are received from various project participants. This is accomplished in two ways, with neither method being totally satisfactory.

1. Scan the document, and convert it to a readable computer file.
2. Transmit the document electronically, via a modem.

Most firms are not electronically transmitting documents, except for facsimile transmissions (faxes). Today's computers have fax capabilities and can both transmit and receive faxes. Many firms use this type of message delivery to augment the standard fax machine. E-mail is another alternate to receiving written documents, in a form that can be stored electronically. The idea of scanning and converting all correspondence and documents seems too time-consuming for today's practical purposes. Construction has not yet evolved into a paperless business but the documentation and control of the general contractor's correspondence is an important part of the management picture. The control process suggested for storing various paper documents includes the use of logs for tracking both incoming and outgoing correspondence and for attaching an issue bar code to each.

Most firms are using generic tools to accomplish many of the defined tasks relating to contract and document control. Figure 12–3 illustrates the common software tools used for project management tasks. Opposition to converting some manual work to computerized forms is sometimes justified. One of the most criticized aspects of using computerized systems is the time required to learn the specifics. The training issue can be somewhat minimized by using only programs with graphical interfaces. A graphical interface provides the user with more of a "point and click" (open and close) method of working with the program. The OS-2, OS-2 Warp, Windows, Windows '95 and Macintosh Operating Systems all provide a graphical interface. Icons are used to graphically depict applica-

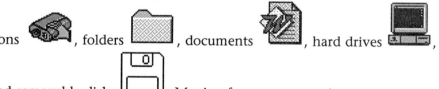

tions   , folders   , documents   , hard drives   ,

and removable disks   . Moving from one operating system to another, the user will note that different icons are used for different depictions and that not all operating systems use icons for everything. A mouse is used to move the cursor on the screen to areas that need to be viewed, and the screen depicts how the actual printed document will appear. This is called WYSIWYG, or "What You See Is What You Get." Figure 12–4 shows a screen from Microsoft Word, using a pull-down menu.

Windows are used to open applications and documents, with a button activating many of the functions needed. All of these operating systems use a type of pull-down menu for further help in using the

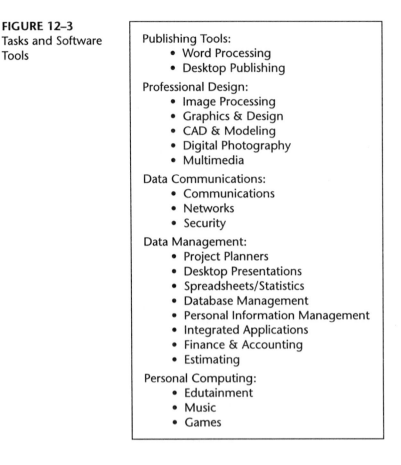

FIGURE 12–3
Tasks and Software Tools

Publishing Tools:
- Word Processing
- Desktop Publishing

Professional Design:
- Image Processing
- Graphics & Design
- CAD & Modeling
- Digital Photography
- Multimedia

Data Communications:
- Communications
- Networks
- Security

Data Management:
- Project Planners
- Desktop Presentations
- Spreadsheets/Statistics
- Database Management
- Personal Information Management
- Integrated Applications
- Finance & Accounting
- Estimating

Personal Computing:
- Edutainment
- Music
- Games

application. As icons and menus become more standardized from application to application, learning time is decreased.

Word Processing Software

Most management and project administration functions rely on some form of communications, retrieval, and use of information. Today, the primary form of written communication is done using **word processors,** which are versatile programs and include features formerly found only in desktop publishing applications. Documents can easily display graphics, photos, and short video clips. All of these items can be displayed on computer monitors, but the process of printing them becomes somewhat more difficult. High-quality color output is not used in day-to-day project administration, due to its high cost. Most written communications use a basic format, although digital photos are being used more and more for projects. Two advantages of using word processing programs are

1. The word processing program can store project participants' names and addresses, as well as form letters, both of which can be merged into a single letter or transmittal for mailing and/or faxing.

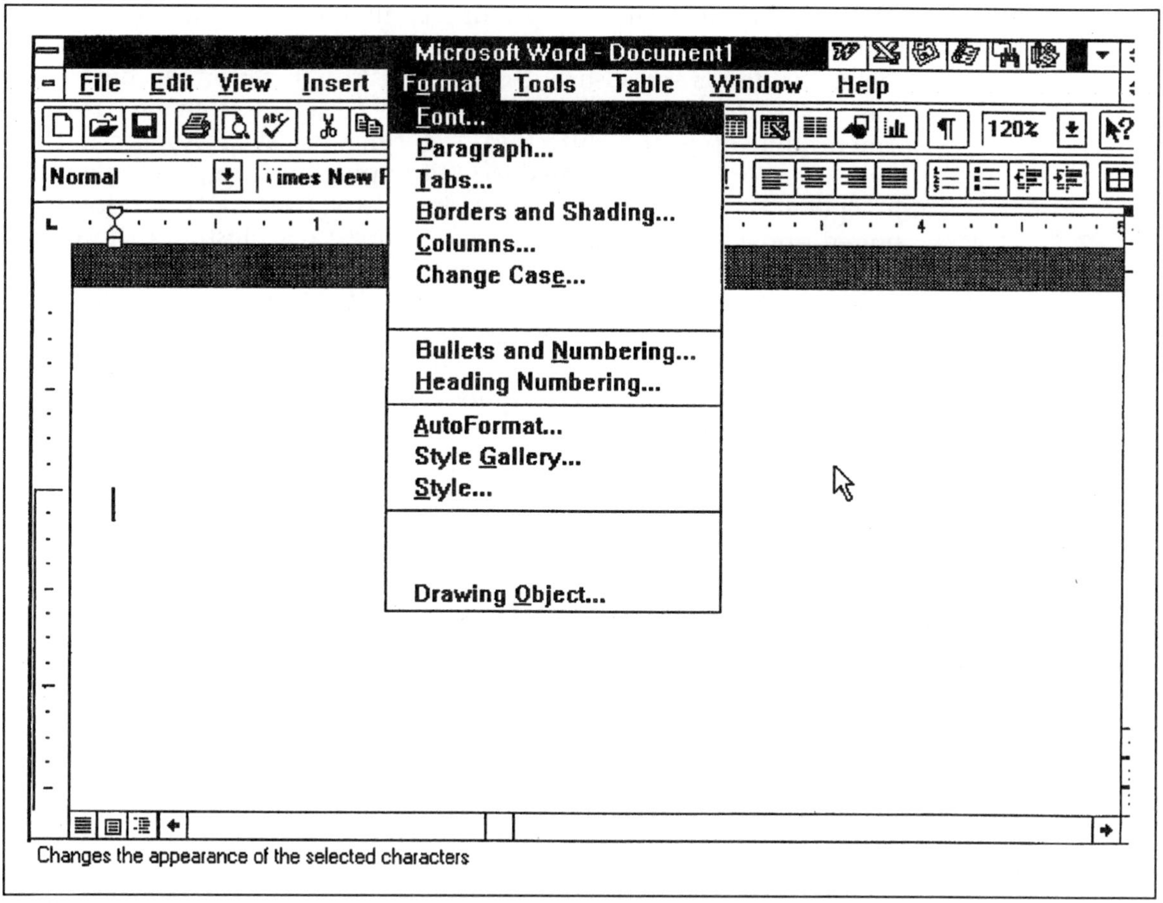

FIGURE 12–4 Screen From Microsoft Word Showing Pull-Down Menus

2. The word processing program has the ability to electronically store and retrieve a document, linking its contents to issues, changes, and other areas that require complete records to discuss and resolve many situations.

The list of word processing programs is long, but a sample follows:

Microsoft Word (UNIX, OS-2, Windows, DOS, Mac, Windows '95)
MacWrite Pro (Mac)
WriteNow (Mac)
WordPerfect (UNIX, OS-2, Windows, DOS, Mac)
Ami Pro (WordPro, OS-2, Windows)

Spreadsheet Software

The second largest type of generic application used by project management professionals for daily operations is the spreadsheet. The spreadsheet allows users to develop, analyze, and change mathematical or statistical information. The spreadsheet allows users to use a variety of mathematical applications in one document. One of the simple documents that uses a spreadsheet-based application to perform its function is the application for payment used on most projects (see chapter 14). The American Institute of Architects (AIA) documents, G702 "Application and Certificate for Payment," and G703 "Continuation Sheet," as shown later in Figures 14–14 and 14–15, and various derivatives are typical forms required and used on many projects. Figure 12–5 illustrates the use of spreadsheets for the computation of the Application for Payment, while Figure 12–6 shows a spreadsheet for a submittal log (see chapter 3).

When duplicating a copyrighted form, the user must obtain copyright releases to reproduce and use it. Most firms have developed their own in-house form that does not look like the AIA form but incorporates the same standard requirements, which are not copyrighted. Most owners and architects are receptive to changing the look of the form if the content and totals appear on the contractor's form.

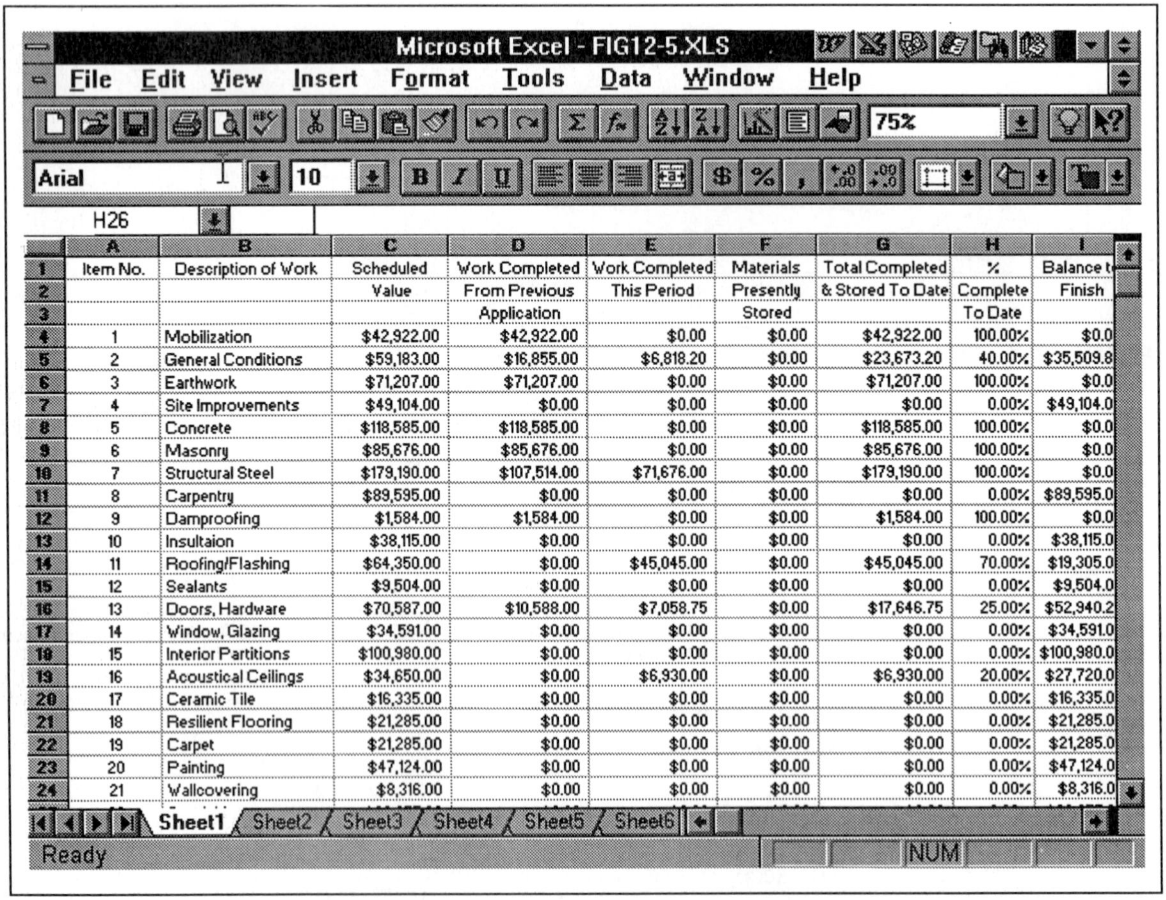

Item No.	Description of Work	Scheduled Value	Work Completed From Previous Application	Work Completed This Period	Materials Presently Stored	Total Completed & Stored To Date	% Complete To Date	Balance to Finish
1	Mobilization	$42,922.00	$42,922.00	$0.00	$0.00	$42,922.00	100.00%	$0.0
2	General Conditions	$59,183.00	$16,855.00	$6,818.20	$0.00	$23,673.20	40.00%	$35,509.8
3	Earthwork	$71,207.00	$71,207.00	$0.00	$0.00	$71,207.00	100.00%	$0.0
4	Site Improvements	$49,104.00	$0.00	$0.00	$0.00	$0.00	0.00%	$49,104.0
5	Concrete	$118,585.00	$118,585.00	$0.00	$0.00	$118,585.00	100.00%	$0.0
6	Masonry	$85,676.00	$85,676.00	$0.00	$0.00	$85,676.00	100.00%	$0.0
7	Structural Steel	$179,190.00	$107,514.00	$71,676.00	$0.00	$179,190.00	100.00%	$0.0
8	Carpentry	$89,595.00	$0.00	$0.00	$0.00	$0.00	0.00%	$89,595.0
9	Damproofing	$1,584.00	$1,584.00	$0.00	$0.00	$1,584.00	100.00%	$0.0
10	Insultaion	$38,115.00	$0.00	$0.00	$0.00	$0.00	0.00%	$38,115.0
11	Roofing/Flashing	$64,350.00	$0.00	$45,045.00	$0.00	$45,045.00	70.00%	$19,305.0
12	Sealants	$9,504.00	$0.00	$0.00	$0.00	$0.00	0.00%	$9,504.0
13	Doors, Hardware	$70,587.00	$10,588.00	$7,058.75	$0.00	$17,646.75	25.00%	$52,940.2
14	Window, Glazing	$34,591.00	$0.00	$0.00	$0.00	$0.00	0.00%	$34,591.0
15	Interior Partitions	$100,980.00	$0.00	$0.00	$0.00	$0.00	0.00%	$100,980.0
16	Acoustical Ceilings	$34,650.00	$0.00	$6,930.00	$0.00	$6,930.00	20.00%	$27,720.0
17	Ceramic Tile	$16,335.00	$0.00	$0.00	$0.00	$0.00	0.00%	$16,335.0
18	Resilient Flooring	$21,285.00	$0.00	$0.00	$0.00	$0.00	0.00%	$21,285.0
19	Carpet	$21,285.00	$0.00	$0.00	$0.00	$0.00	0.00%	$21,285.0
20	Painting	$47,124.00	$0.00	$0.00	$0.00	$0.00	0.00%	$47,124.0
21	Wallcovering	$8,316.00	$0.00	$0.00	$0.00	$0.00	0.00%	$8,316.0

FIGURE 12–5   Screen From Microsoft Excel Showing "Application for Payment" Spreadsheet

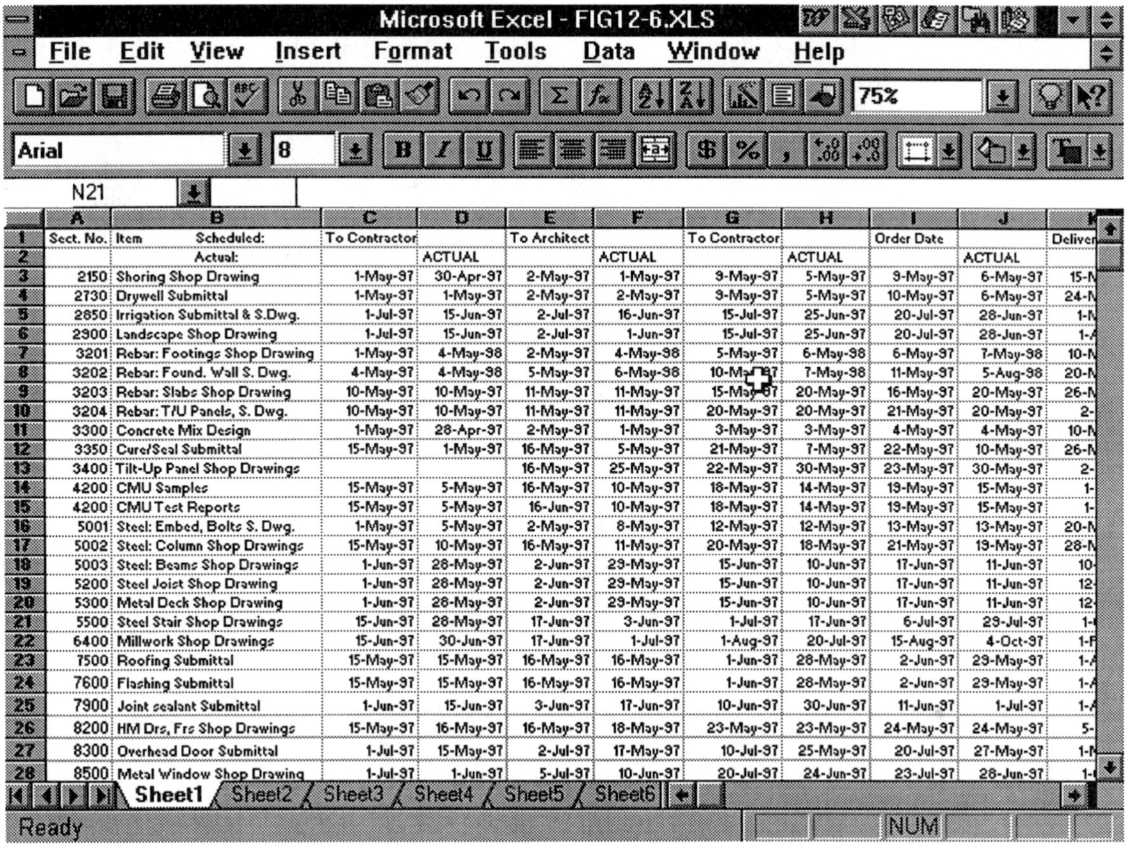

Microsoft Excel - FIG12-6.XLS

File   Edit   View   Insert   Format   Tools   Data   Window   Help

Arial   8   75%

N21

Sect. No.	Item	Scheduled: Actual:	To Contractor	ACTUAL	To Architect	ACTUAL	To Contractor	ACTUAL	Order Date	ACTUAL	Deliver
2150	Shoring Shop Drawing	1-May-97	30-Apr-97	2-May-97	1-May-97	9-May-97	5-May-97	9-May-97	6-May-97	15-N
2730	Drywell Submittal	1-May-97	1-May-97	2-May-97	2-May-97	3-May-97	5-May-97	10-May-97	6-May-97	24-N
2850	Irrigation Submittal & S.Dwg.	1-Jul-97	15-Jun-97	2-Jul-97	16-Jun-97	15-Jul-97	25-Jun-97	20-Jul-97	28-Jun-97	1-N
2900	Landscape Shop Drawing	1-Jul-97	15-Jun-97	2-Jul-97	1-Jun-97	15-Jul-97	25-Jun-97	20-Jul-97	28-Jun-97	1-A
3201	Rebar: Footings Shop Drawing	1-May-97	4-May-98	2-May-97	4-May-98	5-May-97	6-May-98	6-May-97	7-May-98	10-N
3202	Rebar: Found. Wall S. Dwg.	4-May-97	4-May-98	5-May-97	6-May-98	10-May-97	7-May-98	11-May-97	5-Aug-98	20-N
3203	Rebar: Slabs Shop Drawing	10-May-97	10-May-97	11-May-97	11-May-97	15-May-97	20-May-97	16-May-97	20-May-97	26-N
3204	Rebar: T/U Panels, S. Dwg.	10-May-97	10-May-97	11-May-97	11-May-97	20-May-97	20-May-97	21-May-97	20-May-97	2-
3300	Concrete Mix Design	1-May-97	28-Apr-97	2-May-97	1-May-97	3-May-97	3-May-97	4-May-97	4-May-97	10-N
3350	Cure/Seal Submittal	15-May-97	1-May-97	16-May-97	5-May-97	21-May-97	7-May-97	22-May-97	10-May-97	26-N
3400	Tilt-Up Panel Shop Drawings			16-May-97	25-May-97	22-May-97	30-May-97	23-May-97	30-May-97	2-
4200	CMU Samples	15-May-97	5-May-97	16-May-97	10-May-97	18-May-97	14-May-97	19-May-97	15-May-97	1-
4200	CMU Test Reports	15-May-97	5-May-97	16-Jun-97	10-May-97	18-May-97	14-May-97	19-May-97	15-May-97	1-
5001	Steel: Embed, Bolts S. Dwg.	1-May-97	5-May-97	2-May-97	8-May-97	12-May-97	12-May-97	13-May-97	13-May-97	20-N
5002	Steel: Column Shop Drawings	15-May-97	10-May-97	16-May-97	11-May-97	20-May-97	18-May-97	21-May-97	19-May-97	28-N
5003	Steel: Beams Shop Drawings	1-Jun-97	28-May-97	2-Jun-97	23-May-97	15-Jun-97	10-Jun-97	17-Jun-97	11-Jun-97	10-
5200	Steel Joist Shop Drawing	1-Jun-97	28-May-97	2-Jun-97	23-May-97	15-Jun-97	10-Jun-97	17-Jun-97	11-Jun-97	12-
5300	Metal Deck Shop Drawing	1-Jun-97	28-May-97	2-Jun-97	23-May-97	15-Jun-97	10-Jun-97	17-Jun-97	11-Jun-97	12-
5500	Steel Stair Shop Drawings	15-May-97	28-May-97	17-Jun-97	3-Jun-97	1-Jul-97	17-Jun-97	6-Jul-97	29-Jul-97	1-
6400	Millwork Shop Drawings	15-May-97	30-Jun-97	17-Jun-97	1-Jul-97	1-Aug-97	20-Jul-97	15-Aug-97	4-Oct-97	1-F
7500	Roofing Submittal	15-May-97	15-May-97	16-May-97	16-May-97	1-Jun-97	28-May-97	2-Jun-97	23-May-97	1-A
7600	Flashing Submittal	15-May-97	15-May-97	16-May-97	16-May-97	1-Jun-97	28-May-97	2-Jun-97	23-May-97	1-A
7900	Joint sealant Submittal	1-Jun-97	15-Jun-97	3-Jun-97	17-Jun-97	10-Jun-97	30-Jun-97	11-Jun-97	1-Jul-97	1-A
8200	HM Drs, Frs Submittal	15-May-97	16-May-97	16-May-97	18-May-97	23-May-97	23-May-97	24-May-97	24-May-97	5-
8300	Overhead Door Submittal	1-Jul-97	15-May-97	2-Jul-97	17-May-97	10-Jul-97	25-May-97	20-Jul-97	27-May-97	1-N
8500	Metal Window Shop Drawing	1-Jul-97	1-Jun-97	5-Jul-97	10-Jun-97	20-Jul-97	24-Jun-97	23-Jul-97	28-Jun-97	1-

Sheet1 / Sheet2 / Sheet3 / Sheet4 / Sheet5 / Sheet6

Ready   NUM

FIGURE 12–6   Screen From Microsoft Excel Showing "Submittal Log" Spreadsheet

Spreadsheets automatically perform mathematical functions, without error, assuming that the formulas, data fields, and input are correct. When using a standardized form, typing is not required (with the exception of updates, such as additional change orders), however updating forms is a relatively simple process, showing the instant totals and amounts to be reimbursed each month. Even creating "what if" scenarios can be easily done using a spreadsheet.

Database Software

Another category of software that is becoming more critical to the construction professional is the **database**. Databases allow the user to store, group, and quickly retrieve data. The data can take many different forms. In construction, accounting and cost control forms utilize databases to generate jobsite cost items. Quantities, labor hours, equipment usage, change order costs, and other pertinent cost items can be entered into databases. The summarizing and reporting ability of most databases makes an excellent tool for cost control areas. Storage of cost items for estimating purposes is another form of database that can be used. Contractors use in-house prepared cost databases and commercially prepared databases that are available from numerous vendors. Some contractors prefer the in-house type, as it relates to actual historical costs from previous projects. Because this method can be fairly costly to compile and maintain, some contractors prefer to use the commercial type. The commercial cost databases, however do not accurately reflect a particular contractor's costs. Both databases must be kept current, either by manually entering current data or by purchasing periodic updates. Some interpretation of conditions and applications is necessary when using either type of database. Many spreadsheets can also perform many of the same functions as a database, but generally relational databases are easier and more functional than are spreadsheet databases.

Computer Hardware

Computers are being used on the jobsite, as well as in the contractor's office. It is important to select the appropriate amount and type of hardware to match the field staff and its needs. Computers and office equipment are part of the jobsite overhead cost and should be selected with care, minimizing unnecessary equipment. The following list describes typical computer hardware and software that can be used at the jobsite.

Computers:
486 based Intel computers (IBM compatibles), or
Pentium based Intel Computer (IBM compatibles)
Laptops, using the aforementioned processors
Power PC from Macintosh and IBM

Printers:
For everyday use—Ink jet printers
Laser printers

Combination:
Laser printers/fax machines/copy machines

Fax modems—28800 V .42bis MINPS for E-mail, file transfers, faxing from a computer

Document and Contract Control Software

Super Prolog Plus/Prolog Manager and *Expedition* are the two leading software programs in the construction industry that use an integrated approach to contract control. *Expedition* and *Super Prolog Plus/Prolog Manager* perform many of the same functions for the contractor's project management staff and are used by all project team members. As one becomes familiar with these two programs, *Super Prolog Plus/Prolog Manager*'s menus and icons feel more directed to the general contractor's personnel, but the engineer/architect/designer, construction manager, general contractor, owner, and subcontractor can all utilize this software to their advantage. This software contains everything that is needed for the entire documentation process. Team members may interpret documents differently, but the information in the letter or on the form will meet the needs of all of the involved parties. Contract control software attempts to provide an integrated approach to documentation, with better software programs being able to link pertinent issues with different types of documents.

The *Super Prolog Plus/Prolog Manager* and *Expedition* programs allow the team member to answer many critical questions in administrating the main contract with the owner, supplying information to and from the architect, and maintaining the control needed on the subcontractor's contract. Both programs allow project costs to be tracked. One of the manuals that accompanies *Expedition,* called *Perspectives,* uses critical questions to outline the uses of *Expedition* for each team member. The questions deal with submittals; management of design submissions; procurement tracking; problem and issue identification; tracking of costs, changes, and budgets; schedule and progress recording; and quality and safety. All information is then generated to the proper team members. The same critical questions are addressed in *Super Prolog Plus/Prolog Manger* as well.

Contract management software must have the ability to file information and summarize it or report it back in many different ways. It must be able to generate forms and correspondence or have the ability to import this information from other spreadsheets, databases, and word processing programs. The key to contract management issues is the ability to store and retrieve information that is pertinent to issues that must be resolved.

In general, most computer software products that are designed for filing, organizing, and retrieving documents are a derivative of database software. With the ability to create forms, generate correspondence, and perform mathematical calculations with costs. Features that project administration software should contain is discussed next.

Flexibility.   Software flexibility enables a construction firm or owner to manage projects in compliance with management styles, rules, and regulations. Each project a contractor begins has its own particular set of work criteria. Some projects do not have the manpower to update and maintain a complex database, while others require more documentation because of complexities, potential litigations, and so on. Systems must be flexible and easy to maintain.

Integration Capabilities.   Integration involves the compatibility of different forms of documents, such as databases, written documents, and spreadsheets, to be used, accessed, and combined in the same software application. Being able to use and move information from one source to another is an important feature.

Access and Retrieval Capabilities.   During the course of the construction process, it is important to be able to quickly access and retrieve documents and information related to critical issues. Search, find, and sort features are desirable in project administration software.

Security Features.   Recommendations for a computerized system must take into consideration the access rights and key lock/password security the software provides. Many construction firms prefer restricted access to their project files.

Networking Capabilities.   As projects grow in size and complexity, the number of users on a project administration system increases. The ability of a system to be networked to several workstations and provide multiple points of data retrieval and entry also is important.

Standardized Formats.   A document control system should have, in addition to integration, flexibility, network ability, and access protection, a means of creating a company standardized format. Standardization allows for quicker training of personnel and a more efficient use of the system. When systems are universally used, tracking by management can be done on a much larger scale.

Illustrations of screens from *Super Prolog Plus/Prolog Manager* are used in the examples that follow and in the discussion of software uses for the general contractor. It should be noted that many of the forms shown will look like those used throughout this book. *Expedition* and *Super Prolog Plus/Prolog Manager* programs do not provide the flexibility to mirror contractor forms, but by using add-on applications some level of form duplication can occur. The examples of *Super Prolog Plus/Prolog Manager* will assume a traditional contractual relationship, shown in Figure 12–7, which also illustrated the paper flow that will result from using contract management software.

One of the first orders of business once a contract has been signed with the owner is the submission of product data, shop drawings, and material samples to the architect, as discussed in chapter 3. *Expedition* and *Super Prolog Plus/Prolog Manager* allow the contractor to enter all key owner

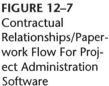

FIGURE 12–7
Contractual Relationships/Paperwork Flow For Project Administration Software

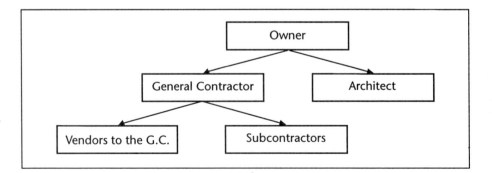

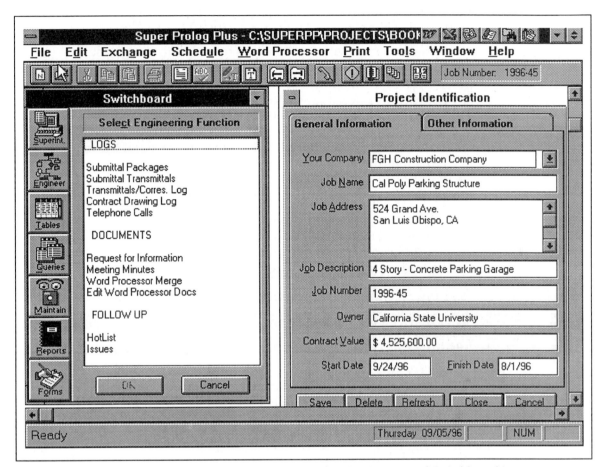

FIGURE 12–8   Screen From Meridian's Prolog Manager Showing Menus and Switchboard Items

contract provisions at the start of the project. This information will later be used for cost and schedule tracking, along with billings and changes to the contract (both cost and schedule). Many contractors use a spreadsheet for this tracking activity, but generally this approach to tracking lacks the integration and ability to link issues, key criteria when evaluating construction software.

The contract summary for *Super Prolog Plus/Prolog Manager* is displayed in the "Project Summary Window" shown in Figure 12–8. This program creates logs (under the Log menu) for correspondence, daily reports, issues, punch lists, and submittals. *Super Prolog Plus/Prolog Manager* and *Expedition* use a similar menu approach, but the former utilizes an icon menu bar and has a "switchboard." The switchboard is an icon-based pallet that moves in and out of the different application parts of the program and in between the different areas that the documents may originate.

A typical submittal tracking system will answer and track the following questions:

- What submittals are required by the specifications?
  (type of submittal, i.e. shop drawings, brochure, samples, etc.)
- When are these submittals needed to maintain the schedule?
  (dates required based on critical, noncritical nature)
- Who are our submittals being generated by?
  (which subcontractor, vendor, or in-house person?)
- Where are the submittals today, or who has the submittal?
  (subcontractor, vendor, general contractor, architect)
- Is the submittal timely, or impacting the purchase of the item?
  (schedule and procurement)
- What is the submittal status from the design professionals?
  (approved, rejected, approved as noted)

Figure 12–9 illustrates the *Super Prolog Plus/Prolog Manager* window that relates to submittals and submittal control.

Both *Super Prolog Plus/Prolog Manager* and *Expedition* allow users to expand from the submittal logs to a more detailed description if they so de-

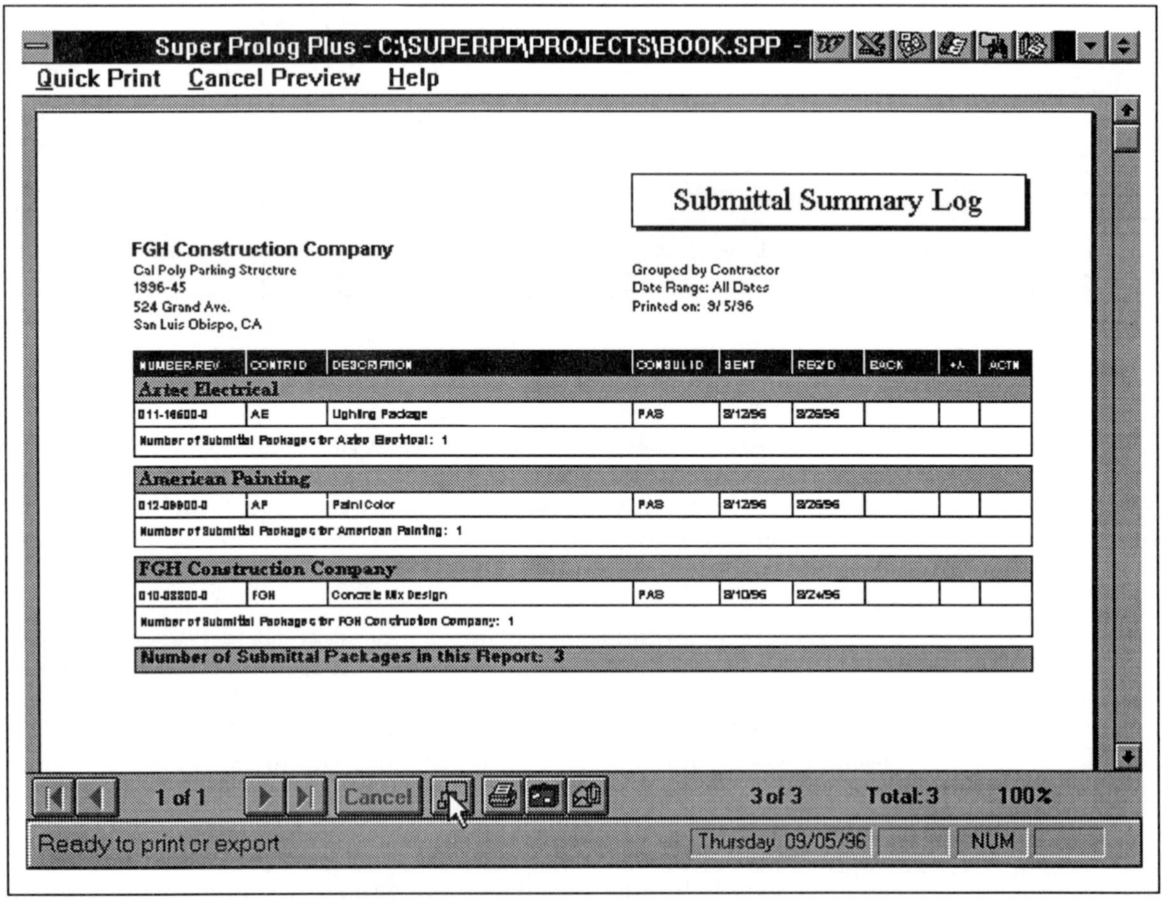

FIGURE 12–9   Screen From Meridian's Prolog Manager Showing "Submittal Summary Log" Example

sire. The submittals in *Expedition* also can be linked to the schedule if the general contractor is using the *P3* or *Finest Hour* schedule from Primavera Systems, Inc. *Super Prolog Plus/Prolog Manager* activities can be directly imported and exported to *MS Project*. Both programs can import and export in ASCII format, allowing a greater range of information.

Another module of *Expedition* and *Super Prolog Plus/Prolog Manager* is the Meeting Minutes log. In Figure 12–10, *Super Prolog Plus/Prolog Manager*, note all of the critical parts required for the Meeting Minutes. *Super Prolog Plus/Prolog Manager* and *Expedition* can now prepare a transmittal (Figure 12–11) to all attendees, with addresses attached, link the critical issues for later retrieval, and file them in a location that can be retrieved when needed. The ability to easily create transmittals and other forms of correspondence is key to controlling and retrieving documents.

Report generation by the contractor's document control system is an important tool that should be integrated into the software. *Expedition* allows a company to establish standard report forms such as daily progress

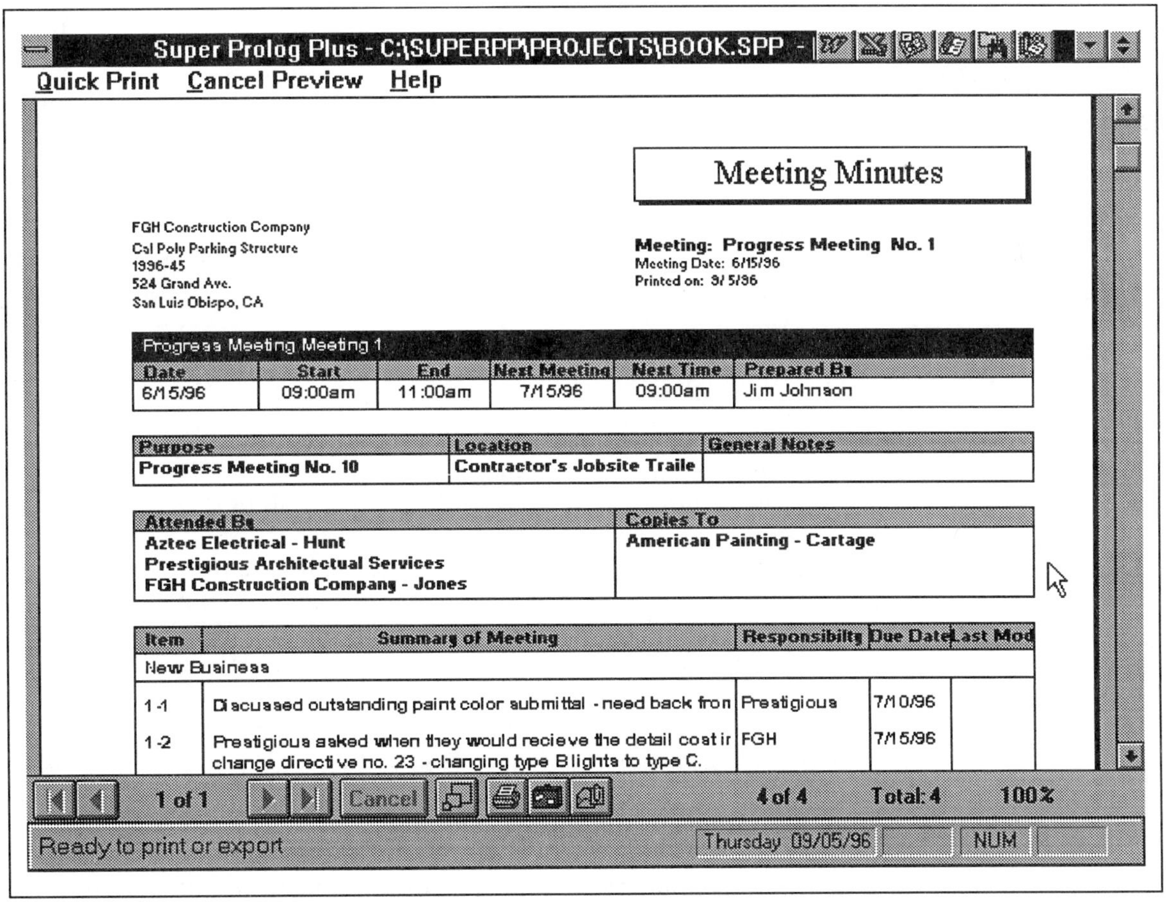

FIGURE 12–10   Screen From Meridian's Prolog Manager Showing "Meeting Minutes" Example

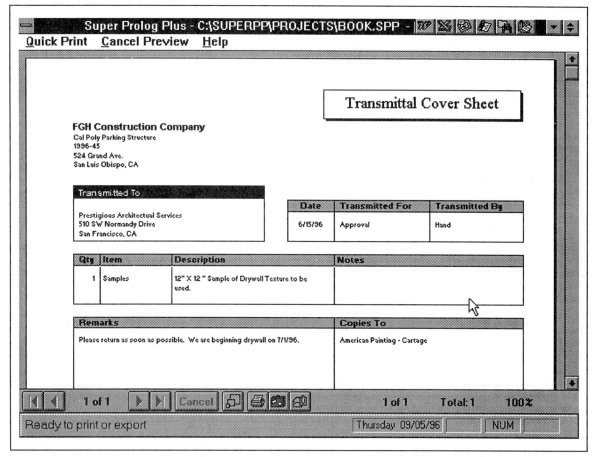

FIGURE 12–11   Screen From Meridian's Prolog Manager Showing "Transmittal" Example

reports, delivery records, recording visitors, and other structured form letters. *Super Prolog Plus/Prolog Manager* and *Expedition* call the latter, which are primarily reminder letters, "dunning letters." All of these types of correspondence and reports can be logged. "Correspondence Sent" Logs are automatically generated for correspondence as it is sent. Another important feature, mentioned earlier, is the exporting and importing data ability. *Expedition* can export data to *dBASE, Lotus 1-2-3,* or ASCII files and can exchange data with *P3* and *Finest Hour. Expedition* also will import data from Lotus.WK1, dBase.DBF, or ASCII files.

In addition to the other document control systems within *Expedition* and *Super Prolog Plus/Prolog Manager,* a cost module is available for control of contract costs and purchase orders. Purchase orders can be written within *Expedition* and *Super Prolog Plus/Prolog Manager,* with automatic distribution of costs to the costing windows and cost worksheet. This allows for the complete management of purchase orders from purchase to delivery. Contracts, subcontracts, and costs related to these different contracts are also easily processed and managed from these programs. Main contract changes and the costs associated with them can be attached and controlled from both programs. Subcontractor contract control is also another strong feature needed in the area of document control. As changes are processed, documentation and cost control can be maintained using the contract management modules.

Super Prolog Plus/Prolog Manager and *Expedition* also can be used by the other construction team members. Most of the operations previously described are done by all members in one form or another. In the near future, document and contract control software will be commonly used on jobsites. As these software packages continue to develop, their operations will become more seamless and will be easily integrated into other parts of the electronic jobsite.

There are, of course, many other software packages available to contractors in the project administration area. Careful consideration is necessary for contractors to obtain the software package that is most compatible with their operation.

Planning and Scheduling Software

Two additional software programs used by the construction industry are Microsoft's *MS Project for Windows* software and *P3 for Windows* from Primavera Systems. Both products are consistent with standard planning and scheduling practices. *MS Project* has a similar approach to the high-end *P3* product. *Sure Track for Windows* is similar to *P3*, without the reporting capabilities. *MS Project* has *Windows* and *Macintosh* versions available. This software integrates well with the other popular Microsoft products of *Word* (word processing) and *Excel* (spreadsheet). Primavera *P3* still holds the distinction of being the scheduling software that is most likely to be specified by some of the larger contracts that use construction management firms as owner's representatives. The learning curve from one scheduling package to another is very short, with most of these products having graphical interfaces. Sometimes different words are used for the same program activity or screen result, but these differences are easily learned.

A quick review follows regarding how *MS Project* and *P3 for Windows* work. The main user screen of project management (scheduling software) software is the project screen or a split screen of task or activities and bars in the form of a bar/Gantt chart, as shown in Figure 12–12. This is the most readable type of schedule and the most common. Today's software uses networks as a basis for the bar chart, and the network chart, as shown in Figure 12–13, can be used to first determine schedule logic. In addition to activity duration, resources such as manpower, costs, equipment, and so on can be attached to the activity. Different dependencies also can be placed between activities on the network, and typical start-to-start, start-to-finish, and finish-to-start can all be depicted by the network. Lags and lead times are available to provide the user with a more realistic technique

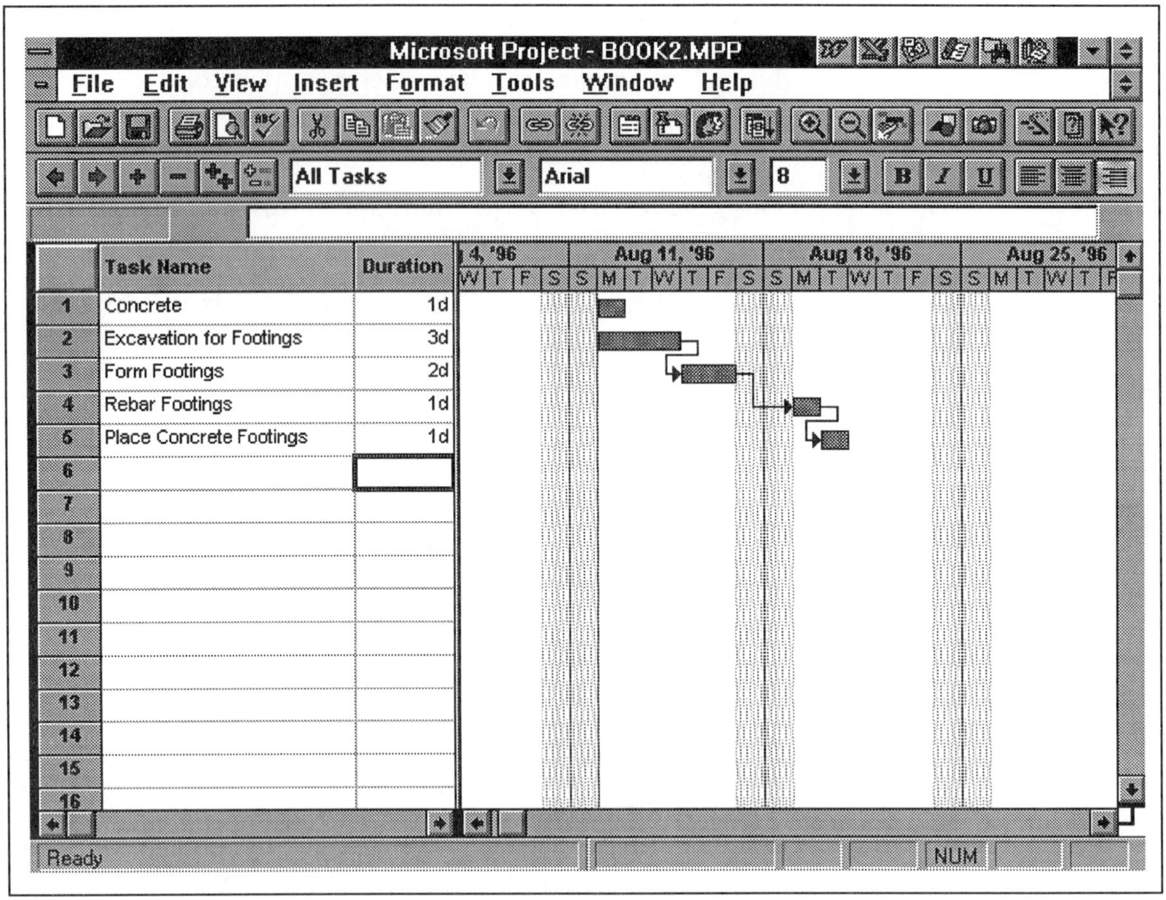

FIGURE 12–12   Screen From Microsoft Project Showing Gantt Chart

for improved scheduling. All scheduling software today will provide the user early and late start dates, along with the critical path, milestones, and end dates. Schedules also must contain the ability to be timely updated as actual progress is achieved in the field. Good software will allow management to compare planned activity duration and actual work activities. A critical element in planning and scheduling software is the ability to provide different looks for different analyses. *MS Project,* like *P3,* provides a Work Breakdown Structure (WBS). This WBS allows users multiple summary levels, methods of searching, selecting, and sorting schedules. Another helpful feature of most planning and scheduling software is an ability to use different calendars for different projects, if needed. Calendars can be configured with holidays or other non-work days, depending on project needs.

Multiple levels of searching, selecting, and sorting occur on a con-

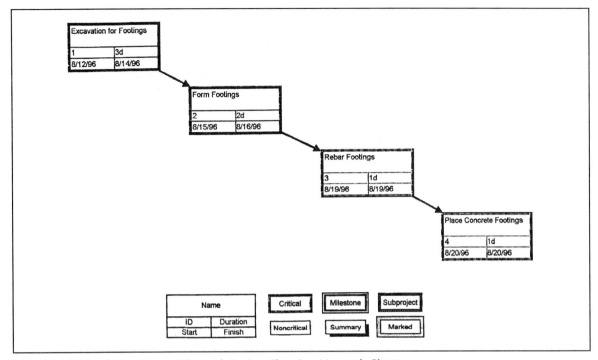

FIGURE 12–13   Report From Microsoft Project Showing Network Chart

struction project, and most software allows the activity Identification Number (ID) to be used as a form of code for this use. Along with searching, selecting, and sorting, these programs permit the filtering of information. This filtering can provide information in summary forms or only items that need to be viewed and analyzed. Figures 12–14 and 12–15 show the same information, but at different summary levels.

Today's planning and scheduling tools have a virtually unlimited number of activities that can be placed in the network. This was not always true with the past generation of this type of software, which was limited by the number of activities it could handle and the time it took to run (calculate) the network. This meant that many complex issues were inaccurately described with less detail, and important submittal dates and delivery activities were left off. Present project management recognizes the need for submittal and delivery information in planning today's projects, with the need to include these items as a true schedule activity.

In the current construction industry, planning and scheduling software is used as billing tools. Costs are loaded as resources, then as tasks are completed, a summary of dollars per task or activity can be viewed and printed, creating a "value" of work completed. Many contractors prefer to separate the billing process from the scheduling process. The items and amounts to be billed on schedule values do not necessarily match a detail

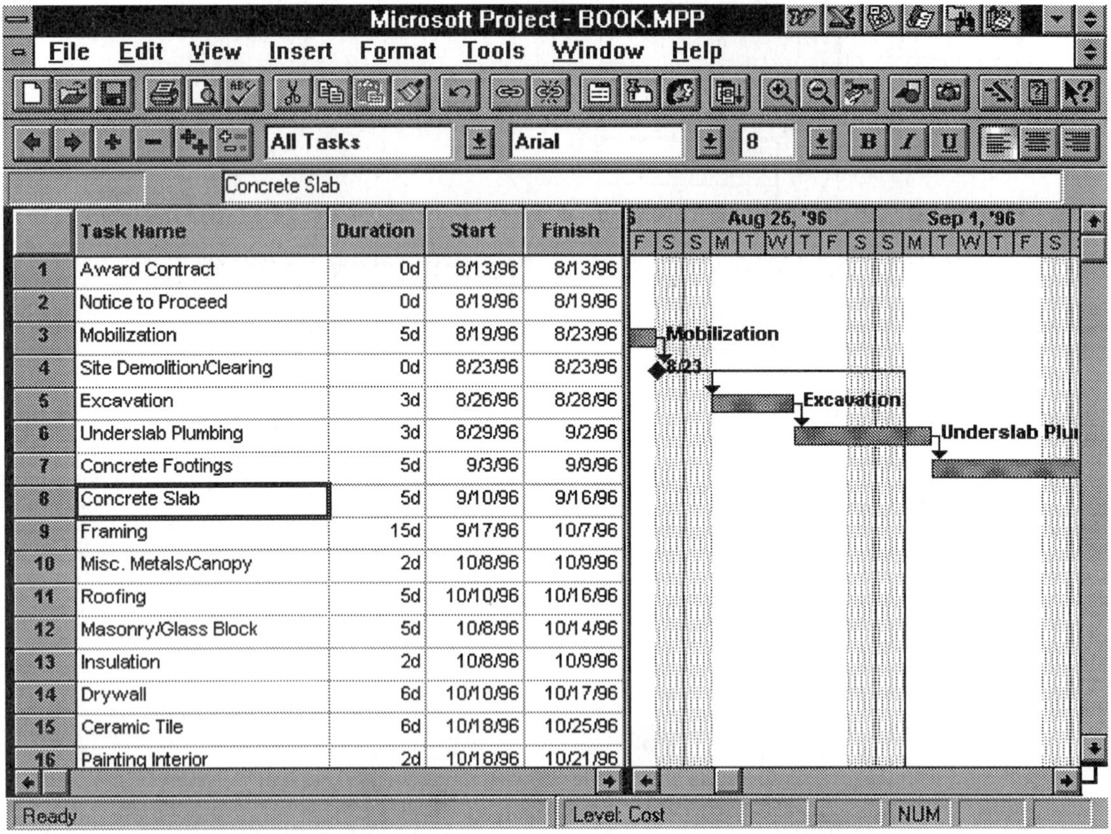

FIGURE 12–14   Screen From Microsoft Project Showing Time Scaled-Networked Gantt Chart

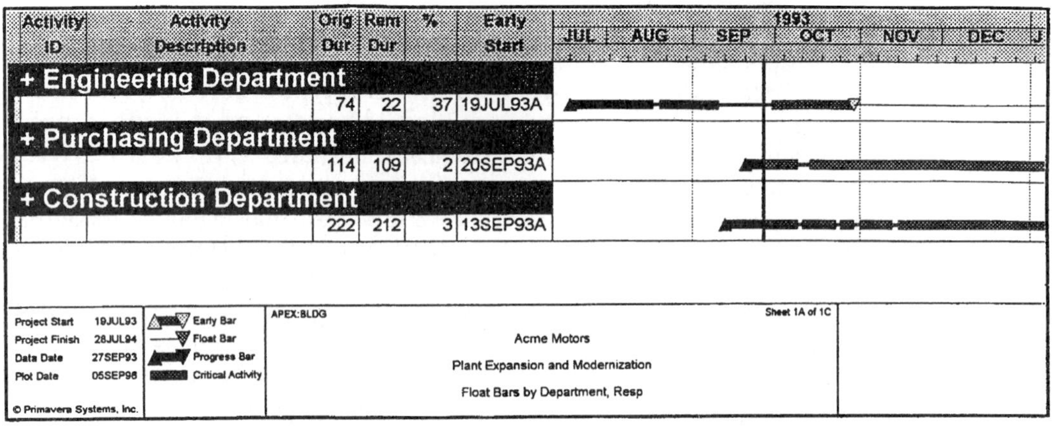

FIGURE 12–15   Report From Primavera Project Planner Showing Time Scaled Gantt Chart

schedule. Costs are applied to schedule activities as a resource, much like labor or equipment.

Additionally, as cost based on activities is analyzed, summaries of costs will produce cash flow curves. Figure 12–16 on page 322 illustrates a cost summary based on scheduled activities showing an early and last curve of cost that can be quickly generated into analysis form using *P3* software.

Electronic Photographic Documentation

Computers are beginning to play a role in the area of project photographic documentation. The recent advent of digital cameras and the ability to capture frames off video makes the use of digital images very viable on today's jobsite. Digital cameras traditionally were expensive and hard to use, but several computer and camera manufacturers are currently manufacturing them. They are now easy to use and competitively priced. Photos from digital cameras can be directly loaded into a Windows- or Macintosh-based computer for storage. These photos can be compressed in size for storage, sent by modem to other locations to help visualize problems, or placed on electric white boards for work group collaboration and problem solving by the architect, subconsultants, the general contractor, and subcontractors. Depending on the particular camera, in-camera storage of images varies. Some cameras can only carry a limited number of images, for example, eight. This problem can be eliminated by using a laptop for downloading images to the hard drive, or by using video-capture technology. Video-capture technology works by capturing images from a video source such as a television, VCR, 8 mm, or VHS camcorder. This type of image capturing requires the use of a low-light camcorder and some type of video capture board (if not built into the computer). Resolution from the camcorder is not as high as a digital camera (Figure 12–17 on page 323), but is quite adequate for documentation and problem solving. This second technique also leaves the contractor with a videotape of the jobsite.

Some of the uses of computerized photographs include:

- incorporation of photos in daily reports
- incorporation of photos in emergency reports, which can immediately be electronically transmitted
- incorporation of photos in schedules, meeting minutes, inspection reports, and progress reports
- incorporation in punch lists
- incorporation in dispute documentation
- photographs of jobsite personnel in personnel charts

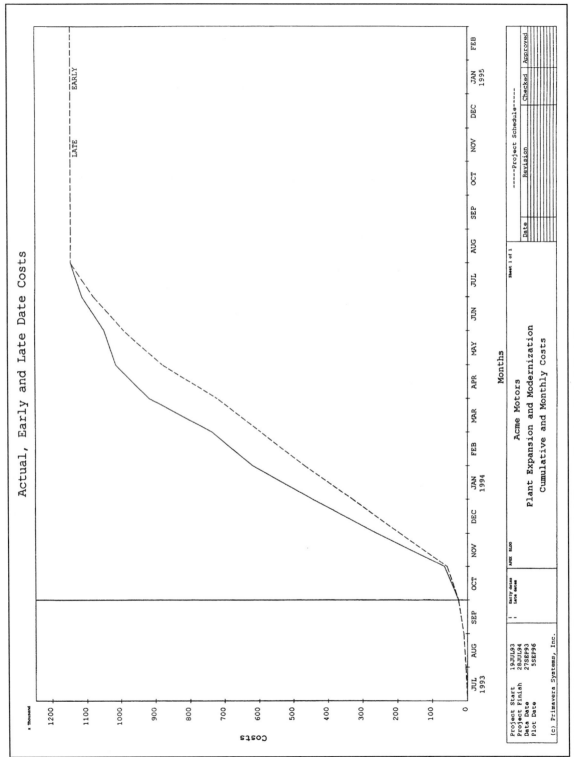

FIGURE 12–16 Report From Primavera Project Planner Showing Project Cash Flow

FIGURE 12–17
"Quick-Take" Digital
Camera

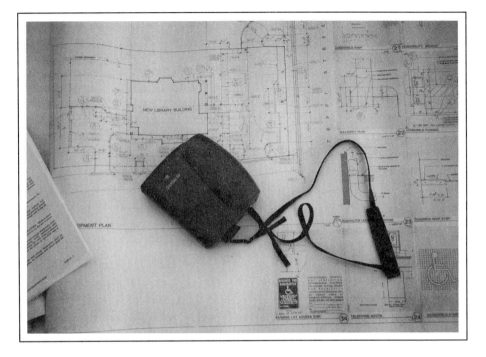

Summary

The construction industry has embraced computer technology. Computerized accounting was the first computer application adopted in the contractor's office, leading to computerized cost control and cost accounting. This was quickly followed by project planning and scheduling software. As hardware prices dropped, word processing was adopted by most businesses. The use of computers to aid in the estimating process has been slower in coming, but today's contractors are using some form of computerized estimating. Some use spreadsheets for estimating summaries, while larger companies use complete software packages. The remaining major areas of computer adoption are communications and networking, contract and document management, and photographic documentation. These areas are being implemented slowly but steadily. The widespread use of graphical interfaces has made computer training and their use much easier. This ease of use will be the catalyst for the continuing addition of computers on the jobsite, and with the continued standardization of programs, the transition from one jobsite to the next will be a smooth process.

CHAPTER 13
CHANGES AND CLAIMS

CHAPTER OUTLINE

Changes in the Construction Project

Changes will occur in every construction project. These changes can be small, such as clarifications to drawings or specifications, or quite large, dramatically affecting the scope of a project. Changes are a normal part of the construction process, as many factors must be considered during its duration. The owner, architect, and contractor must be aware of each change process and be equipped to deal with these changes professionally, objectively, and quickly. The contractor's field personnel must understand the process, be able to assess each situation accurately, and be able to work with the owner and architect regarding each change.

Changes can be made at no cost and with no change in duration, perhaps clarifying a construction detail or an installation method. Changes also can result in greater or lesser costs and require an addition or a reduction in contract time. A change order is a settlement for including a change in the contract, thus a change order normally costs the owner more than the work that was included in the original contract amount. The optimum price on construction work for the owner is contained in the original bid, if construction documents are completed within industry standards. Owner decisions on items that affect the scope of the contract should be made prior to the completion of construction documents and receipt of bids. Building a project from change orders is much like building an automobile from the auto parts store—the cost would be prohibitive. Change orders are not a means for the contractor to make enormous profits, but they reflect the cost of changing the installation during the project, with a specified mark-up on each item. The costs are fairly high, as most change order work must be done quickly and amended to the original work package. The specified mark-up for overhead and profit, particularly in public projects, may be more than the contractor used to bid the project under competitive market conditions. The change order should be viewed as a means for the owner to pay for project elements that were not included in the original documents, rather than as an opportunity for the contractor to make unreasonable profits.

Changes can occur for many reasons. A discussion follows.

Owner Directed Change of Scope. Extra work is added to or deducted from the project and it is clearly a change of scope to the contract. Owners will occasionally add or deduct portions of work that qualify as a change in scope. A change in the scope of the work is when the amount of work is changed from that contained in the contract, as indicated by the contract documents. A change order is almost always authorized for this kind of change, although the cost of the change will influence the owner's decision to accept it.

Constructive Change. The architect or owner representative causes the contractor to perform work outside his contract. This change could occur from a simple defective specification to a directed change in the contractor's method of accomplishing the work at hand. Construction document errors and omissions can fall into this category.

Consequential Change.   This occurs as a consequence of the preceding two changes. It is the impact cost that occurs on the other work that is being accomplished. Sometimes delays, resequencing and rework of other work, or the cost of extended overhead will occur. This type of change is generally the hardest to price, understand, and explain and sell to the owner and architect on the project.

Differing Site Conditions.   This also can be an area for a contract change. This change usually applies to differing subsurface conditions than are indicated by contract documents or are available from geotechnical reports. Occasionally, actual conditions will not appear as assumed by the designer and as indicated in the construction documents. For instance, many times in renovation projects the designer does not have all of the previous construction details and plans of the original project that existed prior to the redesign. Unknown, or latent, conditions can appear after the start of construction which can cause a change to the contract.

Jobsite Discovery of Hazardous Materials.   This often causes work to stop. Compliance to federal and state codes must be reviewed and adhered to when handling hazardous material for removal, storing, or dumping. Figure 13–1 is an example of a typical clause in the General Conditions to the Contract regarding the discovery of hazardous materials.

In this case, the contractor would notify the owner of any discovery of hazardous materials, as described in the specification. The owner then needs to decide on the best way of handling the material. A change order might be written to the contractor to remedy the problem. If the owner chooses the option to hire an outside contractor, as allowed in the speci-

FIGURE 13–1
Hazardous Material Clause, General Conditions of the Contract

4.5 Hazardous Materials

A. **Owner** shall be responsible for any Asbestos, PCBs, Petroleum, Hazardous Waste, or Radioactive Material uncovered or revealed at the site which was not shown or indicated in the Contract Documents to within the scope of the Work and which may present a substantial danger to persons or property exposed thereto in connection with the work at the site. The **Owner** will not be responsible for any such material brought to the site by the **Contractor**, Subcontractor, suppliers, or anyone else for whom the **Contractor** is responsible.

fication, the original contractor may request a change order for the costs and time associated with the necessary delay for remediation work.

Code Revisions.   Changes can occur as a result of outside agencies, such as the local building code authority, reviewing the project after the construction contract has been awarded. If the contract has been signed before the final plan check by the inspection agencies, change orders must be written to modify the change of scope caused by the review. The construction documents are often reviewed prior to pricing the project by contractors, but occasionally some reviews are made subsequent to issuance of the construction contract.

Vendor Coordination.   Additional problems can occur near the end of a project when the contractor is requested to coordinate with vendors that supply and install outside equipment. Often, initial drawing and submittal coordination is not adequate to finalize the installation and operation of the equipment. Most of this equipment being discussed is noted as N. I. C. (not in contract) or O. F. C. I. (owner furnished, contractor installed). Additional installations, such as mechanical piping, might be necessary to facilitate equipment installation.

Product Substitution.   Occasionally products specified are not available as market conditions shift. New products are developed, product lines change, and vendors and manufacturers drop one product for another. Sometimes the new product completely fills the requirements specified by the original item, but often it will take longer to obtain, cost more, and require a longer installation time, which can require changes in time and/or money as they can be a possible change of scope.

Change Orders.   The change of scope or the addition of extra work is generally a tangible type of change. Owners can basically see the extra materials being added to the project and understand that they are paying for and receiving something additional. The extra work will require a change in the documents or a design change. As a project progresses, owners' needs may change and they may want to change or add to the scope of the project. Construction contracts contain provisions allowing owners to make changes to the work. A contract clause providing owners with this right might be worded as shown in Figure 13–2.

It should be noted that the change is authorized by a written notice

FIGURE 13–2
Changes in the Work Clause, General Conditions of the Contract

---

Article 10 - Changes in the Work

10.1 General
     A. Without invalidating the Agreement and without notice to any surety, the owner may at any time or from time to time, order additions, deletions, or revisions in the work; these will authorized by a written Field Order and /or a Change Order issued by the Architect.

and not verbal communication, so both parties are provided with a clear definition of the work being required and who authorized the work. It is extremely important that the contractor know who can authorize a change order. This is an important issue that should be clarified at the pre-construction meeting.

Some contracts include unit prices for work that may be encountered during the project. For instance, the construction documents may request a unit price for solid rock removal beyond a specific quantity. Assume that the specification calls for 100 cubic yards of solid rock removal to be included within the contract price and requests a unit price for material beyond the 100 cubic yards. The specification would describe how the solid rock is to be measured, by bank cubic yards or loose cubic yards. The amount of rock removal over 100 cubic yards would be priced by the stipulated amount per cubic yard and incorporated in a change order.

In a unit price contract, where quantities are specified, compensation for variance with the specified quantities may be available. The contract clause in Figure 13–3 can allow adjustment for quantity variance in a unit price contract.

An allowable quantity variations clause in the contract gives the contractor additional protection for this type of change. This clause permits the renegotiation of a price when the quantity changes substantially, which in this case is defined as an increase or decrease in excess of 25 percent. It should be noted that not all unit price contracts allow for adjustment of quantities. Careful examination of the contract documents is always necessary to determine the methods allowed within the contract.

Change orders for design deficiencies, omission in the project documents, or errors in the specifications and drawings are fairly common. The process of visualizing, designing, and communicating all of the intricacies and details for a construction project is quite complicated. Architects and engineers, for the most part, produce complete and accurate construction documents. Some items, however, must be clarified or modified to meet project needs.

Occasionally during some projects, architects, engineers, and owners direct the contractor to alter the construction plan. This direction may be made to accelerate a certain portion of the project, delay the project,

FIGURE 13–3
Changes in the Work Clause, General Conditions of the Contract, Unit Price Contract

Article 10 - Changes in the work

10.2 Allowable Quantity Variations
    A. In the event of an increase or decrease in bid item quantity of a unit price contract, the total amount of work actually done or materials or equipment furnished will be paid for according to the unit price established for such work under the Contract Documents, wherever such unit price has been established; provided, that an adjustment in the contract Price may be made for changes which result in an increase or decrease in excess of 25 percent of the estimated quantity of any unit price bid item of work.

change a method of installation, or several other modifications of the contractor's plan or procedures. As these changes directly affect the cost of the construction bid by the contractor, compensation by change order may be necessary. The contractor's "means and methods" are considered to be under the contractor's control. **Means and methods** can be defined as the way in which the contractor constructs the designed product. Interference by the owner in these means and methods will result in extra costs. This type of change usually involves a dispute between the owner and contractor, resulting in difficulties in settlement by change order. The following example illustrates an owner-directed change concerning means and methods.

EXAMPLE

In a building project, the contractor's excavation subcontractor decided on using a track-type hydraulic excavator for excavation of a sewer line trench. The excavation subcontractor's bid was based on the productivity of the hydraulic excavator, which was owned by the excavation contractor. There were no restrictions in the construction documents on the type of equipment, access, or protection of existing surrounding areas. The means and methods of the trench excavation were to be the contractor's responsibility. As construction proceeded, the owner requested that the contractor use a smaller piece of excavation equipment, to minimize damage to the surrounding existing environment. Because the smaller piece of equipment was not owned by the excavation contractor, the rental rate exceeded the rate for the owned piece of equipment. The productivity of the smaller equipment was lower than the hydraulic excavator, requiring more time for the activity and resulting in higher labor costs. The excavation subcontractor requested a change order for an extra amount for a directed change in means and methods. Careful analysis and comparison was necessary to prove to the owner that the cost for the change was more than anticipated in the bid, as illustrated in Figure 13–4.

A **change order** is an adjustment to the contract amount and/or to the contract duration, approved, accepted, and signed by the owner, architect, and contractor. An adjustment to the contract duration is needed if the contract's schedule is impacted by the additional work or delay caused in considering the change order. Some change orders will not affect the contract duration. Many contracts contain a liquidated damages clause, which stipulates a specific amount of reimbursement to the owner for completion of the contract beyond the contract completion date. A

Item	Hydraulic Excavator	Rubber-tired Backhoe
Equipment	16 hrs @ $80/hr = $1280.	24 hrs @ $100/hr = $2400.
Labor	16 hrs @ $30/hr = $ 480.	24 hrs @ $ 30/hr = $ 720.
	16 hrs @ $20/hr = $ 320.	24 hrs @ $ 20/hr = $ 480.
Total Cost	$2080.	$3600.
Net Additional Cost		$1520.
Plus: 15% Allowable overhead & profit		$ 228.
Additional Cost Impact		$1748.

FIGURE 13–4
Cost Comparison, Excavation Change Order Example

FIGURE 13–5
Contract Comple-
tion Clause, General
Conditions of the
Contract

Article 2 - Contract times

The work shall be completed within 500 successive days from the commencement date state in the Notice to Proceed

Article 3 - Liquidated Damages

Owner and the contractor recognize that time is of the essence of this agreement and that the owner will suffer financial loss if the work is not completed within the time specified in Article 2 herein, plus any extensions thereof allowed in accordance with Article 12 of the General Conditions. They also recognize the delays, expense, and difficulties involved in proving in a legal proceeding the actual loss suffered by the owner if the work is not completed on time. Accordingly, instead of requiring any such proof, the owner and the contractor agree that as liquidated damages for delay (but not as a penalty) the contractor shall pay the owner $ 6,000.00 for each day that expires after the time specified in Article 2 herein.

liquidated damage is an amount stipulated in the contract document that states that the owner receives compensation for impacted costs of late completion of a project. Liquidated damages are usually stated in a monetary amount per day, such as $1,000. Liquidated damages are not considered a penalty to the contractor for late completion of the project, but rather compensation to the owner for impacted costs of not being able to use the facility when expected and contracted. If a change order affects the contractor's schedule, an extension of the appropriate number of days will be requested to ensure that liquidated damages are not imposed for the impacted period of time. Many contractors also will request payment for "impacted overhead," or jobsite overhead, for the extended contract period.

The General Conditions of the Contract clause, shown in Figure 13–5, is an example of a contract completion clause, including liquidated damages.

This example emphasizes that cost and time can be significant when related to liquidated damages. Considering the cost of jobsite overhead with liquidated damages, time is obviously an important factor in the construction project.

The Change Order Process

The change order process is normally detailed in the construction contract. The term **change order** can be defined as the specific document signed by the contractor and owner that authorizes an addition, a deletion, or a revision in the scope of the work. The change order usually will adjust the contract amount and/or the contract completion date. A no-cost change order can be used to change the scope of work only, to document changes from the contract documents. The sequence of events that follows is typical of the change order process, although some construction contracts will require specific time frames and slightly different sequences.

1. The contractor notices a difference between actual conditions and the conditions shown in the construction documents. The contractor notifies the architect via a Request For Information (RFI) of the varying condition and requests direction.
2. The architect must respond to the contractor regarding what needs to be done. This is normally done in Contract Clarification.
3. The contractor must determine if this clarification is considered additional work. If it is, the contractor must price the work as quickly as possible. The architect may prepare a proposal request, followed by the official Change Order Proposal (COP), which is completed by the contractor with the pricing and request for time extension.
4. If the architect agrees that the price is appropriate for the work, and if the owner agrees that a change order is necessary, the architect can prepare a change order. Work is not authorized to proceed until the change order is signed by both the owner and contractor. As the processing of a change order takes considerable time, the architect might issue a change directive, or field order for the contractor to proceed with the work immediately. This change directive specifies a price for the work, or a method to determine the price, such as cost-plus-a-fee. The owner must to sign the change directive. A change order may be a summary of several change directives.

The procedure described is a simplification and indicates accord by all parties. Construction contracts have a more involved procedure that considers all options. Figure 13–6 indicates the change order process under the AIA Document 201.

The process detailed in Figure 13–6 and in the General Conditions of the Contract, such as AIA Document A201, is critical to follow for successful change orders. The type of notification and conformance to the time frame are essential to a change order. Time is of the essence in this process. Authorization must be given for extra work to enable compensation for that work. It is important that all of the notifications, requests for proposals, pricing, and other change order paper work be executed promptly and accurately. Multiple changes occurring simultaneously impact field personnel with considerable project pressures, in both managing the work and documenting the changes. Despite the pressures, proper notification and authorization must be made for each change. Managing the project during these periods of pressure is still necessary. Pursuit of compensation for changes occasionally takes precedence over normal project management, resulting in lower productivity and profits for the

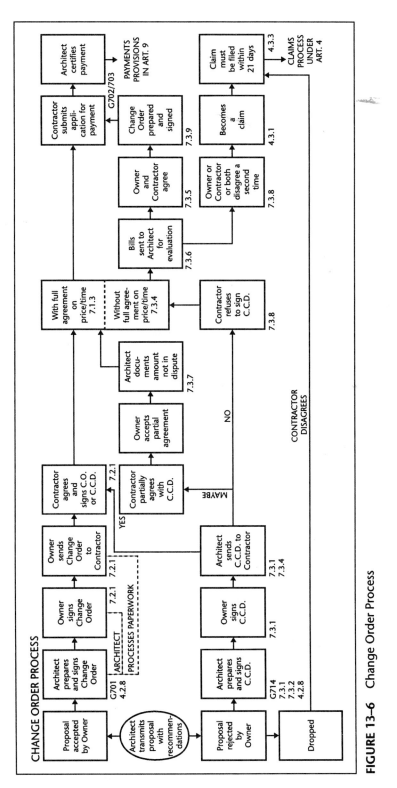

FIGURE 13–6   Change Order Process

contractor. A balance must be maintained during these periods, possibly with the addition of field personnel during these critical periods.

The change order process can be initiated by any of the parties involved, but must be made in writing. These notifications must be made within a certain time period and in a specified format. As the process continues, the owner and architect have opportunities to agree or disagree with the proposal from the contractor. If the owner and architect agree with the proposal by the contractor, the change order is processed. If the owner and architect disagree with the contractor's proposal, the contractor has three options: to revise the proposal, to withdraw the proposal, or to pursue the proposal as submitted. As this process continues, the change proposal becomes a claim. A **claim** is an unresolved change request. A claim, as defined by AIA Document A201, Section 4.3—Claims and Disputes, is a "demand or assertion by one of the parties seeking, as a matter of right, adjustment or interpretation of Contract terms, payment of money, extension of time, or other relief with respect to the terms of the Contract." During the process, the claim can be settled, resulting in a change order, without further processing. Figure 13–7 is an example of the written notification required by the contract to start the process.

This written claim must be made in a timely manner. The exact time period will be specified in the General Conditions to the Contract. AIA Document A201 requires that the claim be made within twenty-one days after recognizing the change. Most contractors have formal letters of notification containing specific language, so the contractor is not waiving any contractual rights, but fulfilling the notification time period as noted by the general conditions. Typical language in the body of the letter could read:

> You are hereby notified that the contractor has uncovered a change of scope which directly changes the contract work. This change potentially will have costs and time impact on our contract. Please acknowledge this change of scope by issuing a formal Change Order per contract documents.

The change order or claims process will not stop the work from continuing unless it greatly impacts the proceeding work. Under section 4.3.4, Continuing Contract Performance, of AIA Document A201, General Conditions of the Contract, the contractor is contractually obligated to proceed with the completion of the contract and the owner is obligated to continue making payments to the contracts. This area is often used to put pressure on both parties, the contractor threatening to stop work and the owner threatening to stop all payment. Contractually, neither party can refuse to uphold their part of the contract. Occasionally, both parties to the contract slow their obligations and use many tactics to influence the other.

A **Construction Change Directive (CCD)** can be used by the architect to keep work going and initiate a change order. The Construction Change Directive is defined as a written notice directing a change in the

FGH Construction Company
5390 Walnut Avenue
San Francisco, California
93422-0027

Phone: (415) 555-2346   Fax: (415) 555-2300

To: _____   Date: _____
_____   Job No.: _____
_____
_____

Subject: Notification of Change or Claim for: _____

Attn.: _____

Gentlemen:
The following <u>Request for Information, No. 45</u> has been determined by FGH Construction Company
to be beyond the scope of our contract. You are hereby notified that this problem may create a
suspension /delay of the work, increase scheduled time to complete the project, and/or cause additional
cost to our work. We reserve the right to request additional time and costs for this work. This
work could potentially have an adverse effect on other work being performed or that will be performed.

Description of Occurrence or Request for Information:
_____
_____
_____
_____
_____

Date of Occurrence   On: _____
                          From: _____   To: _____

You are hereby notified of our intention to seek recovery of all extra costs, including but not limited
to General Condition Expenses if this delay effects scheduled completion. It is our intent to minimize
the effect of this change.

Please issue the appropriate paperwork to complete this change. Thank you.

Sincerely,

Bill Jones
Project Engineer

cc: Frank Canteen
         Project Manager
         File CP - _____

Enclosures: _____

FIGURE 13–7   Change or Claim Notification

work. This document can be used before a written change order has been fully agreed to by all parties. Many times a written Change Order Proposal (COP) from the contractor is sent to the architect following the notice of claim and used to write the CCD or the initial proposal to the owner. The Construction Change Directive could look like the example shown in Figure 13–8.

It should clearly describe the additional work. It also should specify a method of determining payment by one of the following methods: a stipulated price; by unit price, including a method of measuring the units attributed to the change; or cost-plus-a-fee, when the method of measuring the cost and the fee are stipulated. Each of these price alternatives relates to the extent of the work known at the time of writing the Contract Change Directive.

The information required for the Change Order Proposal (COP) is specifically defined by the General Conditions and often modified by the Supplementary Conditions. AIA Document A201 Section 7.3.6 generally defines the information requirement needed in the COP. A more complete specification or the modification found in the Supplementary Conditions could read as shown in Figure 13–9 on page 338.

The fee for overhead and profit allowed in change orders is usually stated in the Contract Documents. The fee allowable may vary for different cost segments, as per the following example:

EXAMPLE

Actual overhead and profit percentages to be added to the following direct cost categories.

Labor	20 percent
Materials	15 percent
Equipment	15 percent
Subcontractor	10 percent

The type of costs for the change order are usually stipulated in the Contract Documents, excluding off-site overhead, which is to be covered by the fee.

The Change Order Proposal will vary from one construction company to another but the specific form will contain most of the same information. Figure 13–10 on page 339 is an example of a Change Order Proposal (COP) that could be adopted as a company standard.

All necessary subcontract work is included in the Change Order Proposal. The contractor must solicit the necessary pricing from each subcontractor and coordinate the proper pricing. The contractor must determine which subcontracts are impacted by the change and request pricing for that work. As many changes affect virtually all of the work on the project, careful examination of the impact of each change is necessary. Determination of the work required by the contract and the additional work is usually required as well. A letter must be sent to each affected subcontractor, including an accurate description of the scope of the change; an accurate description of the scope of change to the particular subcontract; the form of proposal required; the allowable mark-up

CONSTRUCTION CHANGE DIRECTIVE

CONSTRUCTION MANAGER-ADVISER EDITION

AIA DOCUMENT G714/CMa

(Instructions on reverse side)

OWNER	☐
CONSTRUCTION MANAGER	☐
ARCHITECT	☐
CONTRACTOR	☐
FIELD	☐
OTHER	☐

PROJECT:
(Name and address)

DIRECTIVE NO.:

DATE:

TO CONTRACTOR:
(Name and address)

PROJECT NOS.:

CONTRACT FOR:

CONTRACT DATE:

You are hereby directed to make the following change(s) in this Contract:

PROPOSED ADJUSTMENTS

1. The proposed basis of adjustment to the Contract Sum or Guaranteed Maximum Price is:

   ☐ Lump Sum (increase) (decrease) of $ _____ .

   ☐ Unit Price of $ _____ per _____ .

   ☐ as provided in Subparagraph 7.3.6 of AIA Document A201/CMa, 1992 edition.

   ☐ as follows:

2. The Contract Time is proposed to (be adjusted) (remain unchanged). The proposed adjustment, if any, is (an increase of _____ days) (a decrease of _____ days).

Signature by the Contractor indicates the Contractor's agreement with the proposed adjustments in Contract Sum and Contract Time set forth in this Construction Change Directive.

CONTRACTOR

Address

_____

BY _____

DATE _____

When signed by the Owner, Construction Manager and Architect and received by the Contractor, this document becomes effective IMMEDIATELY as a Construction Change Directive (CCD), and the Contractor shall proceed with the change(s) described above.

OWNER	CONSTRUCTION MANAGER	ARCHITECT
Address	Address	Address
BY _____	BY _____	BY _____
DATE _____	DATE _____	DATE _____

CAUTION: You should use an original AIA document which has this caution printed in red. An original assures that changes will not be obscured as may occur when documents are reproduced.

AIA DOCUMENT G714/CMa • CONSTRUCTION CHANGE DIRECTIVE • CONSTRUCTION MANAGER-ADVISER EDITION • 1992 EDITION • AIA® • ©1992 • THE AMERICAN INSTITUTE OF ARCHITECTS, 1735 NEW YORK AVENUE, N.W., WASHINGTON, D.C. 20006 5292 • WARNING: Unlicensed photocopying violates U.S. copyright laws and will subject the violator to legal prosecution.

G714/CMa-1992

FIGURE 13–8 Construction Change Directive

FIGURE 13-9
Change Order Pro-
posal Requirements,
Supplementary
Conditions to the
Contract

Article 14 - Changes in the work:

....

1. Additive changes:
   a. Direct labor costs including foreman.
   b. Direct cost of materials to be entered into the work.
   c. ....
2. Deductive changes:

   ....
3. Additive changes and deductive changes together:

   ....

4. Changes under $ 500.00:
   If the description of the change in the work is, in the opinion of the architect, definitive enough for the architect to determine fair value and the total of the change does not exceed $ 500.00, no breakdown is required.

5. Changes between $ 500.00 and $ 1,500.00:
   If the description of the change in the work is, in the opinion of the architect, definitive enough for the contracting authority to determine fair value the breakdown shall consist of the following:
   a. Lump sum Labor
   b. Lump sum Material
   c. Lump sum equipment usage
   d. Appropriate overhead and profit as defined in 1.e. and 1.f.

specified in the contract documents and the subcontract agreement; and other factors affecting the change. A formal request for cost information must be worded carefully to avoid committing the general contractor to a change until approved by the owner. Figure 13–11 on page 340 is a sample letter requesting a change proposal from a subcontractor.

When using unit-price or cost-plus determination of the change order price, measurement procedures are necessary. Unit-price determination may require verification of the units affected, such as a truck count verified by the owner's inspector. Daily work reports, listing labor hours and material received, may be required for cost-plus changes. The daily work reports will probably also require verification from the owner's inspector. The form shown in Figure 13–12 on page 341 could be used for cost-plus or "force account" work.

Payment for additional work is not allowed until the official change order is processed. Work description and the firm price for the work must be determined, and the change order has to be signed by both the owner and contractor. This firm price is then included in the contract amount in the progress payment request, with payments related to the completion of the change order work.

Change Order Proposal
No. _____

Project: _____
Date: _____

To: _____
_____
_____

From: _____
_____
_____

FGH Construction Co., wishes to submit the following Change Order Proposal for the change in scope of the contract work as described:

_____
_____
_____
_____

Cost of the work:

Description	Labor	Mat.	Equip.	Other	Sub	Total

Labor Burden _____%   _____   _____
          Subtotal   _____
Bond Premium   _____%   _____
Liability Insurance   _____%   _____
Subtotal   _____

Overhead   _____
Profit   _____

Grand Total   _____

Schedule Extension:

          Calendar Days to be added to the contract time: _____

FGH Construction will proceed on this change when authorized by _____ in writing. This information is time sensitive and is valid for 10 days without review.

Signed: _____
Title: _____
FGH Construction Co.

FIGURE 13–10   Example Change Order Proposal

FGH Construction Company
5390 Walnut Avenue
San Francisco, California
93422-0027

Phone: (415) 555-2346   Fax: (415) 555-2300

To: _____   Date: _____
_____   Job No.: _____
_____
_____

Subject: CP - __
            Change Order Proposal for: _____

Attn.: _____

Gentlemen:

Please submit to the above address your firm change cost proposal encompassing your itemized Labor/Material/Jobsite Overhead/Markup as required by the specification. This cost proposal should include all additions and deletions.

For your portions of the work according to the following:

_____
_____
_____
_____
_____

In the interest of saving time, submit this proposal first by fax no later than _____ with your written confirmation to follow by mail.

This change cost proposal request is not authorization to proceed at this time so it is imperative that we have your proposal for review and approval by the above date. Proposals are subject to review and approval by _____ (owner/architect) when it is incorporated into FGH Construction Company's proposal on the above mentioned project for this proposed change.

Please refer to Number CP - _____ on all future correspondence related to this proposed cost change proposal.

Sincerely

Bill Jones
Project Engineer

cc: Frank Canteen
         Project Manager
         File CP - _____

         Enclosures: _____

FIGURE 13–11   Sample Letter to Subcontractor Requesting Change Proposal Information

FGH Construction Company

5390 Walnut Avenue
San Francisco, California
93422-0027

Phone: (415) 555-2346   Fax: (415) 555-2300

Daily Time and Material Work Form

Subcontractor: _____   Date: _____
_____   Job No.: _____
_____   CP No.: _____

Description of Work: _____
_____
_____
_____

Work In Progress: _____
Work Completed: _____

Payment for the above described work will be based upon actual cost for labor, materials, equipment, and jobsite overhead (discribed in specifications and subcontract)

Labor:

Name	Craft	Journeyman Apprentice Foreman	Man-Hours

Materials Used:

_____
_____
_____

Equipment Used:

_____
_____
_____

_____   _____
FGH Const. - Superintendent   Subcontractor

FIGURE 13–12   Force Account Work Form

Time Extension

The time extension requested by the contractor depends on a variety of factors. Extra work usually requires extra time to the contract. Analysis of the project schedule is necessary to show the impact of this additional work. The project may be delayed in determining what change is necessary, which also should be included in the requested time extension. Most time extension requests are fairly subjective, as the true impact of the change is rarely clearly determined. Discussion and negotiation of the time extension is a frequent occurrence in the change order process.

Some time extensions are due to delays in the project caused by outside influences. Most contracts allow extension of contract time for delays due to "labor disputes, fire, unusual delay in deliveries, unavoidable casualties, or other causes beyond the contractor's control." Delay to the project due to unusual weather often is compensated by a time extension. Comparison of the average weather conditions to the actual weather conditions is made to determine the appropriate time extension. The severity of the weather also is examined. As mentioned in chapter 4, daily documentation of weather and construction activity is essential in determining appropriate time extension. Delays in the project from unexpected sources also will extend the contractor's jobsite overhead, resulting in a request for compensation for this extended overhead cost.

Documentation of Changes

A file should be established for each possible change, as defined by the change order proposal. This file should include all relevant documentation to the change, including photographs, daily reports, cost records, subcontractor correspondence, excerpts from meeting minutes, correspondence with the architect or owner, drawings details, applicable excerpts from the specification and contract, and other applicable information. Some computerized project documentation systems have a search function for all items related to a particular issue. This information should be assembled and kept for future reference.

Several change orders may be in processing at the same time during the project. As this involves many steps, it is extremely important to track all change proposals and change orders. A change order log records the dates of the steps in processing the change order. This log can be used to determine the current location of the change order. Clear knowledge of the responsibilities of the process helps speed up the system. Recording the dates of the processing also is necessary in claims procedures. Project meetings should involve updating the log, resulting in an updated version of the change order log being distributed with meeting minutes.

Each contractor will include different information in the change order log. Some typical information that might be included follows:

- change order proposal number
- change order number (after a formal change is issued)

- description
- original date of the C. O. P.
- who initiated the change (individual and company)
- price associated with the C. O. P.
- number of calendar days for extension of contract requested
- subcontractors that are affected, and dates associated with pricing
- date that formal change order was issued for signatures
- status (approved or rejected), and date status was determined
- reprocessing dates, if required
- actual amount of the change order
- date change order approved

An example of a change order log is contained in chapter 4. The contractor must be able to use this log as a tool to facilitate the processing of the change orders. The log also will serve as a record of the processing. The information contained in the log for use in processing may be a bit different than the information used as a record. Change order logs can be formatted and developed on a computer spreadsheet. By using a computerized version of the change order log, different types of information can be made and completed information can be left off if not required. As part of the change order control process, the general contractor must have a formal system for subcontract change control. On large projects, separate logs are kept of each of the major subcontractors within each correspondence file. This log must track all correspondence that relates to the change.

The formal change order should contain a complete description of the change, including additional drawings, sketches, and specification descriptions, as necessary. A detailed description of the change order will be referenced during the project, defining the work to be accomplished. AIA Document G701 is used as the formal change order when using an AIA Contract. This form, when filled out, keeps a continual cost update of the contract. This same information will show up on the application for payment. A special emphasis should note the line for change of contract time. "The Contract Time will be (increased) (decreased) (unchanged) by (____) days. The Date of Substantial Completion as of the date of this Change Order therefore is _____." This document is also signed by all three team members—the architect, contractor, and owner. AIA Form G701, Change Order, is shown in Figure 13–13.

Implementation of Change Orders

The contractor's project team has not completed their work even after receiving a signed change order. Issuance of change orders to the subcontracts affected must also be accomplished. Many times the actual subcontractor change order still must be finalized and negotiated. If the contractor has asked for cost and time input from the subcontractor at each negotiation stage, from the original change order proposal to the

CHANGE ORDER

CONSTRUCTION MANAGER-ADVISER EDITION

AIA DOCUMENT G701/CMa

(Instructions on reverse side)

OWNER	☐
CONSTRUCTION MANAGER	☐
ARCHITECT	☐
CONTRACTOR	☐
FIELD	☐
OTHER	☐

PROJECT:
(Name and address)

CHANGE ORDER NO.:

INITIATION DATE:

TO CONTRACTOR:
(Name and address)

PROJECT NOS.:

CONTRACT FOR:

CONTRACT DATE:

The Contract is changed as follows:

Not valid until signed by the Owner, Construction Manager, Architect and Contractor.

The original (Contract Sum) (Guaranteed Maximum Price) was . . . . $

Net change by previously authorized Change Orders . . . . $

The (Contract Sum) (Guaranteed Maximum Price) prior to this Change Order was . . . . $

The (Contract Sum) (Guaranteed Maximum Price) will be (increased) (decreased) (unchanged) by
this Change Order . . . . $

The new (Contract Sum) (Guaranteed Maximum Price) including this Change Order will be . . . . $

The Contract Time will be (increased) (decreased) (unchanged) by . . . .   (   ) days

The date of Substantial Completion as of the date of this Change Order therefore is . . . .

NOTE: This summary does not reflect changes in the Contract Sum, Contract Time or Guaranteed Maximum Price which have been authorized by Construction Change Directive.

CONSTRUCTION MANAGER _____   ARCHITECT _____

ADDRESS _____   ADDRESS _____

BY _____   DATE _____   BY _____   DATE _____

CONTRACTOR _____   OWNER _____

ADDRESS _____   ADDRESS _____

BY _____   DATE _____   BY _____   DATE _____

AIA   CAUTION: You should use an original AIA document which has this caution printed in red. An original assures that changes will not be obscured as may occur when documents are reproduced.

FIGURE 13–13   Change Order, Construction Manager-Adviser Edition

finalized change order completion, then a finalized subcontract change order will not be hard to accomplish. This becomes a difficult process if the contractor has negotiated different terms with the owner than proposed by the subcontractor.

Summary

Changes to the contract require careful processing by the contractor. Prompt recognition of the change and its impact is important in avoiding unnecessary costs and delays. Cooperation between the architect, owner, contractor, and subcontractor also is necessary to continue construction without impact by changes.

Specific procedures are contained in each contract relating to the process flow of change requests, change orders, and claims. It is essential that these procedures are followed, complying with the requirements and time frame. In-depth knowledge of the process is necessary for all contractor's personnel involved with the project.

A variety of forms and correspondence is necessary in the change order process. Some contract document forms, such as the AIA documents, provide a complete set of forms for the process, including proposals, directives, and change orders. Proper use of these documents provides a uniform and systematic way of dealing with changes to the project. Documentation of each change order enables the contractor to deal completely with each change. This documentation also is necessary if the change order is not accepted and if a claim is pursued.

The contractor must manage the subcontractors' proposals and changes to subcontract amounts, as well as the changes to the contract with the owner. Careful record keeping is necessary to keep track of all of these changes.

Changes can be managed without producing claims, which are unresolved change orders. Pursuit of claims can be expensive, involving arbitration or litigation. Honest, prompt, efficient, and cooperative processing of change orders will bring the project to a successful close, without further disagreement on contract claims.

CHAPTER 14

PROGRESS PAYMENTS

CHAPTER OUTLINE   The Schedule of Values

Unit Price Contracts

Project Cash Flow Projections

Progress Payment Procedures

Payment Processing

Summary

Proper management of payments for construction work is a very important responsibility of project administration for a contractor. An understanding of the process is necessary to efficiently manage the payment process. Prompt, accurate, and efficient processing of the payment request can literally save the contractor thousands of dollars on each project. A brief payment processing period will reduce interim financing costs incurred by the contractor.

Several methods of payment are available for construction work, depending on the project delivery method, the owner's financial situations, and the arrangements made for payment and payment frequency. The most common method of payment in commercial building construction, in the traditional owner-contractor contract, is by monthly **progress payments**, which are partial payments for work completed during a portion, usually a month, of the construction period. By agreement, this payment frequency can be modified.

Payment for the project at the completion of the project is also done in some arrangements. This arrangement can be termed **turnkey**, as the project is delivered to the owner and payment is made for the project. In this type of arrangement, interim financing costs are included in the project cost. Speculative construction also normally requires payment after the project is completed. Although these arrangements require payment after the construction, accounting for the percentage complete at periodic periods is still necessary, usually to justify financing payments.

As the monthly progress payment is the most common method of payment in commercial building construction, this chapter will primarily deal with procedures concerning these periodic payments

Progress payments reflect payment for the work in place and material delivered during the applicable period. Under the progress payment method, an estimate of the work complete is made and payment is made for that amount. Most contracts deduct an amount for **retainage** from the amount earned during the period. Retainage is an amount held by the owner until the final completion of the contract. It can be held for a number of reasons, depending on the owner and the owner's contracting rules. In some cases, retainage is held by the owner to complete the project, if necessary. Some owners, particularly public owners, retain an amount as a reserve fund to cover materialman and subcontractor liens on the project. Varying amounts of retainage are held by owners, usually 5 percent or 10 percent. This amount is expressed in the contract documents and is established in the agreement between the owner and contractor. Some contracts do not retain funds, as the contract's performance and payment bond is intended for the same purpose as retainage. When retainage is held, it is normally released in the final payment, when the contract requirements are completed.

Material delivered to the jobsite, even when not actually incorporated into the work, is usually included in the progress payment. Special care is normally taken to inventory the material stored at the jobsite. Some owners will also include material stored away from the jobsite in the progress payment, with special provisions. It is necessary to assure the owner that the material is insured and will be installed during the project. Some reluctance may occur on the owner's part in paying for material stored off-site that could be used for other projects, such as non-fabricated and generic material and lumber. The owner's primary concern with stored material is whether it will actually be incorporated into the project.

Varying methods of inspection and verification are done by the owner to determine accuracy of the payment request. This verification process can vary from a meticulous measurement of work in place and material stored to an approximation of the percentage completed for the activities. Owners and their agents, such as architects, engineers, and construction managers, are concerned that the proper amount be paid, without overpayment.

The actual payment period varies from contract to contract. The frequency of payment is generally on a one month, or 30-day, basis, but the actual cutoff dates can vary. Generally, the payment period extends from the first to the last day of the month. There are, however, many variations regarding the exact start and cutoff date, depending on the owner's processing requirements. If the owner requires the verified payment on the first day of the month, the actual cutoff date may be the twentieth day of the month to facilitate compilation and review the progress payment request. This type of payment period is relatively easy to incorporate into the project procedures, however each supplier and subcontractor must be made aware of the cutoff dates used for the project.

The Schedule of Values

The **schedule of values** is a written list in tabular form of the value amounts relating to the activities of work, associated with appropriate monetary amounts for each activity. The schedule of values is not time-related, but is merely a list of the value of the construction activities, upon which progress payments are based.

The items for the schedule of values are often prescribed in the Project Manual and are selected to accurately monitor the progress of the project. The activities listed should be adequately detailed to facilitate estimation of the percentage of completion of the item. Too many items in the schedule of values unnecessarily complicate the payment process. Some projects use the activities in the construction progress schedule as the schedule of values. This method creates a large number of payment activities. It is the authors' opinion that use of the construction schedule activities as the schedule of values for payment requests emphasizes the payment schedule, rather than using it as a planning tool during construction. A separate schedule of values separates the management of payment request processing and construction scheduling.

The schedule of values normally will list construction activities that are identifiable as a system or installation. The list in Figure 14–1 shows two fairly typical schedules of values, with varying detail.

Example 1 is a broadscope example of detail in the schedule of values. Example 2 takes the broadscope of items in Example 1 and divides them into further detail, matching actual work activities. Example 2 provides a more accurate activity breakdown for estimating the amount completed in a given period. Estimates of actual progress will be more accurate using the detailed breakdown in Example 2.

The values for these construction activities normally include all material, labor, equipment, subcontracts, and distributed overhead and profit for each item.

Several items are optional for inclusion in the schedule of values, depending on the customs of the owner and architect and engineer. Some owners prefer that all overhead items be distributed into the construction activities. Others realize that some direct overhead items should be included in the schedule of values, as they relate to specific expenses incurred by the contractor. Some of these items include:

- *Mobilization:* A number of costs are incurred by the contractor at the very start of a project. Some relate to temporary facilities, including transportation and setup of office and storage facilities for the contractor and subcontractors; temporary security fencing; installation of temporary utilities, such as power, water, and telephone; and access routes for the jobsite. Some of these initial costs include immediate payments required for performance and payment bonds, builder's risk insurance, and building permits. Some contracts treat the building permit as a reimbursable cost, paid by the owner when incurred by the contractor. Some owners refuse to pay for the performance bond initially and feel that it should be distributed evenly throughout the contract.

FIGURE 14–1
Comparison of Detail in a Schedule of Values

Example 1		Example 2	
Sitework	$ 50,000.00	Mass Excavation	$15,000.00
		Structural Excavation	$20,000.00
		Landscaping	$ 5,000.00
		Asphalt Paving	$ 7,000.00
		Concrete Walks	$ 3,000.00
Concrete	$125,000.00	Concrete Walls	$75,000.00
		Concrete Slabs	$35,000.00
		Precast Concrete	$15,000.00
Masonry	$ 80,000.00	Masonry	$80,000.00

- *General conditions:* General conditions costs normally relate to direct (jobsite) overhead. The following items could be included in the general conditions value: cost of supervisory personnel, such as superintendent and field engineers, which would not be included in the work activities; monthly costs for temporary facilities; cleaning costs, both in progress and final; jobsite trucks and equipment; and miscellaneous direct overhead. These costs would be paid in monthly progress payments, as the costs would be incurred.
- *Punch list:* Many owners and architects and engineers include an additional item for punch list items. The **punch list** is a list of items that must be completed or repaired prior to final completion. Without the punch list value, all of the money in the project is claimed without anything left to cover the cost of completing the punch list. The project retainage often is not available to finish the punch list items, as it may be reserved for liens or reinvested in an interest-bearing escrow account. The punch list items are actually part of the work activities and the amount to cover the list should be deducted from the work activity values. Many owners will assign a percentage of the contract amount to the punch list, usually from 1 to 5 percent.

The contractor's estimate is normally not arranged into the necessary items for a schedule of values. The estimate probably has more items than are required for the schedule of values. The contractor's estimate, which is confidential to the contractor, includes explicit items for indirect overhead and profit, included in the line items in the schedule of values. It is not customary for the contractor to reveal the amount of profit or indirect overhead in the project. As the estimate detail is different than the detail in the schedule of values, computation of the schedule of values is necessary. This computation involves the combination of estimate items relating to the schedule of values items; the separation of some lump sum items, particularly subcontractor bids, into two or more schedule of value items; and the distribution of overhead and profit into the schedule of value items. Figures 14–2 through 14–7 illustrate the translation of the estimate into the schedule of values.

Figure 14–2: Example Estimate Summary Sheet. This is a typical contractor's estimate summary sheet, following the CSI MasterFormat, and showing the costs for labor, material, equipment, and subcontracts for each item. It is more detailed than required for a schedule of values. The overhead and profit are shown as line items. The information from this summary sheet should be reorganized into a schedule of values.

Figure 14–3: Example Schedule of Values Computations. This figure shows the combination of items from the estimate summary sheet (Figure 14–2) into items for the schedule of values. The requirements for the schedule of values on this particular project requires a punch list item of 1 percent of the contract amount. One percent of each schedule of value item has been deducted and combined into a punch list line item.

FIGURE 14–2
Example Estimate
Summary Sheet

Example Estimate Summary Sheet

Section	Description	Labor	Material	Equipment	Sub-Contract	Total
00600	Bonds					$15,600
00650	All-Risk Insurance					$1,600
01050	Field Engineering	$2,000				$2,000
01450	Superintendent	$52,200				$52,200
01500	Temporary Facilities	$2,000	$2,000	$5,000	$1,000	$10,000
01710	Final Cleaning	$2,000	$300	$250		$2,550
02200	Earthwork				$45,000	$45,000
02510	Asphalt Paving				$27,600	$27,600
02900	Landscaping				$22,000	$22,000
03100	Concrete Formwork	$2,000	$1,300		$25,000	$28,300
03200	Concrete Reinforcing		$9,000		$10,500	$19,500
03300	Cast-In-Place Concrete	$4,000	$22,400	$2,500		$28,900
04000	Masonry				$65,000	$65,000
05100	Structural Steel	$4,500	$67,000		$45,000	$116,500
05200	Metal Joists		$34,000			$34,000
05300	Metal Deck		$23,000			$23,000
05520	Metal Handrails	$3,500	$4,000			$7,500
06100	Rough Carpentry	$1,200	$1,300			$2,500
06200	Finish Carpentry	$23,000	$65,000			$88,000
07150	Dampproofing	$600	$1,000			$1,600
07200	Insulation				$22,000	$22,000
07250	Fireproofing				$16,500	$16,500
07500	Roofing & Flashing				$65,000	$65,000
07900	Joint Sealants				$9,600	$9,600
08100	Metal Doors & Frames	$7,500	$34,000			$41,500
08350	Folding Doors	$2,500	$4,300			$6,800
08500	Metal Windows				$25,600	$25,600
08700	Finish Hardware		$23,000			$23,000
08800	Glazing				$9,340	$9,340
09250	Metal Studs/Gyp.Bd.				$102,000	$102,000
09310	Ceramic Tile				$16,500	$16,500
09510	Acoustical Ceilings				$35,000	$35,000
09600	Floor Covering				$43,000	$43,000
09900	Painting				$56,000	$56,000
10100	Tackboards	$1,000	$3,000			$4,000
10150	Toilet Compartments	$2,000	$4,000			$6,000
10400	Signage	$2,000	$5,600			$7,600
10520	Fire Extinguishers	$500	$4,500			$5,000
10800	Toilet Accessories	$2,400	$4,350			$6,750
15000	Mechanical				$123,000	$123,000
16000	Electrical				$104,000	$104,000
	Subtotal					$1,321,540
	3 % Indirect Overhead					$39,646
						$1,361,186
	5 % Profit					$68,059
	Total Bid					$1,429,246

FIGURE 14–3
Example Schedule of
Values Computations

Example Project, Schedule of Values Computations

SOV Item	Estimate Item	Amount		Less 1%
Mobilization	Bonds	$15,600		
	All-Risk Insurance	$1,600		
	Temporary Facilities	$10,000		
			$27,200	$26,928
General Conditions	Field Engineering	$2,000		
	Superintendent	$52,200		
	Final Cleaning	$2,550		
			$56,750	$56,183
Earthwork	Earthwork	$45,000	$45,000	$44,550
Site Improvements	Asphalt Paving	$27,600		
	Landscaping	$22,000		
			$49,600	$49,104
Concrete	Concrete Formwork	$28,300		
	Concrete Reinforcing	$19,500		
	Cast-in-place Concrete	$28,900		
			$76,700	$75,933
Masonry	Masonry	$65,000	$65,000	$64,350
Structural Steel	Structural Steel	$116,500		
	Metal Joists	$34,000		
	Metal Deck	$23,000		
	Metal Handrails	$7,500		
			$181,000	$179,190
Carpentry	Rough Carpentry	$2,500		
	Finish Carpentry	$88,000		
			$90,500	$89,595
Dampproofing	Dampproofing	$1,600	$1,600	$1,584
Insulation	Insulation	$22,000		
	Fireproofing	$16,500		
			$38,500	$38,115
Roofing/Flashing	Roofing/Flashing	$65,000	$65,000	$64,350
Sealants	Sealants	$9,600	$9,600	$9,504
Doors, Hardware	Metal Doors & Frames	$41,500		
	Folding Doors	$6,800		
	Finish Hardware	$23,000		
			$71,300	$70,587
Windows, Glazing	Metal Windows	$25,600		
	Glazing	$9,340		
			$34,940	$34,591
Interior Parititons	Metal studs/Gyp.Bd.	$102,000	$102,000	$100,980
Acoustical Ceilings	Acoustical Ceilings	$35,000	$35,000	$34,650
Ceramic Tile	Ceramic Tile	$16,500	$16,500	$16,335
Resilient Flooring	Floor Covering (50%)	$21,500	$21,500	$21,285
Carpet	Floor Covering (50%)	$21,500	$21,500	$21,285
Painting	Painting (85%)	$47,600	$47,600	$47,124
Wallcovering	Painting (15%)	$8,400	$8,400	$8,316
Specialties	Tackboards	$4,000		
	Toilet Compartments	$6,000		
	Signage	$7,600		
	Fire Extinguishers	$5,000		
	Toilet Accessories	$6,750		
			$29,350	$29,057
Plumbing	Mechanical (40%)	$49,200	$49,200	$48,708
HVAC	Mechanical (45%)	$55,350	$55,350	$54,797
Fire Protection	Mechanical (15%)	$18,450	$18,450	$18,266
Electrical Power	Electrical (25%)	$26,000	$26,000	$25,740
Electrical Circuits	Electrical (20%)	$20,800	$20,800	$20,592
Lighting Fixtures	Electrical (35%)	$36,400	$36,400	$36,036
Fire Alarm	Electrical (20%)	$20,800	$20,800	$20,592
Punch List				$13,215
				$1,321,540

The total for this worksheet indicates the project costs, without mark-up added at this point.

Figure 14–4: Example Schedule of Values Computations With Evenly Distributed Overhead and Profit. This figure is a progression from Figure 14–3, adding distribution of the overhead and profit evenly and proportionately to each schedule of values item. The right column, labeled "Value," will be the value shown on the schedule of values. The total of this column is the contract amount.

Figure 14–5: Schedule of Values, Evenly Distributed Overhead. This is the schedule of values that would be submitted to the owner for inclusion in the progress payment documentation. The architect and owner typically would not be aware of the computations accomplished to arrive at this schedule of values. (See Figure 14–5 on page 356.)

Figure 14–6: Schedule of Values Computations With Front-Loaded Distribution of Profit and Overhead. Many contractors prefer to distribute their profit and overhead to activities that will be completed in the early phases of the project. In this computation, profit and overhead are distributed only to Mobilization, Earthwork, Concrete, and Masonry items. (See Figure 14–6 on page 357.)

Figure 14–7: Schedule of Values, Front-loaded Distribution of Profit and Overhead. This would be the schedule of values submitted to the owner (see Figure 14–7 on page 358). In comparison with the schedule of values shown in Figure 14–5, the following conclusions can be made:

> The total amount is the same for either schedule of values. Mobilization, Earthwork, Concrete, and Masonry activities are higher on the front-loaded schedule of values. All other items are slightly higher on the even distribution than in the front-loaded version.

Preparation of the schedule of values from the estimate can include the following steps:

1. Combination of items: The schedule of values may consolidate several line items from the estimate for clarity in the schedule of values. For example, the schedule of values may list "Concrete" as a single item. In the contractor's estimate, several items relate to concrete. In the estimate summary shown in Figure 14–2, three line items relate to "Concrete":

Section 03100 Concrete Formwork	$28,300.00
Section 03200 Concrete Reinforcing	$19,500.00
Section 03300 Cast-in-Place Concrete	$28,900.00

   These three line items from the estimate total $76,700 for "Concrete."

2. Separation of lump sum items: Some items, particularly subcontractor bids, combine several items into a lump sum in the estimate. The schedule of values may require further breakdown of these items. For example, the schedule of values requires an item for Resilient Flooring and another item for Carpet. The estimate has one item, a subcon-

FIGURE 14–4
Example Schedule of Values Computations, Evenly Distributed Markup

Example Project, Schedule of Values Computations				Evenly Distributed Overhead		
SOV Item	Estimate Item	Amount	Less 1%	% of Total OH	Distributed	Value
Mobilization	Bonds	$15,600				
	All-Risk Insurance	$1,600				
	Temporary Facilities	$10,000				
		$27,200	$26,928	2.04%	$2,195	$29,123
General Conditions	Field Engineering	$2,000				
	Superintendent	$52,200				
	Final Cleaning	$2,550				
		$56,750	$56,183	4.25%	$4,579	$60,761
Earthwork	Earthwork	$45,000	$45,000	3.37%	$3,631	$48,181
Site Improvements	Asphalt Paving	$27,600				
	Landscaping	$22,000				
		$49,600	$49,104	3.72%	$4,002	$53,106
Concrete	Concrete Formwork	$28,300				
	Concrete Reinforcing	$19,500				
	Cast-in-place Concrete	$28,900				
		$76,700	$75,933	5.75%	$6,189	$82,122
Masonry	Masonry	$65,000	$65,000	4.87%	$5,245	$69,595
			$64,350			
Structural Steel	Structural Steel	$116,500				
	Metal Joists	$34,000				
	Metal Deck	$23,000				
	Metal Handrails	$7,500				
		$181,000	$179,190	13.56%	$14,604	$193,794
Carpentry	Rough Carpentry	$2,500				
	Finish Carpentry	$88,000				
		$90,500	$89,595	6.78%	$7,302	$96,897
Dampproofing	Dampproofing	$1,600	$1,600	0.12%	$129	$1,713
Insulation	Insulation	$22,000				
	Fireproofing	$16,500				
		$38,500	$38,115	2.88%	$3,106	$41,221
Roofing/Flashing	Roofing/Flashing	$65,000	$65,000	4.87%	$5,245	$69,595
Sealants	Sealants	$9,600	$9,600	0.72%	$775	$10,279
			$9,504			
Doors, Hardware	Metal Doors & Frames	$41,500				
	Folding Doors	$6,800				
	Finish Hardware	$23,000				
		$71,300	$70,587	5.34%	$5,753	$76,340
Windows, Glazing	Metal Windows	$25,600				
	Glazing	$9,340				
		$34,940	$34,591	2.62%	$2,819	$37,410
Interior Parititons	Metal studs/Gyp.Bd.	$102,000	$102,000	7.64%	$8,230	$109,210
			$100,980			
Acoustical Ceilings	Acoustical Ceilings	$35,000	$35,000	2.62%	$2,824	$37,474
			$34,650			
Ceramic Tile	Ceramic Tile	$16,500	$16,500	1.24%	$1,331	$17,666
			$16,335			
Resilient Flooring	Floor Covering (50%)	$21,500	$21,500	1.61%	$1,735	$23,020
			$21,285			
Carpet	Floor Covering (50%)	$21,500	$21,500	1.61%	$1,735	$23,020
			$21,285			
Painting	Painting (85%)	$47,600	$47,600	3.57%	$3,841	$50,965
			$47,124			
Wallcovering	Painting (15%)	$8,400	$8,400	0.63%	$678	$8,994
			$8,316			
Specialties	Tackboards	$4,000				
	Toilet Compartments	$6,000				
	Signage	$7,600				
	Fire Extinguishers	$5,000				
	Toilet Accessories	$6,750				
		$29,350	$29,057	2.20%	$2,368	$31,425
Plumbing	Mechanical (40%)	$49,200	$49,200	3.69%	$3,970	$52,678
			$48,708			
HVAC	Mechanical (45%)	$55,350	$55,350	4.15%	$4,466	$59,262
			$54,797			
Fire Protection	Mechanical (15%)	$18,450	$18,450	1.38%	$1,489	$19,754
			$18,266			
Electrical Power	Electrical (25%)	$26,000	$26,000	1.95%	$2,098	$27,838
			$25,740			
Electrical Circuits	Electrical (20%)	$20,800	$20,800	1.56%	$1,678	$22,270
			$20,592			
Lighting Fixtures	Electrical (35%)	$36,400	$36,400	2.73%	$2,937	$38,973
			$36,036			
Fire Alarm	Electrical (20%)	$20,800	$20,800	1.56%	$1,678	$22,270
			$20,592			
Punch List			$13,215	1.00%	$1,075	$14,290
			$1,321,540	100%	$107,704	$1,429,246
Markup to Distribute	$107,706					

FIGURE 14–5
Schedule of Values,
Evenly Distributed
Overhead

Example Project, Schedule of Values
Evenly Distributed Overhead

Item	Value
Mobilization	$29,123
General Conditions	$60,761
Earthwork	$48,181
Site Improvements	$53,106
Concrete	$82,122
Masonry	$69,595
Structural Steel	$193,794
Carpentry	$96,897
Dampproofing	$1,713
Insulation	$41,221
Roofing/Flashing	$69,595
Sealants	$10,279
Doors, Hardware	$76,340
Windows, Glazing	$37,410
Interior Parititons	$109,210
Acoustical Ceilings	$37,474
Ceramic Tile	$17,666
Resilient Flooring	$23,020
Carpet	$23,020
Painting	$50,965
Wallcovering	$8,994
Specialties	$31,425
Plumbing	$52,678
HVAC	$59,262
Fire Protection	$19,754
Electrical Power	$27,838
Electrical Circuits	$22,270
Lighting Fixtures	$38,973
Fire Alarm	$22,270
Punch List	$14,290
	$1,429,246

tractor bid for "floorcovering," including both resilient flooring and carpet. This lump sum item must be divided into proportionate amounts for each item. The subcontractor would be asked to divide the bid into the two items. Contractors often will estimate the percentage of each item, based on unit price and quantity, to determine this division.

3. Distribution of overhead: Several different methods exist for distribution of overhead and nonallocated items to the items of the schedule of values. The method used may be determined by the contractor, regulations, or local custom. Equal distribution of the overhead to the items, based on a percentage, is a method of allocating overhead. Figure 14–4 illustrates a proportionate allocation of overhead, with each item receiving an overhead allocation proportionate to its relative value.

4. Front-loading: Another method of allocating overhead is commonly referred to as **front-loading**. Front-loading is the allocation of overhead to items that will be completed early in the project. There are different degrees of front-loading, depending on the contractor and contract conditions. This method is commonly used by contractors. As the owner is very reluctant to pay for anything that is not earned,

FIGURE 14–6
Schedule of Values,
Computation, Front-
loaded Distribution
of Markup

Example Project, Schedule of Values Computations   Front-Load Overhead

SOV Item	Estimate Item	Amount	Less 1%	% of OH	Distribution	Value
Mobilization	Bonds	$15,600				
	All-Risk Insurance	$1,600				
	Temporary Facilities	$10,000				
		$27,200	$26,928	15%	$15,994	$42,922
General Conditions	Field Engineering	$2,000				
	Superintendent	$52,200				
	Final Cleaning	$2,550				
		$56,750	$56,183			$56,183
Earthwork	Earthwork	$45,000	$45,000	$44,550	25%	$26,657	$71,207
Site Improvements	Asphalt Paving	$27,600				
	Landscaping	$22,000				
		$49,600	$49,104			$49,104
Concrete	Concrete Formwork	$28,300				
	Concrete Reinforcing	$19,500				
	Cast-in-place Concrete	$28,900				
		$76,700	$75,933	40%	$42,652	$118,585
Masonry	Masonry	$65,000	$65,000	$64,350	20%	$21,326	$85,676
Structural Steel	Structural Steel	$116,500				
	Metal Joists	$34,000				
	Metal Deck	$23,000				
	Metal Handrails	$7,500				
		$181,000	$179,190			$179,190
Carpentry	Rough Carpentry	$2,500				
	Finish Carpentry	$88,000				
		$90,500	$89,595			$89,595
Dampproofing	Dampproofing	$1,600	$1,600	$1,584			$1,584
Insulation	Insulation	$22,000				
	Fireproofing	$16,500				
		$38,500	$38,115			$38,115
Roofing/Flashing	Roofing/Flashing	$65,000	$65,000	$64,350			$64,350
Sealants	Sealants	$9,600	$9,600	$9,504			$9,504
Doors, Hardware	Metal Doors & Frames	$41,500				
	Folding Doors	$6,800				
	Finish Hardware	$23,000				
		$71,300	$70,587			$70,587
Windows, Glazing	Metal Windows	$25,600				
	Glazing	$9,340				
		$34,940	$34,591			$34,591
Interior Partitions	Metal studs/Gyp.Bd.	$102,000	$102,000	$100,980			$100,980
Acoustical Ceilings	Acoustical Ceilings	$35,000	$35,000	$34,650			$34,650
Ceramic Tile	Ceramic Tile	$16,500	$16,500	$16,335			$16,335
Resilient Flooring	Floor Covering (50%)	$21,500	$21,500	$21,285			$21,285
Carpet	Floor Covering (50%)	$21,500	$21,500	$21,285			$21,285
Painting	Painting (85%)	$47,600	$47,600	$47,124			$47,124
Wallcovering	Painting (15%)	$8,400	$8,400	$8,316			$8,316
Specialties	Tackboards	$4,000				
	Toilet Compartments	$6,000				
	Signage	$7,600				
	Fire Extinguishers	$5,000				
	Toilet Accessories	$6,750				
		$29,350	$29,057			$29,057
Plumbing	Mechanical (40%)	$49,200	$49,200	$48,708			$48,708
HVAC	Mechanical (45%)	$55,350	$55,350	$54,797			$54,797
Fire Protection	Mechanical (15%)	$18,450	$18,450	$18,266			$18,266
Electrical Power	Electrical (25%)	$26,000	$26,000	$25,740			$25,740
Electrical Circuits	Electrical (20%)	$20,800	$20,800	$20,592			$20,592
Lighting Fixtures	Electrical (35%)	$36,400	$36,400	$36,036			$36,036
Fire Alarm	Electrical (20%)	$20,800	$20,800	$20,592			$20,592
Punch List				$13,215		$1,077	$14,292
				$1,321,540			$1,429,246

Markup	$107,706
Less markup to Punchlist	-$1,077
Markup to Distribute	**$106,629**

FIGURE 14–7
Schedule of Values, Front-loaded Distribution of Profit and Overhead

Schedule of Values,
Front-Loaded

Item	Value
Mobilization	$42,920
General Conditions	$56,183
Earthwork	$71,207
Site Improvements	$49,104
Concrete	$118,585
Masonry	$85,676
Structural Steel	$179,190
Carpentry	$89,595
Dampproofing	$1,584
Insulation	$38,115
Roofing/Flashing	$64,350
Sealants	$9,504
Doors, Hardware	$70,587
Windows, Glazing	$34,591
Interior Parititons	$100,980
Acoustical Ceilings	$34,650
Ceramic Tile	$16,335
Resilient Flooring	$21,285
Carpet	$21,285
Painting	$47,124
Wallcovering	$8,316
Specialties	$29,057
Plumbing	$48,708
HVAC	$54,797
Fire Protection	$18,266
Electrical Power	$25,740
Electrical Circuits	$20,592
Lighting Fixtures	$36,036
Fire Alarm	$20,592
Punch List	$14,292
Total	$1,429,246

considerable caution is used by owners, architects and engineers, and construction managers to avoid front-loading. Regulations on public work may prohibit front-loading.

Many contractors feel the anticipated profit and indirect overhead should be available for sufficient funds to manage the project. Contractors normally use a single mark-up for all costs, including their work and the subcontract. This mark-up is fairly low, considering that the majority of costs in a project are subcontracts. Obviously, less overhead and cost is required for the subcontracts than work with the contractor's own forces. The majority of the contractor's work is completed early in the project, such as concrete work and the structure of the building. It is logical then that the profit and overhead be allocated early in the project.

As front-loading is a common method of allocating overhead and profit in the schedule of values, it has been included in this discussion. The authors do not advocate front-loading or proportionate allocation. Local customs, regulations, and owners' attitudes vary on this subject, and contractors must determine their own course of action within context. A spread of overhead and profit into several progress payments is normally acceptable. Allocation of all overhead and profit into the first one or two progress payments is normally not acceptable.

The owner and architect review the schedule of values prior to its use in progress payments. The owner may reject the schedule of values submitted and request a revision by the contractor. Common reasons for rejection of the schedule of values include excessive front-loading, insufficient detail of construction activities, and noncompliance with contract requirements.

The total of the schedule of values is the contract amount. As change orders are added to the contract, they are added to the schedule of values, usually as lump sums.

Unit Price Contracts

Some contracts require that the bid be presented in unit prices for particular items of work. These types of contracts usually are in engineering construction and are found frequently in public works contracts. In these contracts, the unit price breakdown is used as the schedule of values. Each unit price includes labor, material, equipment, subcontracts, direct overhead, indirect overhead, and distributed profit. A finite quantity is usually furnished in the bid forms, and the contractor supplies the unit price. A variety of strategies is used in allocating overhead and profit to the unit prices, including front-loading.

Figure 14–8 is an example of a typical unit price breakdown contained in the bid form. The contractor normally would furnish the unit price and extension. These unit prices would be used as the schedule of values. Quantities of each item complete are measured and tabulated, then multiplied by the unit price to determine the amount due for each work activity.

Item #	Item	Quantity	Unit	Unit Price	Extension
1	Mobilization	1	L.S.	$20,000.00	$ 20,000.00
2	Trench Excavation	2,340	Cu. Yd.	$ 5.40	$ 12,636.00
3	Clear & Grub	15,555	Sq. Yd.	$ 1.36	$ 21,154.80
4	Mass Excavation	5,670	Cu. Yd.	$ 1.75	$ 9,922.50
5	Structual Excavation	12,965	Cu. Yd.	$ 2.90	$ 37,598.50
6	Structural Backfill	2,570	Cu. Yd.	$ .75	$ 1,927.50
7	3/8" Gravel	488	Tons	$ 7.67	$ 3,742.96
Total					$106,982.26

FIGURE 14–8
Example of Unit Price Breakdown

Project Cash Flow Projections

Project cash flow projections relate the schedule of values to the construction schedule, projecting the progress payments through the duration of the project. The cash flow projection approximates the progress payments for each payment period during the construction contract. This projection is used by the owner to make financial arrangements, which creates available funds for the payments and optimizes investment opportunities. The contractor also can use the cash flow projection for anticipating revenue for future periods.

Because the cash flow projection is based on the construction schedule, the projections are estimates only, as many fluctuations exist in the construction schedule. Several factors can affect the actual amount of progress payments at each period, producing a variance from the projection: late or early delivery of materials; late or early completion of work activities; revision in sequencing of work activities; and the disproportionate completion of work activities in their scheduled time period.

An approximate range can be established for the cash flow curve, using early start-early finish and late start-late finish dates defining the limits of the range. Some contractors prefer to present a range to the owner, as the cash flow projection is approximate. Other cash flow projections use a single curve, with a disclaimer that the amounts are approximate. Updating the cash flow projection is necessary as the project progresses to provide current and more accurate projections.

Because the cash flow projection relates to the amounts that will be requested for each progress payment, some factors differ from those relating to the construction schedule. The actual payments may not be equally distributed in the work activity period.

EXAMPLE

Structural steel erection for the example project has a duration of three months, starting on June 1. During this three-month period, there are three payment requests: July 1, August 1, and September 1. Labor and equipment are fairly consistent throughout the period, but material is not evenly distributed. A summary of estimated costs for this work follows:

Item	Material	Labor	Equipment	Total
Structural Steel	$103,400.00	$28,300.00	$24,000.00	$155,700.00
Metal Joists	$ 32,400.00			$ 32,400.00
Metal Deck	$ 13,650.00			$ 13,650.00
Total Cost	$149,450.00	$28,300.00	$24,000.00	$201,750.00

Some cash flow projections will take the total, $201,750, and divide it evenly into each of the three payment periods, or $67,250.

Realistically, the material will not be received at the jobsite evenly throughout the period. The chart that follows reflects a realistic projection of structural steel costs.

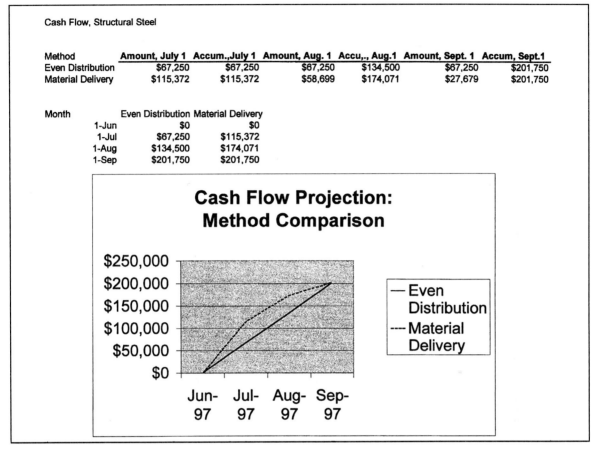

Cash Flow, Structural Steel

Method	Amount, July 1	Accum.,July 1	Amount, Aug. 1	Accu,., Aug.1	Amount, Sept. 1	Accum, Sept.1
Even Distribution	$67,250	$67,250	$67,250	$134,500	$67,250	$201,750
Material Delivery	$115,372	$115,372	$58,699	$174,071	$27,679	$201,750

Month	Even Distribution	Material Delivery
1-Jun	$0	$0
1-Jul	$67,250	$115,372
1-Aug	$134,500	$174,071
1-Sep	$201,750	$201,750

FIGURE 14–9   Cash Flow Projection, Method Comparison

EXAMPLE continued

Item	July 1 %	July 1 Amount	Aug. 1 %	Aug. 1 Amount	Sept. 1 %	Sept. 1 Amount
Str. Steel (material)	50%	$ 51,700	40%	$ 41,360	10%	$10,340
Metal Joists	100%	$ 32,400				
Metal Deck	100%	$ 13,650				
Labor	34%	$ 9,622	33%	$ 9,339	33%	$ 9,339
Equip.	33.33%	$ 8,000	33.33%	$ 8,000	33.33%	$ 8,000
Totals	57%	$115,372	29%	$ 58,699	14%	$27,679

Quite a difference exists between the evenly distributed projection and the realistic delivery projection. Figure 14–9 compares these two methods, which illustrate that by anticipating material delivery, the cash flow projection for that activity will be a larger payment request earlier in the process.

Because cash flow projections are used by the owner and contractor for financial arrangements, the projected payments should reflect material delivery and other factors that would influence the payment amount.

Figure 14–10 is a tabular computation of the cash flow for each month of the example project, from July 1 through May 1, with the actual project starting on June 5 and ending on April 12. The schedule of values, on the left side of the spreadsheet, is from that shown in Figure 14–7. The Gantt chart for the construction schedule is shown in Figure 14–11 on page 365. The percentage complete for each payment period is the cumulative total. What follows are examples of how the percentages in each period are determined, relating to the construction schedule in Figure 14–11.

EXAMPLE

1. Mobilization is completed between the start of the project and July 1. The percent complete then for July 1 and all subsequent periods is 100%. 100% of the value of $42,922 is entered in the amount column for July 1.
2. General conditions are spread evenly in ten periods, with 10% earned during the first period. Most items in the general conditions value, such as superintendent, temporary office rent, telephone charges, and so on are fairly constant for each period. The value of $56,183 is multiplied by 10%, with the resulting amount of $5,618 for July 1. For August 1, the total percentage earned is 20%; 10% for the first period and 10% for the second period.
3. The construction schedule shows the Earthwork item spread between the June and July period. Because about half of the work will be done in June, 50% is entered in the percentage column. The August 1 period, then, shows 100% completion of the activity.
4. Site plumbing is shown on the construction schedule as being completed during June. The schedule of values, in this case, has only one item for plumbing, while four activities for plumbing are shown on the construction schedule. From conversations with the plumbing subcontractor, the following percentage of the plumbing value has been determined for the four activities:

Site Plumbing:	10%
Plumbing rough-in, under slab:	10%
Plumbing rough-in, above grade:	40%
Plumbing fixtures:	40%

Site Plumbing is scheduled for completion during June. Because much of the rough-in material will be delivered with the initial delivery, a total of 20% is estimated for work in place and delivered material during the June period.

The monthly accumulated total is then plotted in a line graph, as shown in Figure 14–12 on page 366. The cash flow curve is normally a "lazy-S" curve. Since this curve is related to specific amounts at the payment dates, the curve is not a smooth line, but segmented at the payment dates.

A front-loaded schedule of values will have a steeper initial section on the cash flow curve than a schedule of values with an evenly distributed overhead. Figure 14–13 on page 367 shows a comparison of cash flow charts for the two different schedule of values methods described in this chapter. This comparison shows a slightly accelerated payment schedule

Item	Value	1-Jul %	1-Jul Amount	1-Aug %	1-Aug Amount	1-Sep %	1-Sep Amount	1-Oct %	1-Oct Amount	1-Nov %
Mobilization	$42,920	100%	$42,920	100%	$42,920	100%	$42,920	100%	$42,920	100%
General Conditions	$56,183	10%	$5,618	20%	$11,237	30%	$16,855	40%	$22,473	50%
Earthwork	$71,207	50%	$35,604	100%	$71,207	100%	$71,207	100%	$71,207	100%
Site Improvements	$49,104									
Concrete	$118,585			60%	$71,151	100%	$118,585	100%	$118,585	100%
Masonry	$85,676					100%	$85,676	100%	$85,676	100%
Structural Steel	$179,190					60%	$107,514	100%	$179,190	100%
Carpentry	$89,595									10%
Dampproofing	$1,584					100%	$1,584	100%	$1,584	100%
Insulation	$38,115									
Roofing/Flashing	$64,350							70%	$45,045	100%
Sealants	$9,504									
Doors, Hardware	$70,587					25%	$17,647	25%	$17,647	50%
Windows, Glazing	$34,591									
Interior Partitions	$100,980									50%
Acoustical Ceilings	$34,650							20%	$6,930	20%
Ceramic Tile	$16,335									
Resilient Flooring	$21,285									
Carpet	$21,285									
Painting	$47,124									
Wallcovering	$8,316									
Specialties	$29,057	20%	$9,742	40%	$19,483	50%	$24,354	60%	$29,225	65%
Plumbing	$48,708							30%	$16,439	40%
HVAC	$54,797									
Fire Protection	$18,266									
Electrical Power	$25,740									
Electrical Circuits	$20,592			20%	$4,118	50%	$10,296	70%	$14,414	90%
Lighting Fixtures	$36,036									
Fire Alarm	$20,592									
Punch List	$14,292									
Total	$1,429,246	7%	$93,883	15%	$220,116	35%	$496,638	46%	$651,335	54%
			1-Jul		1-Aug		1-Sep		1-Oct	

FIGURE 14–10   Tabular Cash Flow Projection

1-Nov Amount	1-Dec %	1-Dec Amount	1-Jan %	1-Jan Amount	1-Feb %	1-Feb Amount	1-Mar %	1-Mar Amount	1-Apr %	1-Apr Amount	1-May %	1-May Amount
$42,920	100%	$42,920	100%	$42,920	100%	$42,920	100%	$42,920	100%	$42,920	100%	$42,920
$28,092	60%	$33,710	70%	$39,328	80%	$44,946	90%	$50,565	100%	$56,183	100%	$56,183
$71,207	100%	$71,207	100%	$71,207	100%	$71,207	100%	$71,207	100%	$71,207	100%	$71,207
									100%	$49,104	100%	$49,104
$118,585	100%	$118,585	100%	$118,585	100%	$118,585	100%	$118,585	100%	$118,585	100%	$118,585
$85,676	100%	$85,676	100%	$85,676	100%	$85,676	100%	$85,676	100%	$85,676	100%	$85,676
$179,190	100%	$179,190	100%	$179,190	100%	$179,190	100%	$179,190	100%	$179,190	100%	$179,190
$8,960	10%	$8,960	90%	$80,636	100%	$89,595	100%	$89,595	100%	$89,595	100%	$89,595
$1,584	100%	$1,584	100%	$1,584	100%	$1,584	100%	$1,584	100%	$1,584	100%	$1,584
	100%	$38,115	100%	$38,115	100%	$38,115	100%	$38,115	100%	$38,115	100%	$38,115
$64,350	100%	$64,350	100%	$64,350	100%	$64,350	100%	$64,350	100%	$64,350	100%	$64,350
	100%	$9,504	100%	$9,504	100%	$9,504	100%	$9,504	100%	$9,504	100%	$9,504
$35,294	70%	$49,411	100%	$70,587	100%	$70,587	100%	$70,587	100%	$70,587	100%	$70,587
	100%	$34,591	100%	$34,591	100%	$34,591	100%	$34,591	100%	$34,591	100%	$34,591
$50,490	70%	$70,686	100%	$100,980	100%	$100,980	100%	$100,980	100%	$100,980	100%	$100,980
$6,930	20%	$6,930	20%	$6,930	100%	$34,650	100%	$34,650	100%	$34,650	100%	$34,650
			100%	$16,335	100%	$16,335	100%	$16,335	100%	$16,335	100%	$16,335
			50%	$10,643	100%	$21,285	100%	$21,285	100%	$21,285	100%	$21,285
			50%	$10,643	100%	$21,285	100%	$21,285	100%	$21,285	100%	$21,285
			40%	$18,850	100%	$47,124	100%	$47,124	100%	$47,124	100%	$47,124
					100%	$8,316	100%	$8,316	100%	$8,316	100%	$8,316
					90%	$26,151	100%	$29,057	100%	$29,057	100%	$29,057
$31,660	65%	$31,660	65%	$31,660	100%	$48,708	100%	$48,708	100%	$48,708	100%	$48,708
$21,919	60%	$32,878	70%	$38,358	80%	$43,838	90%	$49,317	100%	$54,797	100%	$54,797
	80%	$14,613	80%	$14,613	100%	$18,266	100%	$18,266	100%	$18,266	100%	$18,266
	80%	$20,592	100%	$25,740	100%	$25,740	100%	$25,740	100%	$25,740	100%	$25,740
$18,533	90%	$18,533	90%	$18,533	90%	$18,533	100%	$20,592	100%	$20,592	100%	$20,592
					70%	$25,225	100%	$36,036	100%	$36,036	100%	$36,036
							60%	$12,355	100%	$20,592	100%	$20,592
									25%	$3,573	100%	$14,292
$765,388	65%	$933,694	79%	$1,129,556	91%	$1,307,286	94%	$1,346,515	99%	$1,418,527	100%	$1,429,246
1-Nov		1-Dec		1-Jan		1-Feb		1-Mar		1-Apr		1-May

FIGURE 14–10　continued

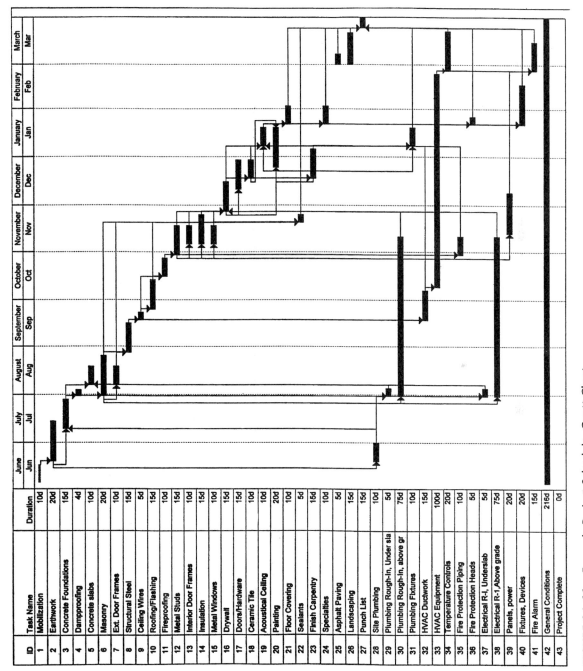

FIGURE 14–11   Example Project Schedule, Gantt Chart

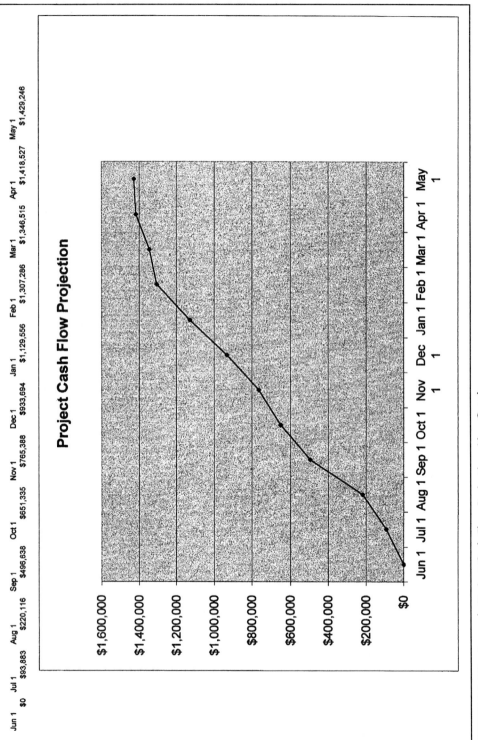

FIGURE 14–12   Example Project Cash Flow Projection, Line Graph

Comparison of SOV Methods

	Jun 1	Jul 1	Aug 1	Sep 1	Oct 1	Nov 1	Dec 1	Jan 1	Feb 1	Mar 1	Apr 1	May 1
Front-Loaded	$0	$93,883	$220,116	$496,638	$651,335	$765,388	$933,694	$1,129,556	$1,307,286	$1,346,515	$1,418,527	$1,429,246
Evenly Distributed	$0	$69,825	$164,255	$421,798	$589,103	$712,452	$894,474	$1,106,298	$1,298,514	$1,340,940	$1,418,529	$1,429,246

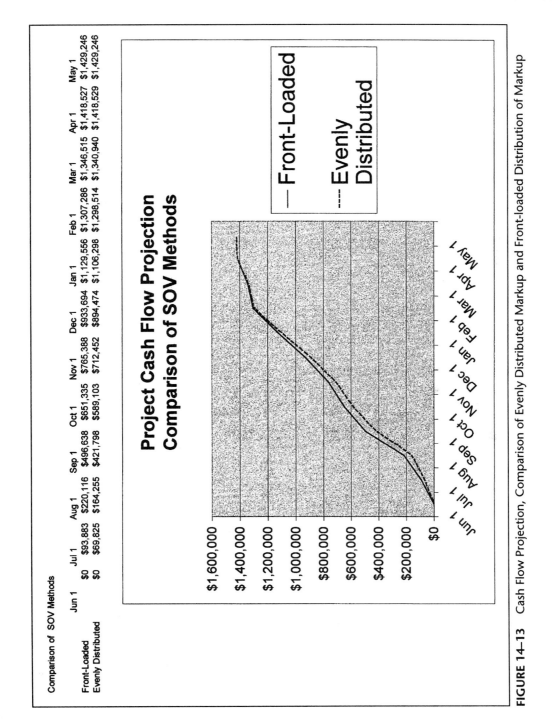

Project Cash Flow Projection
Comparison of SOV Methods

— Front-Loaded

---- Evenly Distributed

FIGURE 14–13   Cash Flow Projection, Comparison of Evenly Distributed Markup and Front-loaded Distribution of Markup

for the front-loaded schedule of values, with similar amounts in the latter periods.

Some of the "bumps" in the cash flow curve, for example, for the September 1 and October 1 periods, indicate additional payment activity, such as payment for material delivered and not installed. This is a fairly standard practice, since contractors prefer to have material delivered as soon as possible, when feasible, rather than delaying the work by waiting for material delivery.

Some computerized scheduling programs generate a cash flow chart from the construction schedule and loaded costs. Careful examination of the cash flow charts in the particular scheduling software is necessary to ensure that pertinent information is generated. Most programs evenly distribute cost when the activity extends into two or more periods. If an uneven distribution of the costs is anticipated, splitting the work activity into segments might be advisable. Using material delivery items and anticipated delivery dates also will help make the cash flow curve more accurate.

Progress Payment Procedures

Each owner will have specific progress payment requirements. Many will use the forms and procedures recommended by the AIA documents. Most public agencies have specific procedures required by law for the municipality, state, or federal agency. Most owners will require a detail sheet for the period and a certificate of payment, which is a summary of payments, and certification by the contractor and the architect and engineer, as representative of the owner.

The AIA forms for payment are G702 and G703 (Figures 14–14 and 14–15). AIA Form G702, "Application and Certificate for Payment," has five major sections:

1. General information: contract information, contractor, architect, owner, payment number, and so on.
2. Contract status: change orders to date, change orders during this period, and net contract change by change orders.
3. Payment computation: total contract, progress to date, retainage, sales tax (if applicable), and payment due.
4. Contractor's certifications: several certifications are necessary with each payment. The contractor must certify that the payment request represents work completed and that the work completed is in accordance with the contract documents. The contractor also must certify that bills were paid relating to previous payments.
5. Architect's certification: certification that the architect has reviewed the work in progress and the payment request.

Reproduced with permission of The American Institute of Architects under license number #96049. This license expires September 30, 1997. FURTHER REPRODUCTION IS PROHIBITED. Because AIA Documents are revised from time to time, users should ascertain from the AIA the current edition of this document. Copies of the current edition of this AIA document may be purchased from The American Institute of Architects or its local distributors. The text of this document is not "model language" and is not intended for use in other documents without permission of the AIA.

APPLICATION AND CERTIFICATE FOR PAYMENT AIA DOCUMENT G702 (Instructions on reverse side) PAGE ONE OF ___ PAGES

TO OWNER:

PROJECT:

APPLICATION NO.:

PERIOD TO:

PROJECT NOS.:

Distribution to:
☐ OWNER
☐ ARCHITECT
☐ CONTRACTOR
☐ ☐

FROM CONTRACTOR:

VIA ARCHITECT:

CONTRACT DATE:

CONTRACT FOR:

CONTRACTOR'S APPLICATION FOR PAYMENT

Application is made for payment, as shown below, in connection with the Contract. Continuation Sheet, AIA Document G703, is attached.

1. ORIGINAL CONTRACT SUM .... $
2. Net change by Change Orders .... $
3. CONTRACT SUM TO DATE (Line 1 ± 2) .... $
4. TOTAL COMPLETED & STORED TO DATE .... $
   (Column G on G703)
5. RETAINAGE:
   a. _____ % of Completed Work $
      (Columns D + E on G703)
   b. _____ % of Stored Material $
      (Column F on G703)
   Total Retainage (Line 5a + 5b or
      Total in Column 1 of G703) .... $
6. TOTAL EARNED LESS RETAINAGE .... $
   (Line 4 less Line 5 Total)
7. LESS PREVIOUS CERTIFICATES FOR PAYMENT .... $
   (Line 6 from prior Certificate)
8. CURRENT PAYMENT DUE .... $
9. BALANCE TO FINISH, INCLUDING RETAINAGE $
   (Line 3 less Line 6)

CHANGE ORDER SUMMARY	ADDITIONS	DEDUCTIONS
Total changes approved in previous months by Owner		
Total approved this Month		
TOTALS		
NET CHANGES by Change Order		

The undersigned Contractor certifies that to the best of the Contractor's knowledge, information and belief the Work covered by this Application for Payment has been completed in accordance with the Contract Documents, that all amounts have been paid by the Contractor for Work for which previous Certificates for Payment were issued and payments received from the Owner, and that current payment shown herein is now due.

CONTRACTOR:

By: _____ Date: _____

State of:
County:
Subscribed and sworn to before
me this _____ day of

Notary Public:
My Commission expires:

ARCHITECT'S CERTIFICATE FOR PAYMENT

In accordance with the Contract Documents, based on on-site observations and the data comprising this application, the Architect certifies to the Owner that to the best of the Architect's knowledge, information and belief the Work has progressed as indicated, the quality of the Work is in accordance with the Contract Documents, and the Contractor is entitled to payment of the AMOUNT CERTIFIED.

AMOUNT CERTIFIED .... $ _____
(Attach explanation if amount certified differs from the amount applied for. Initial all figures on this Application and on the Continuation Sheet that are changed to conform to the amount certified.)
ARCHITECT:

By: _____ Date: _____

This Certificate is not negotiable. The AMOUNT CERTIFIED is payable only to the Contractor named herein. Issuance, payment and acceptance of payment are without prejudice to any rights of the Owner or Contractor under this Contract.

AIA DOCUMENT G702 • APPLICATION AND CERTIFICATE FOR PAYMENT • 1992 EDITION • AIA® • ©1992 • THE AMERICAN INSTITUTE OF ARCHITECTS, 1735 NEW YORK AVENUE, N.W., WASHINGTON, D.C. 20006-5292 • WARNING: Unlicensed photocopying violates U.S. copyright laws and will subject the violator to legal prosecution.

G702-1992

CAUTION: You should use an original AIA document which has this caution printed in red. An original assures that changes will not be obscured as may occur when documents are reproduced.

FIGURE 14–14 AIA Document G702, "Application and Certificate for Payment"

CONTINUATION SHEET   AIA DOCUMENT G703 (Instructions on reverse side)

AIA Document G702, APPLICATION AND CERTIFICATE FOR PAYMENT,
containing Contractor's signed Certification, is attached.
In tabulations below, amounts are stated to the nearest dollar.
Use Column I on Contracts where variable retainage for line items may apply.

APPLICATION NO.:
APPLICATION DATE:
PERIOD TO:
ARCHITECT'S PROJECT NO.:

PAGE   OF   PAGES

A	B	C	D	E	F	G	H	I
			WORK COMPLETED					
ITEM NO.	DESCRIPTION OF WORK	SCHEDULED VALUE	FROM PREVIOUS APPLICATION (D + E)	THIS PERIOD	MATERIALS PRESENTLY STORED (NOT IN D OR E)	TOTAL COMPLETED AND STORED TO DATE (D+E+F)	BALANCE TO FINISH (C – G)	RETAINAGE (IF VARIABLE RATE)
						(G % C)		

G703-1992

FIGURE 14–15   AIA Document G703, Continuation Sheet (Reproduced with permission of The American Institute of Architects under license number #96049. This license expires September 30, 1997. FURTHER REPRODUCTION IS PROHIBITED. Because AIA Documents are revised from time to time, users should ascertain from the AIA the current edition of this document. Copies of the current edition of this AIA document may be purchased from The American Institute of Architects or its local distributors. The text of this document is not "model language" and is not intended for use in other documents without permission of the AIA.)

The Continuation Sheet, AIA Document G703, Figure 14–15, details the progress of each item in the schedule of values, including previous progress, current completed work, current materials stored at the jobsite, total work completed and stored to date, balance to finish, and current retainage. The totals from this sheet are used in the Application and Certification for Payment.

Figures 14–16 and 14–17 illustrate the process of completing Forms G702 and G703 for Payment 1:

1. Figure 14–16 is the detail sheet for the payment (Form G703). The data used in this payment request was derived from the cash flow projection (Figure 14–10). The amounts in the cash flow projection and the payment request rarely coincide exactly, however, this data is appropriate for examples. As mobilization was completed during the month, it is 100% complete, with no balance to finish. Retainage of 5%, or $2,146, will be held for that item. Item 23, Plumbing, is divided equally into work completed and material stored, combined in column G. It is important to separate work completed with material stored for verification of payment. Some owners calculate only work in place as work progress, treating material stored separately. Some owners use different rates of retainage on material stored as well. The retainage amounts for the particular project will be found in the General Conditions to the Contract, the Supplementary Conditions to the Contract, and the Agreement between the owner and the contractor.

2. The totals generated on Form G703 are used to complete Form G702. Since this is the first payment request, no change orders have been processed and the original contract amount is still current. Figure 14–17 is the computation for the progress payment on the Application and Certification for Payment. "Contract Sum to Date" is the contract amount and the same amount from the total of Column C in Figure 14–16. "Total Completed and Stored to Date" is the total of Column G in Figure 14–16. The retainage breakdown takes retainage separately on work completed and material stored on site. In this case, the retainage is the same percentage, 5%, for both. "Total Retainage" is the same as Column I on Figure 14–16. The retainage is then deducted from the amount earned to determine the total amount due the contractor to date. Previous payments, then, are deducted from the amount due the contractor to determine the amount due for the current payment period. In this case, no previous payments have been made, since this is the first payment request. The amount remaining to be paid, plus retainage, is also computed.

CONTINUATION SHEET

AIA DOCUMENT G703 (Instructions on reverse side)

APPLICATION NO.:
APPLICATION DATE:
PERIOD TO:
ARCHITECT'S PROJECT NO.:

PAGE   OF   PAGES

AIA Document G702, APPLICATION AND CERTIFICATE FOR PAYMENT,
containing Contractor's signed Certification, is attached.
In tabulations below, amounts are stated to the nearest dollar.
Use Column I on Contracts where variable retainage for line items may apply.

A	B	C	D	E	F	G		H	I
			WORK COMPLETED		MATERIALS PRESENTLY STORED (NOT IN D OR E)	TOTAL COMPLETED AND STORED TO DATE (D+E+F)			
ITEM NO.	DESCRIPTION OF WORK	SCHEDULED VALUE	FROM PREVIOUS APPLICATION (D + E)	THIS PERIOD			% (G ÷ C)	BALANCE TO FINISH (C − G)	RETAINAGE (IF VARIABLE) RATE)
1	Mobilization	$42,922		$42,922		$42,922	100.00%	$0	$2,146
2	General Conditions	$56,183		$5,618		$5,618	10.00%	$50,565	$281
3	Earthwork	$71,207		$35,604		$35,604	50.00%	$35,603	$1,780
4	Site Improvements	$49,104							
5	Concrete	$118,585							
6	Masonry	$85,676							
7	Structural Steel	$179,190							
8	Carpentry	$89,595							
9	Dampproofing	$1,584							
10	Insulation	$38,115							
11	Roofing/Flashing	$64,350							
12	Sealants	$9,504							
13	Doors, Hardware	$70,587							
14	Windows, Glazing	$34,591							
15	Interior Partitions	$100,980							
16	Acoustical Ceilings	$34,650							
17	Ceramic Tile	$16,335							
18	Resilient Flooring	$21,285							
19	Carpet	$21,285							
20	Painting	$47,124							
21	Wallcovering	$8,316							
22	Specialties	$29,057							
23	Plumbing	$48,708		$4,871	$4,871	$9,742	20.00%	$38,966	$487
24	HVAC	$54,796							
25	Fire Protection	$18,265							
26	Electrical Power	$25,740							
27	Electrical Circuits	$20,592							
28	Lighting Fixtures	$36,036							
29	Fire Alarm	$20,592							
30	Punch List (1%)	$14,292							
	Total	$1,429,246		$89,015	$4,871	$93,886	6.57%	$125,134	$4,694

G703-1992

FIGURE 14-16 Example Computation, Payment 1, Continuation Sheet (Reproduced with permission of The American Institute of Architects under license number #96049. This license expires September 30, 1997. FURTHER REPRODUCTION IS PROHIBITED. Because AIA Documents are revised from time to time, users should ascertain from the AIA the current edition of this document. Copies of the current edition of this AIA document may be purchased from The American Institute of Architects or its local distributors. The text of this document is not "model language" and is not intended for use in other documents without permission of the AIA.)

APPLICATION AND CERTIFICATE FOR PAYMENT AIA DOCUMENT G702 (Instructions on reverse side) PAGE ONE OF   PAGES

TO OWNER:	PROJECT:	APPLICATION NO.:	Distribution to:
		PERIOD TO:	☐ OWNER
		PROJECT NOS.:	☐ ARCHITECT
			☐ CONTRACTOR
FROM CONTRACTOR:	VIA ARCHITECT:	CONTRACT DATE:	☐
			☐

CONTRACT FOR:

CONTRACTOR'S APPLICATION FOR PAYMENT
Application is made for payment, as shown below, in connection with the Contract.
Continuation Sheet, AIA Document G703, is attached.

1. ORIGINAL CONTRACT SUM.................$ __1,429,246.00__
2. Net change by Change Orders ....$ __0.00__
3. CONTRACT SUM TO DATE (Line 1 ± 2).......$ __1,429,246.00__
4. TOTAL COMPLETED & STORED TO DATE ......$ __93,886.00__
   (Column G on G703)
5. RETAINAGE:
   a. __5__ % of Completed Work $ __4,450.75__
      (Columns D + E on G703)
   b. __5__ % of Stored Material $ __243.55__
      (Column F on G703)
   Total Retainage (Line 5a + 5b or
   Total in Column I of G703) ....$ __4,694.30__
6. TOTAL EARNED LESS RETAINAGE............$ __89,191.70__
   (Line 4 less Line 5 Total)
7. LESS PREVIOUS CERTIFICATES FOR PAYMENT
   (Line 6 from prior Certificate) .... __0.00__
8. CURRENT PAYMENT DUE ....$ __89,191.70__
9. BALANCE TO FINISH, INCLUDING RETAINAGE
   (Line 3 less Line 6) $ __1,340,054.30__

CHANGE ORDER SUMMARY	ADDITIONS	DEDUCTIONS
Total changes approved in previous months by Owner		
Total approved this Month		
TOTALS		
NET CHANGES by Change Order		

The undersigned Contractor certifies that to the best of the Contractor's knowledge, information and belief the Work covered by this Application for Payment has been completed in accordance with the Contract Documents, that all amounts have been paid by the Contractor for Work for which previous Certificates for Payment were issued and payments received from the Owner, and that current payment shown herein is now due.

CONTRACTOR:

By: _____   Date: _____

State of:
County of:
Subscribed and sworn to before
me this   day of
Notary Public:
My Commission expires:

ARCHITECT'S CERTIFICATE FOR PAYMENT

In accordance with the Contract Documents, based on on-site observations and the data comprising this application, the Architect certifies to the Owner that to the best of the Architect's knowledge, information and belief the Work has progressed as indicated, the quality of the Work is in accordance with the Contract Documents, and the Contractor is entitled to payment of the AMOUNT CERTIFIED.

AMOUNT CERTIFIED ....$_____
(Attach explanation if amount certified differs from the amount applied for. Initial all figures on this Application and on the Continuation Sheet that are changed to conform to the amount certified.)
ARCHITECT:
By: _____   Date: _____
This Certificate is not negotiable. The AMOUNT CERTIFIED is payable only to the Contractor named herein. Issuance, payment and acceptance of payment are without prejudice to any rights of the Owner or Contractor under this Contract.

FIGURE 14–17   Example Computation, Payment 1, Application and Certificate for Payment

Figures 14–18, 14–19, and 14–20 illustrate a later progress payment, Progress Payment Application Number 4, October 1:

1. More items are active in Figure 14–18, Form G703. Three change order items have been added, as the three change orders have been formally approved. The total of Column C, Scheduled Value, reflects the current amended contract price.
2. Figure 14–19 illustrates the Change Order Summary found on Form G702, Application and Certification for Payment. Change orders previously approved (only Change Order 1) amount to an additional $15,300. Change orders 2 and 3 are added to the contract amount for this Application for Payment, as they were approved during the payment period. Figure 14–18 shows that Change Order 2 was completed during this period, but that Change Order 3 must still be completed.

CONTINUATION SHEET   AIA DOCUMENT G703 (Instructions on reverse side)

AIA Document G702, APPLICATION AND CERTIFICATE FOR PAYMENT,
containing Contractor's signed Certification, is attached.
In tabulations below, amounts are stated to the nearest dollar.
Use Column 1 on Contracts where variable retainage for line items may apply.

APPLICATION NO.:
APPLICATION DATE:
PERIOD TO:
ARCHITECT'S PROJECT NO.:

PAGE   OF   PAGES

A ITEM NO.	B DESCRIPTION OF WORK	C SCHEDULED VALUE	D WORK COMPLETED FROM PREVIOUS APPLICATION (D + E)	E WORK COMPLETED THIS PERIOD	F MATERIALS PRESENTLY STORED (NOT IN D OR E)	G TOTAL COMPLETED AND STORED TO DATE (D+E+F)	% (G ÷ C)	H BALANCE TO FINISH (C - G)	I RETAINAGE (IF VARIABLE RATE)
1	Mobilization	$42,922	$42,922			$42,922	100.00%	$0	$2,146
2	General Conditions	$56,183	$16,855	$5,618		$22,473	40.00%	$33,710	$1,124
3	Earthwork	$71,207	$71,207			$71,207	100.00%	$0	$3,560
4	Site Improvements	$49,104				$0		$49,104	$0
5	Concrete	$118,585	$118,585			$118,585	100.00%	$0	$5,929
6	Masonry	$85,676	$85,676			$85,676	100.00%	$0	$4,284
7	Structural Steel	$179,190	$107,514	$71,676		$179,190	100.00%	$0	$8,960
8	Carpentry	$89,595				$0		$89,595	$0
9	Dampproofing	$1,584	$1,584			$1,584	100.00%	$0	$79
10	Insulation	$38,115				$0		$38,115	$0
11	Roofing/Flashing	$64,350		$19,305	$25,740	$45,045	70.00%	$19,305	$2,252
12	Sealants	$9,504				$0		$9,504	$0
13	Doors, Hardware	$70,587	$10,588		$7,059	$17,647	25.00%	$52,940	$882
14	Windows, Glazing	$34,591				$0		$34,591	$0
15	Interior Partitions	$100,980				$0		$100,980	$0
16	Acoustical Ceilings	$34,650		$6,930		$6,930	20.00%	$27,720	$347
17	Ceramic Tile	$16,335				$0		$16,335	$0
18	Resilient Flooring	$21,285				$0		$21,285	$0
19	Carpet	$21,285				$0		$21,285	$0
20	Painting	$47,124				$0		$47,124	$0
21	Wallcovering	$8,316				$0		$8,316	$0
22	Specialties	$29,057				$0		$29,057	$0
23	Plumbing	$48,708	$24,354	$4,871		$29,225	60.00%	$19,483	$1,461
24	HVAC	$54,796		$10,959	$5,480	$16,439	30.00%	$38,357	$822
25	Fire Protection	$18,265				$0		$18,265	$0
26	Electrical Power	$25,740				$0		$25,740	$0
27	Electrical Circuits	$20,592	$10,296	$4,118		$14,414	70.00%	$6,178	$721
28	Lighting Fixtures	$36,036				$0		$36,036	$0
29	Fire Alarm	$20,592				$0		$20,592	$0
30	Punch List (1%)	$14,292				$0		$14,292	$0
31	Change Order 1	$15,300	$15,300			$15,300	100.00%	$0	$765
32	Change Order 2	$22,450		$22,450		$22,450	100.00%	$0	$1,123
33	Change Order 3	$5,600				$0		$5,600	$0
	Total	$1,472,596	$504,881	$145,927	$38,279	$689,087	46.79%	$783,509	$34,454
		$1,472,596				$689,087		$1,472,596	$34,454

AIA DOCUMENT G703 • CONTINUATION SHEET FOR G702 • 1992 EDITION • AIA® • ©1992 • THE AMERICAN INSTITUTE OF ARCHITECTS, 1735 NEW YORK AVENUE, N.W., WASHINGTON, D.C. 20006-5292 • WARNING: Unlicensed photocopying violates U.S. copyright laws and will subject the violator to legal prosecution.

G703-1992

CAUTION: You should use an original AIA document which has this caution printed in red. An original assures that changes will not be obscured as may occur when documents are reproduced.

FIGURE 14–18 Example Computation, Payment 4, Continuation Sheet (Reproduced with permission of The American Institute of Architects under license number #96049. This license expires September 30, 1997. FURTHER REPRODUCTION IS PROHIBITED. Because AIA Documents are revised from time to time, users should ascertain from the AIA the current edition of this document. Copies of the current edition of this AIA document may be purchased from The American Institute of Architects or its local distributors. The text of this document is not "model language" and is not intended for use in other documents without permission of the AIA.)

APPLICATION AND CERTIFICATE FOR PAYMENT AIA DOCUMENT G702 (Instructions on reverse side) PAGE ONE OF   PAGES

TO OWNER: PROJECT: APPLICATION NO.:   Distribution to:
                                                                 PERIOD TO: ☐ OWNER
                                                                 PROJECT NOS.:   ☐ ARCHITECT
                                                                                           ☐ CONTRACTOR
FROM CONTRACTOR: VIA ARCHITECT: CONTRACT DATE: ☐
                                                                                           ☐

CONTRACT FOR:

CONTRACTOR'S APPLICATION FOR PAYMENT

Application is made for payment, as shown below, in connection with the Contract.
Continuation Sheet, AIA Document G703, is attached.

1. ORIGINAL CONTRACT SUM...................$_____
2. Net change by Change Orders..............$_____
3. CONTRACT SUM TO DATE (Line 1 ± 2).......$_____
4. TOTAL COMPLETED & STORED TO DATE......$_____
   (Column G on G703)
5. RETAINAGE:
   a. _____% of Completed Work $_____
      (Columns D + E on G703)
   b. _____% of Stored Material $_____
      (Column F on G703)
   Total Retainage (Line 5a + 5b or
   Total in Column I of G703)................$_____
6. TOTAL EARNED LESS RETAINAGE............$_____
   (Line 4 less Line 5 Total)
7. LESS PREVIOUS CERTIFICATES FOR PAYMENT
   (Line 6 from prior Certificate).................$_____
8. CURRENT PAYMENT DUE ....│ $_____ │
9. BALANCE TO FINISH, INCLUDING RETAINAGE
   (Line 3 less Line 6) $_____

CHANGE ORDER SUMMARY	ADDITIONS	DEDUCTIONS
Total changes approved in previous months by Owner	15,300	
Total approved this Month	28,050	
TOTALS	43,350	
NET CHANGES by Change Order	43,350	

The undersigned Contractor certifies that to the best of the Contractor's knowledge, information and belief the Work covered by this Application for Payment has been completed in accordance with the Contract Documents, that all amounts have been paid by the Contractor for Work for which previous Certificates for Payment were issued and payments received from the Owner, and that current payment shown herein is now due.

CONTRACTOR:

By: _____   Date: _____

State of:
County of:
Subscribed and sworn to before
me this _____ day of _____

Notary Public: _____
My Commission expires: _____

ARCHITECT'S CERTIFICATE FOR PAYMENT

In accordance with the Contract Documents, based on on-site observations and the data comprising this application, the Architect certifies to the Owner that to the best of the Architect's knowledge, information and belief the Work has progressed as indicated, the quality of the Work is in accordance with the Contract Documents, and the Contractor is entitled to payment of the AMOUNT CERTIFIED.

AMOUNT CERTIFIED ....$_____
(Attach explanation if amount certified differs from the amount applied for. Initial all figures on this Application and on the Continuation Sheet that are changed to conform to the amount certified.)
ARCHITECT:

By: _____   Date: _____
This Certificate is not negotiable. The AMOUNT CERTIFIED is payable only to the Contractor named herein. Issuance, payment and acceptance of payment are without prejudice to any rights of the Owner or Contractor under this Contract.

FIGURE 14–19   Example Computation, Payment 4, Change Order Summary

3. The payment computation for Payment 4, Figure 14–20, is similar to the application for Payment 1. The "Contract Sum to Date" has been changed due to the three change orders. Previous certificates for payment have been deducted. The previous certificates for payment amount is the net amount received from previous payments and does not include retainage.

Some states require payment of sales tax on construction projects. This sales tax computation modifies the computation for payment amount. As each state has different requirements concerning the tax and its collection, the reader is encouraged to contact the appropriate state agency to determine where sales tax applies and its common method of collection.

APPLICATION AND CERTIFICATE FOR PAYMENT AIA DOCUMENT G702 (Instructions on reverse side) PAGE ONE OF   PAGES

TO OWNER: PROJECT: APPLICATION NO.:   Distribution to:
                                                                     PERIOD TO: ☐ OWNER
                                                                     PROJECT NOS.:   ☐ ARCHITECT
                                                                                                 ☐ CONTRACTOR
FROM CONTRACTOR: VIA ARCHITECT: CONTRACT DATE: ☐
                                                                                                 ☐

CONTRACT FOR:

CONTRACTOR'S APPLICATION FOR PAYMENT

Application is made for payment, as shown below, in connection with the Contract.
Continuation Sheet, AIA Document G703, is attached.

1. ORIGINAL CONTRACT SUM .... $ 1,429,246.00

2. Net change by Change Orders .... $ 43,350.00

3. CONTRACT SUM TO DATE (Line 1 ± 2) .... $ 1,472,596.00

4. TOTAL COMPLETED & STORED TO DATE ...... $ 689,087.00
   (Column G on G703)

5. RETAINAGE:
   a. __5__ % of Completed Work $ 32,540.40
      (Columns D + E on G703)
   b. __5__ % of Stored Material $ 1,913.95
      (Column F on G703)
   Total Retainage (Line 5a + 5b or
   Total in Column I of G703) .... $ 34,454.35

6. TOTAL EARNED LESS RETAINAGE .... $ 654,632.65
   (Line 4 less Line 5 Total)

7. LESS PREVIOUS CERTIFICATES FOR PAYMENT
   (Line 6 from prior Certificate) .... $ 479,636.95

8. CURRENT PAYMENT DUE .... $ 174,995.70

9. BALANCE TO FINISH, INCLUDING RETAINAGE
   (Line 3 less Line 6) $ 817,963.35

CHANGE ORDER SUMMARY	ADDITIONS	DEDUCTIONS
Total changes approved in previous months by Owner		
Total approved this Month		
TOTALS		
NET CHANGES by Change Order		

The undersigned Contractor certifies that to the best of the Contractor's knowledge, information and belief the Work covered by this Application for Payment has been completed in accordance with the Contract Documents, that all amounts have been paid by the Contractor for Work for which previous Certificates for Payment were issued and payments received from the Owner, and that current payment shown herein is now due.

CONTRACTOR:

By _____   Date: _____

State of:
County of:
Subscribed and sworn to before
me this   day of

Notary Public:
My Commission expires:

ARCHITECT'S CERTIFICATE FOR PAYMENT

In accordance with the Contract Documents, based on on-site observations and the data comprising this application, the Architect certifies to the Owner that to the best of the Architect's knowledge, information and belief the Work has progressed as indicated, the quality of the Work is in accordance with the Contract Documents, and the Contractor is entitled to payment of the AMOUNT CERTIFIED.

AMOUNT CERTIFIED .... $ _____

(Attach explanation if amount certified differs from the amount applied for. Initial all figures on this Application and on the Continuation Sheet that are changed to conform to the amount certified.)

ARCHITECT:

By: _____   Date: _____

This Certificate is not negotiable. The AMOUNT CERTIFIED is payable only to the Contractor named herein. Issuance, payment and acceptance of payment are without prejudice to any rights of the Owner or Contractor under this Contract.

FIGURE 14–20   Example Computation, Payment 4, Application and Certificate for Payment

Payment Processing

After the contractor completes the application for payment, the architect and engineer review the application and certify the payment request if correct. Some architects and engineers request a rough payment to review, then the final application is prepared with the architect's comments. This procedure usually saves time, avoiding rejection of the application for payment. Most contracts specify how long the owner can take to process the progress payment after receipt of the certified application. This period usually is thirty days, but can vary depending on specific factors relating to the owner and project financing. Figure 14–21 illustrates the progress payment process. Many owners, even public owners, try to

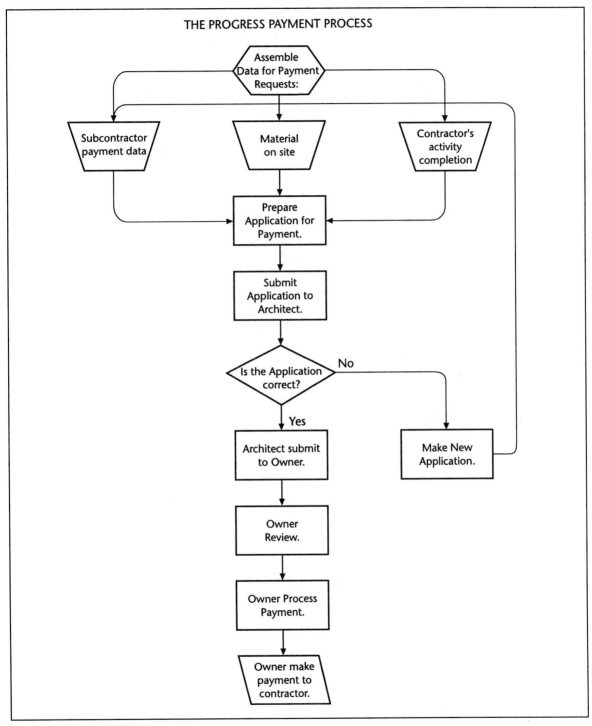

FIGURE 14–21   The Progress Payment Process

expedite progress payments to the contractor. A good relationship between the owner and contractor can facilitate prompt payment of applications.

Summary

Progress payments are an important part of contract administration for the contractor. Proper preparation of the information necessary for payment processing can help the contractor financially complete the project.

At the start of the project, the contractor prepares the schedule of values and a cash flow projection. The schedule of values is a list of the amounts for work activities, which is the basis for progress payments. Preparation of the schedule of values from the estimate can include a combination of estimate items, a separation of estimate items, and distribution of mark-up (indirect overhead and profit).

The cash flow projection is an estimate of progress payments throughout the project, according to the construction schedule. Some construction scheduling software provides cash flow projections. Cash flow projections should accurately estimate progress payments, since the owner and contractor make financial arrangements based on the cash flow projection.

Most contracts use a standard format for progress payments. These forms use a detail sheet to show the completion of each payment item, material stored, amount remaining, and proportionate amount of retainage held for each item. The Application and Certification for Payment includes certifications for both the contractor and architect, a change order summary, and a computation of the current progress payment.

CHAPTER 15

PROJECT CLOSEOUT

CHAPTER OUTLINE   The Closeout Process

Financial Resolution of the Project

Summary

The contractor's primary goal is to complete a successful project. The **project closeout** is an important element in achieving this goal and is defined as the process of completing a construction project, including completion of contractual requirements, approvals, financial resolution, and required documentation. Completing a construction project can take a long time due to waiting for delivery of minor items, repair work, or completion of required paperwork. During this period, the contractor still has overhead costs such as temporary jobsite facilities and field staff. Quick and efficient completion of closeout requirements minimizes the additional amount spent by the contractor on continuing overhead and enables personnel to pursue other projects. The owner is also usually anxious to complete the project as quickly as possible. The owner remembers the completion phase of the project and the last impression of the contractor will affect the project's timeliness and ultimate success.

Project closeout consists of the activities that complete the project requirements. This phase of the project naturally includes the physical completion of the project, with the remaining items completed and the defective areas repaired. The project closeout phase also involves completion of paperwork and other contractual requirements, such as operation and maintenance manuals and instructions. Financial resolution of the project is necessary during the closeout phase as well.

This phase of the construction process is perhaps the most difficult to successfully coordinate. As the project approaches completion, the remaining items usually consist of the following:

1. Completion of seemingly minor details of work. Many of these remaining details are neglected by subcontractors when they leave the project.
2. Delivery of late material and equipment. Many of these items are not essential to the operation of the facility.
3. Replacement of defective materials, equipment, or parts of assemblies. Replacement materials normally are ordered prior to the closeout period, but they may not have arrived or been installed.
4. Repair of defective workmanship. Most defective workmanship relates primarily to finishes, as workmanship repairs to structural assemblies or the building envelope probably are completed prior to the project closeout phase.
5. Testing and approval of building systems, including code inspections, fire alarm and building safety, testing of HVAC systems, temperature control systems, and other specialized systems.

The aforementioned items require identification and usually a special follow-up with subcontractors and suppliers. As the majority of the subcontractor's work is complete, and most of the payment has been received, the subcontractor usually is not highly motivated to return to the project to complete minor items and repair defective work. Suppliers, as well as subcontractors, have priorities other than pursuing replacement parts or equipment. To accomplish this work, the contractor must ensure that subcontractors and suppliers complete contractual requirements. Some methods that help minimize the time required to complete these items include:

1. The contractor's field personnel should be aware of the quality of the subcontractors' work during its progress. A list of remaining items and defective installations should be presented to the subcontractor, emphasizing that the items must be completed prior to the subcontractor leaving the jobsite. Although this method alone does not ensure that the subcontractor will complete the work satisfactorily before leaving the jobsite, it does help.

2. All defective equipment delivered to the jobsite should be tracked for: date delivered to jobsite; date rejected; description of defective parts or equipment needing replacement; responsible party, subcontractor, supplier, or the subcontractor's supplier; date ordered; date scheduled for delivery; follow-up checks by contractor; and date actually delivered to the jobsite. Documentation is necessary to manage the replacement of material or equipment. Written notification to the subcontractor and supplier serves as documentation and also encourages the subcontractor or supplier to order and pursue replacements.

3. At completion, or near completion of the project, the contractor must make for each subcontractor and its crews a list of work that needs to be completed. Many contractors wait for the architect and engineer to make this list, since subcontractors do not enjoy making several trips back to the jobsite to complete work. The earlier the subcontractors receive the list of remaining items, the sooner the problems will be solved. The list of items to the subcontractor should be specific. Some contractors describe the remaining work to the subcontractor as "complete contract requirements" to avoid omitting items that must be completed. A specific list of items, such as "Complete ceramic tile base in restrooms 105 and 106," is much more effective in facilitating successful completion of the work.

4. Follow-up with subcontractors and suppliers usually is necessary. Both written reminders and telephone conversations are necessary to emphasize the importance of completion of the work and to update work and delivery schedules.

5. Many contractors are cautious about releasing all or most of the subcontractors' and suppliers' payments until all work is complete. If no amount is due the subcontractor other than retainage, or a very small amount remains to be paid, subcontractors may have stronger priorities other than completing their work.

As indicated earlier, special care is required for the contractor's personnel to complete the closeout of the project. All of the contract requirements for closeout must be actively pursued, as nothing will automatically fall into place. Because all closeout items, both physical completion and paperwork requirements, require additional effort, they simply will not be completed unless they are pursued.

The Closeout Process

Figure 15–1 illustrates the steps for closing out a project. A distinct order of activities should be observed during this process, as dictated contractually and by custom. The relative amount of completion of work has a large influence on the progress of this sequence.

As the project nears completion, the contractor should compile a list of remaining work to be done, including repairs—this is normally called a punchlist. This list is distributed to all appropriate parties and completed. Most contracts require that this preliminary or contractor's punchlist be completed prior to notification of the architect and engineer. In actual practice, however, this preliminary punchlist may not be actively pursued, since most subcontractors wish to make only one trip back to the jobsite. It is important, however, that this punchlist is addressed and completed, as it actually speeds up the closeout of the project. Most architects and engineers expect the project to be complete upon their arrival, requiring minimal punchlist items. Because the architect's inspection is directly related to substantial completion, most architects will not conduct the punchlist inspection if many and/or large items must still be completed. In this case, architects and engineers bill the contractor for extra inspections, particularly in situations where travel is required.

When the preliminary punchlist is complete, the architect is notified and requested to perform a punchlist on the project. Notification to the architect should always be in writing, as it begins to establish the completion sequence. The architect and consulting engineers will then inspect the project, or the requested portions of the project, and compile the punchlist. Punchlists exist in many forms, but they usually list the room number and the particular problem. The architect probably will not separate items according to subcontract, due to lack of awareness about subcontractor division on the project. Consulting engineers, such as electrical and mechanical engineers, usually will prepare a separate punchlist that relates only to their portion of the work.

Punchlists

Punchlists come in every and any form, typed and handwritten. The owner's maintenance personnel often makes a punchlist, but this should be coordinated with and included in the architect's punchlist. The contractor is responsible for the project, thus must provide specific punchlists to subcontractors and suppliers.

Common types of punchlists include:

- A sheet of paper posted on each door, or in every room, indicating the items that need to be completed in the room, and a sign-off date for each item for both the contractor and architect. This method is effective for having personnel complete items in every room. The sheets of

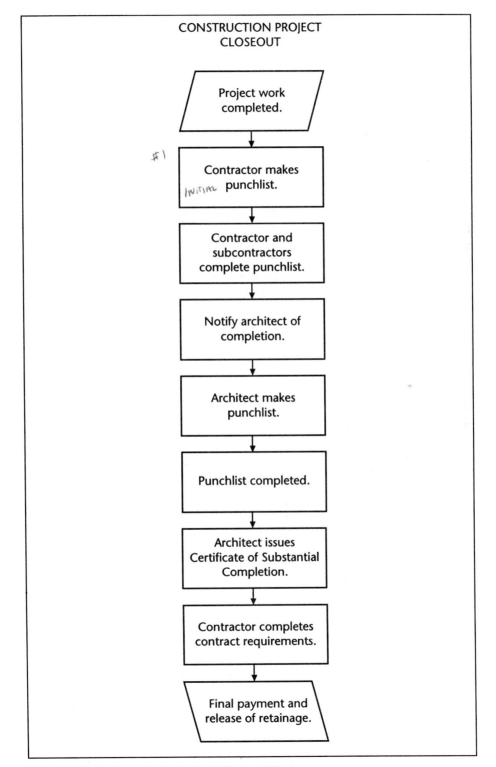

FIGURE 15–1   Steps in the Project Closeout Process

paper, however, are occasionally removed or signed by unauthorized personnel. This punchlist is hard to use as a record and requires separate permanent documentation. Items that apply to the entire project or entire systems do not usually fall into the room-by-room categories and must be tracked as a conventional punchlist. Figure 15–2 illustrates a typical form for this type of punchlist.

• Another form of punchlist is a written list of items from the architect to the contractor. There also are several variations to this type of list. Usually, items are related to the specific room or area. Figure 15–3 illustrates this type of punchlist.

This type of punchlist does not include log information. The contractor probably will want to make a log form on a database for the punchlist, to enable tracking of activities. The use of a database on a computer permits sorting, for example, by room or by subcontractor. The punchlist log in Figure 15–4 contains basic information, but other facts might be included, such as the date the subcontractor was notified, subcontractor identification number, scheduled dates for completion of the items, and punchlist version.

Substantial Completion

After the architect has compiled the punchlist and the items have been corrected, a "Certificate of Substantial Completion" will be issued.

ROOM 1065
OFFICE
POSTED _____ 6/27/97

ITEM	Gen.Cont. Complete	Architect Approval	Remarks
Rubber base	6/29/97	7/5/97 *HDF*	
Replace defective faucet @ sink	6/30/97	7/6/97 *JLT*	
Install Folding Door			Delivery scheduled for 7/10/97
Replace cracked outlet plate	6/29/97	7/6/97 *JLT*	
Touch-up paint, west wall	6/29/97		*Not Acceptable 7/5/97 HDF*

FIGURE 15–2
Punchlist Example, Posted in Each Room

FIGURE 15–3
Punchlist Example,
Comprehensive List

PUNCHLIST
PROJECT: PROMINENT OFFICE BUILDING
JUNE 28, 1997
PUNCHLIST CONDUCTED BY A.L.SMITH
ARCHITECT: SMITH ASSOCIATES, P.S.
CONTRACTOR: XYZ, INC.

I. General Items:
  A. Final Cleaning of all surfaces is necessary, as per section 01710
  B. Install new filters in HVAC equipment
  C. Remove all waste and debris from site.
II. Exterior:
  A. West Elevation:
      1. Install splash block at downspout
      2. Caulk around aluminum storefront
  B. North Elevation:
      1. Touch-up paint on coping
  C. East Elevation:
      1. Adjust operator on overhead door at room 1001
      2. Paint trim around overhead door
  D. South Elevation:
      1. Install sign at entrance
III. Interior:
  A. Room 1065
      1. Install rubber base
      2. Replace defective faucet at sink
      3. Install Folding Door
      4. Replace cracked outlet plate
      5. Touch-up paint, west wall

FIGURE 15–4
Punchlist Log

ITEM #	LOCATION	ITEM	RESPONS	Completed Date	Init	Approval Date	By	REMARKS
I.A.	General	Final Clean	XYZ	7/5/97	RTZ	7/10/97 ALS		
I B.	General	New Filters	ABC Mech.	7/6/97	RTZ	7/10/97	ALS	
I C.	General	Waste Rem.	XYZ	7/6/97	RTZ	7/10/97	ALS	
II A. 1	West ext.	Splash block	XYZ	7/6/97	RTZ	7/10/97	ALS	
II A.2.	West Ext.	Caulking	A-1 Sealants	7/9/97	RTZ	7/10/97	ALS	
II B 1.	North Ext	Paint Coping	Steve's Pntg	7/9/97	RTZ	7/10/97	ALS	
II C 1.	East Ext	Ovhd. Door	Doors, Inc.					Scheduled 7/13/97
II C 2.	East Ext	Paint trim @ ovhd. door	Steve's Painting	7/9/97	RTZ	7/10/97	ALS	
II D 1	South Ext	Sign	DG Specialties					Delivery : 7/15/97
III A 1	Rm 1065	Rub.base	Floors, Inc.	7/8/97	RTZ	7/10/97	ALS	
III A 2	Rm 1065	Repl.Faucet	ABC Mech.					Delivery: 7/15/97
III A 3	Rm 1065	Folding Door	DG Specialties					Delivery: 7/20/97
III A 4	Rm 1065	Replace outlet plate	Sparks Electric	7/6/95	RTZ	7/10/95	ALS	
III A 5	Rm 1065	Touch up Paint, W. wall	Steve's Painting	7/9/95	RTZ	7/10/95	ALS	

Substantial completion has several different meanings to the participants in the construction process, but basically it is defined as the point in the project when the architect has determined that the facility or a portion of the facility is acceptable for owner use and occupancy. The commonly used term for this stage of completion is **beneficial occupancy**, indicating when the facility can be used for its intended purpose. The punchlist has been issued, items have been corrected, and corrections have been approved. Some items may still need to be completed, but these are considered minor. All remaining items are included in the Certificate of Substantial Completion. The date of the substantial completion is extremely important, as it stops the potential of liquidated damages and establishes the contract warranty period. Upon reaching substantial completion, liquidated damages usually cannot be imposed, and if they are in effect, substantial completion will stop the liquidated damages from being further imposed. Each project may have several substantial completions for portions of the work and may also have several warranty periods. The contractor must keep a record of multiple substantial completions when the project has phased completion, as shown in Figure 15–5.

Most contact document systems, such as the AIA and EJCDC documents, use a standard form for the Certificate of Substantial Completion. AIA Document G704, "Certificate of Substantial Completion," is commonly used in commercial building construction (Figure 15–6).

Essential elements of the "Certificate of Substantial Completion" include:

- *Project identification:* Official project name and address, architect's identification number, contract amount, contract date, and names of architect, owner, and contractor.
- *Date of issuance/Date of substantial completion:* The date of issuance is not as important as the date established for substantial completion, which is in a separate location on Form AIA G704. The date of substantial completion establishes the completion date of the project

Area Complete (Description)	Substantial Completion Date	List Completion Date	Warranty Completion Date
Warehouse, Bldg. A	7/31/97	8/15/97	7/31/98
Office, Bldg. A	8/15/97	9/15/97	8/15/98
Building B	10/1/97	11/1/97	10/1/98
Building C	12/1/97	12/30/97	12/1/98

FIGURE 15–5
Substantial Completion Log

CERTIFICATE OF
SUBSTANTIAL COMPLETION

AIA DOCUMENT G704

(Instructions on reverse side)

OWNER ☐
ARCHITECT ☐
CONTRACTOR ☐
FIELD ☐
OTHER ☐

PROJECT:
(Name and address)

PROJECT NO.:

CONTRACT FOR:
CONTRACT DATE:

TO OWNER:
(Name and address)

TO CONTRACTOR:
(Name and address)

Reproduced with permission of The American Institute of Architects under license number #96049. This license expires September 30, 1997. FURTHER REPRODUCTION IS PROHIBITED. Because AIA Documents are revised from time to time, users should ascertain from the AIA the current edition of this document. Copies of the current edition of this AIA document may be purchased from The American Institute of Architects or its local distributors. The text of this document is not "model language" and is not intended for use in other documents without permission of the AIA.

DATE OF ISSUANCE:
PROJECT OR DESIGNATED PORTION SHALL INCLUDE:

The Work performed under this Contract has been reviewed and found, to the Architect's best knowledge, information and belief, to be substantially complete. Substantial Completion is the stage in the progress of the Work when the Work or designated portion thereof is sufficiently complete in accordance with the Contract Documents so the Owner can occupy or utilize the Work for its intended use. The date of Substantial Completion of the Project or portion thereof designated above is hereby established as

which is also the date of commencement of applicable warranties required by the Contract Documents, except as stated below:

A list of items to be completed or corrected is attached hereto. The failure to include any items on such list does not alter the responsibility of the Contractor to complete all Work in accordance with the Contract Documents.

ARCHITECT _____ BY _____ DATE _____

The Contractor will complete or correct the Work on the list of items attached hereto within _____ days from the above date of Substantial Completion.

CONTRACTOR _____ BY _____ DATE _____

The Owner accepts the Work or designated portion thereof as substantially complete and will assume full possession thereof at _____ (time) on _____ (date).

OWNER _____ BY _____ DATE _____

The responsibilities of the Owner and the Contractor for security, maintenance, heat, utilities, damage to the Work and insurance shall be as follows:
(Note—Owner's and Contractor's legal and insurance counsel should determine and review insurance requirements and coverage.)

AIA CAUTION: You should use an original AIA document which has this caution printed in red. An original assures that changes will not be obscured as may occur when documents are reproduced.

AIA DOCUMENT G704 • CERTIFICATE OF SUBSTANTIAL COMPLETION • 1992 EDITION • AIA® • ©1992 • THE AMERICAN INSTITUTE OF ARCHITECTS, 1735 NEW YORK AVENUE, N.W., WASHINGTON, D.C. 20006-5292
WARNING: Unlicensed photocopying violates U.S. copyright laws and is subject to prosecution.

G704-1992

FIGURE 15–6 AIA Document G704, "Certificate of Substantial Completion"

relating to liquidated damages and also is the beginning of the warranty period.

- *Description of the project or portion of the project:* In cases of substantial completion for a portion of the project, the description must be made in adequate detail, including an explanation of the applicable environmental systems. When using a standardized form, an attachment may be necessary for the detailed description of the portion of the work.
- *Definition of "Substantial Completion":* AIA Form G704 defines "Substantial Completion" on the actual form. Although it may not be essential to contain the definition in this certificate, it can be useful since many different opinions exist.
- *List of remaining responsibilities:* A list of remaining items to be "completed or corrected" must be attached to the certificate. Because most punchlist items are completed or corrected prior to the certificate, the attached list should be minimal. Final payment usually will not be made until all times are complete.
- *List of warranty dates:* A list of warranty dates also should be attached to the certificate. Several different warranty dates are possible, however the customary general warranty is one year. Other special warranties such as roofing, equipment, and finish hardware can be specified in the documents.
- *Signatures:* Signatures by the architect, contractor, and owner are necessary to validate this document. The certificate is an agreement by all three parties stating that the work is substantially complete.
- *List of agreements:* An agreed-upon list of responsibilities for heat, utilities, and other services should be attached. Property insurance must be established by the owner, particularly if immediate occupancy is desired. Any future damage to finishes is normally the owner's responsibility.

Following completion of the remaining items and financial resolution of the project, final payment and release of retainage is made to the contractor.

Paperwork Requirements

The old adage that "the job isn't done until the paperwork is done" applies to the construction project. Most contracts contain provisions for lien releases, written warranties, and contract receivables, such as Operating and Maintenance Manuals. Most public work requires further releases, certifications, and affidavits from public agencies.

Prompt pursuit of these finalizing documents is essential to releasing the final payment for the project. The majority of documents are completed by subcontractors and governmental agencies. The contractor must actively pursue all responsible parties to quickly process these documents.

Inspection Agency Releases

Most communities have local building code authorities that must inspect the project through construction at specific phases. Final inspection and a "Certificate of Occupancy" must be issued prior to the owner occupying the facility. Most communities require a final inspection on the entire project and special inspections/certifications in the following areas:

Plumbing	Electrical
HVAC Equipment	Fire Alarm
Elevator	Environmental/Storm Water Drainage
Public Works (roads, etc.)	Health Department (sewage systems)
Planning Compliance	ADA Requirements (handicapped access)
Fire Protection Systems	

Requirements will vary from one community to another. The contractor should devise a checklist that relates to the inspection and certification requirements in the community. These supersede the contractual requirements and probably will not be delineated in the contract documents, other than as a reference to comply to all local codes and regulations.

Documentation of the final inspections should be kept, with a copy of all certifications and final inspections given to the owner. Some inspection certificates, such as plumbing and electrical, are presented to the subcontractor by the inspector. Contractors should obtain copies of these certificates for their own as well as the owner's records.

System Testing and Documentation

During the punchlist and closeout period, certain building systems are tested for compliance to specifications. Most tests are in the mechanical and electrical subcontract areas. Testing may be required in several different ways:

- independent testing, contracted to the owner
- independent testing, contracted to the subcontractor
- testing and start-up by an authorized manufacturer's representative
- testing by the mechanical and/or electrical subconsultant
- testing by the owner's personnel

Figure 15–7 illustrates a sample testing log.

These tests must be accomplished after the system is fully operational. Arrangements should be made with the owner and architect and engineer for their representatives to be present to witness the tests. Depending upon the nature of the test, code officials may wish to witness tests as well. Subcontractor or installer personnel also should be present at this

FIGURE 15–7
Testing Log

Test	System	Date	Method	Results	Tested by	Witnessed
Fire Protect.	Fire Alarm Fire Sprinkler	6/3/98	Alarm, Smoke	OK	John Smith, Fire Marshall	*JLS*
Plumbing Vents, Drains	Plumbing	3/23/98	Hydrostatic Pressure	OK	Fred Johnson Plumbing Inspector	*FEJ*
Pumps	Plumbing, Fire Sprinkler	5/2/98	Pressure, Flow	OK	Ole Olsen, Pump Rep.	*OO*
Fans	HVAC	5/10/98	Speed, Blade angle	OK	N.T. Jones Fan Rep	*NTJ*
Temperatiure Controls	HVAC	5/13/98	Calibration, computer chk	OK	R.T. Andrews Temp.Cont. Rep.	*RTA*
Elevator	Elevator	4/2/98 5/8/98	Complete	No OK	O.McCarthy State Elevator Inspector	*OTM*

time. Documentation of each test should be made, with the following information:

- date and location of test
- system or equipment tested
- method of testing
- individual conducting the test
- results of the test
- witnesses to the test, signed by each

Standardized test forms provide uniform information and are easily referenced. All systems testing should be done prior to occupancy and use of the facility.

HVAC systems require testing and balancing of equipment. Both operations are often combined. Outside consultants may be used to provide this testing and balancing to verify that the system meets the requirements.

Operating and Maintenance Manuals and Instructions

Most contract documents require that the contractor furnish Operating and Maintenance (O & M) Manuals for all equipment on the project. Although these manuals normally include mechanical and electrical information, relevant facts about other operating systems installed by the contractor or other subcontractors should be included as well. Operating doors, finish hardware, folding partitions, and other equipment items could be listed.

The Operating and Maintenance Manuals include manufacturer's instructions about the equipment, including diagrams and parts lists. Some of the data contains catalog information and some is furnished with the equipment. Because installation and maintenance instructions often are destroyed in the uncrating process, special care must be taken to save all information that is sent with equipment. This information must be saved and contained in the appropriate volume of the O & M Manuals.

For larger projects, the contract documents usually must be contained in hard-cover, bound volumes of the O & M Manuals, which are permanent references that should be kept at the facility by maintenance personnel for the life of the building. Documents are bound to prevent removal of information as this information may not be available in later years. Separate volumes usually are made for the mechanical and electrical areas, since they are compiled by separate subcontractors. The subcontractor can compile this information in-house or can ask a consultant to prepare it.

The O & M Manuals should be organized as they relate to project specifications. If the CSI MasterFormat numbering system is used, manuals should be organized in that manner. Contract documents may list a specific organization for these manuals.

Delivery of manuals must be made through the proper channels, with documentation of their receipt. Because these manuals are unique, specifically assembled for the particular project, they are difficult to replace if they are lost or not directed to the proper individual. Follow-up often is necessary to ensure proper distribution.

Instruction sessions also are held for Operating and Maintenance instructions, normally conducted by the subcontractor and manufacturer's representatives for the facility's maintenance personnel. Instructions should be given after systems are complete and tested, but prior to owner occupancy of the building. Maintenance personnel should be familiar with the operation of the equipment prior to occupancy, if possible. Instructional sessions can be held after testing, when the manufacturer's representatives are on the site. Some projects may require several different instructional sessions to cover all of the equipment installed on the project. Documentation of instructional meetings is necessary to indicate who demonstrated the equipment and who received instruction.

Videotaping of instructional periods provides a record that show that instructional sessions were indeed held and also provides the owner with a reference for use by maintenance personnel. Because maintenance personnel changes frequently, the owner can use the videotapes to train new maintenance workers.

Spare Parts and Extra Materials

Most contracts require delivery of extra material and spare parts to the owner at the conclusion of the project. The technical specification for each area of work will specify, if required, the necessary spare parts or material to leave on the project. Material and equipment installed during construction projects can be quickly dated from the manufacturer or supplier. Finish materials usually are in specific dye lots that cannot be easily matched. Technical specifications require extra material, such as carpet, floor tile, or paint, to patch changes or future damage. Some spare parts will be specified to enable the owner to quickly repair equipment items that are worn.

Most spare parts and extra material are from subcontractors. The contractor should prepare a checklist of spare parts and extra material early in the project and refer to it as subcontractors leave the project. Because this material is in order, it will be available as subcontractors finish their segments of the project. The material and spare parts should be collected from subcontractors prior to their leaving the jobsite and should be stored in a safe area. All keys to the equipment should be collected as well, prior to the subcontractor leaving the jobsite. An example of a checklist for spare parts and extra material is shown in Figure 15–8.

Keys, Permanent Cylinders, and Rekeying

Turnover of keys and changing door lock cylinders can be a cumbersome task. Many owners prefer, for security reasons, that the permanent keying of locks be done just prior to occupancy. Some keying systems have removable cores that can be replaced just prior to the owner's occupancy of the building. Under this system, the contractor is supplied "construction cores" by the hardware supplier, providing locking capabilities during construction. At the end of the construction period, the cores are changed by the supplier or the owner's maintenance personnel, providing the final keying and security for the locks.

During the construction period, preferably during the submittal period, the finish hardware supplier requests a keying schedule for the building. This can be a very complicated list, with master and sub-master systems. Depending upon the type of lock furnished, locks may be keyed prior to delivery, or furnished with construction cores. If permanent keying is provided with the locks, the contractor must organize all of the keys to ensure that they are given to the owner. Most security issues impact on the keying of facilities, requiring the contractor to be extremely careful distributing and organizing keys. In some arrangements, the hardware supplier handles the keys and key cabinets with the owner. Regardless of who is responsible for key turnover, the contractor is required to facilitate

Section	Subcontractor	Material Description	Date Received	Location Stored
08700,Fin.Hdwe	National Hdwe.	2 ea. locksets	8/24/97	Trailer
09300, Cer.Tile	United Floors	2 boxes tile	7/27/95	Trailer
09680,Carpet	United Floors	10 SY carpet	10/15/97	Rm. 1011
09900,Painting	City Painting	1 gal/color	10/2/97	Rm. B-0027
09950,Wallcvrg	City Painting	1 bolt/ pattern	10/2/97	Rm. B-0027

FIGURE 15–8
Spare Parts Log

the keying and key turnover prior to final acceptance of the facility. A signed receipt for any keys is necessary, similar to procedures for other receivables to the owner.

Record Drawings

Most contracts require the contractor to prepare record drawings or "as-built" drawings during the project. This process entails recording actual dimensions, location of utilities, and any changes in drawings. These changes and dimensions usually are written on clean sets of drawings during the entire project. Electrical and mechanical contractors normally record relevant information on their own drawings.

Since the as-built set of drawings is usually fairly worn by the end of the project, some contracts request that the contractor transfer changes onto sepia or velum sets of drawings. The contractor normally completes the record drawings and provides the necessary drawings to the owner. Some "as-built" drawings are recorded into CAD media on the jobsite. In this case, the contractor would provide the owner with the revised set on disk, furnishing the hard copy as well.

As with all of the other receivables during the closeout period, signed documentation of receipt of the necessary as-built drawings is necessary. It is important that the contractor keep this documentation safe, in a retrievable location. Owners will occasionally misplace receivables and claim they never received the documents from the contractor, therefore written records are essential.

Warranties and Guarantees

Most contracts require a one-year warranty on all installations in the project. Some items may require two- or five-year warranties. The applicable information is found in the technical specification in each section. Warranties are provided during the closeout period. Most subcontractors furnish a warranty to the contractor for their work. All special warranties beyond the one-year program should be copied and sent to the owner. Since the owner's contract is with the contractor only, the contractor is ultimately responsible for the warranty to the owner.

A checklist on warranties should be made from the information contained in the technical specification, similar to the table shown in Figure 15–9.

Prompt repair of warranty items is essential during the warranty period. Many items, particularly mechanical and electrical, can require emergency repair. The contractor, with the owner's maintenance personnel, should formulate a plan for warranty calls and repairs. Most contractors prefer that mechanical and electrical problems be reported directly to the appropriate subcontractor, with subsequent notice to the contractor that the subcontractor has been contacted. All other calls should be made directly to the contractor. The contractor should discuss and agree with

FIGURE 15–9
Warranty Checklist

Warranty Checklist

Section	Subcontractor	Warranty Required	Date Received	Date to Owner
02850 Irrigation System	City Sprinkler Co.	2-season special warranty	10/10/97	12/1/97
02900 Landscaping	Trees-R-Us	2-season special warranty	10/01/97	12/1/97
03400 Precast Concrete	County Precast, Inc.	1 year standard warranty	9/08/97	
04000 Masonry	ABC Masonry	1 year standard warranty	9/01/97	
07500 Roofing	Anderson Roofing	5 year warranty; bond	10/15/97	12/1/97
08200 Metal Doors / Frames	Doors-R-US	1 year standard warranty	9/15/97	
08700 Finish Hardware	Acme Hardware	1 year standard warranty	9/15/97	

maintenance personnel on the parameters of warranty work. For instance, warranty will include repair to an airhandler, but will not include filters, belts, and other maintenance replacement items.

Affidavits of Payment, Lien Releases, and Consent of Surety

The owner usually will request an affidavit of payment, lien releases, and consent of surety prior to releasing final payment. An "affidavit of payment" certifies that payment of all debts and claims relating to the project has been made, with the noted specific exceptions, which usually is only retainage. If further items remain, the contractor can post a bond to cover payment of the associated debt or claim. The owner must be assured that the contractor has paid all material, labor, taxes, and subcontracts in relation to the project. The owner does not want to be held liable for these debts after releasing payment to the contractor. Retainage, in many cases, is reserved solely for payment of "labor and materialmen" (suppliers) and cannot be released until the owner has some assurance that these entities have been paid.

Lien releases may be requested from the contractor and all subcontractors. The lien release indicates that the potential lien holder has received payment for the work, with the exceptions noted, usually retainage. The owner must be assured that the project will be received free and clear, with no further obligation. The owner normally will require a waiver or release of lien from the contractor, as the legal contract is with the contractor. The owner can ask for lien releases from all subcontractors and suppliers as well, to provide additional assurance that the project will not have further encumbrances.

An accompanying document to the affidavit of payment and lien releases is the "Consent of Surety," when bonding is used on the project. Consent of Surety is executed by the bonding company, indicating approval of the contractor's receiving final payment for the project. This requirement provides the bonding company with the opportunity to audit the contractor's records to ensure that a financial obligation does not exist. Consent of Surety also should be required by the contractor on subcontractors that have been bonded on the project.

AIA standard forms exist for the following documents:

AIA Document G706: Contractor's Affidavit of Payment of Debts and Claims

AIA Document G706 A: Contractor's Affidavit of Release of Liens

AIA Document G707: Consent of Surety Company to Final Payment

AIA Document G707 A: Consent of Surety to Reduction in or Partial Release of Retainage

These forms contain the standard language used in these documents. Custom documents often are used for releases as well. The contractor must use the proper wording and submit the correct forms, as these are important legal documents that establish a basis for final payment and termination of the contract.

Miscellaneous Certifications and Releases

Particularly in public work, the contractor may be required to supply additional certifications, affidavits, and releases from public agencies. These certifications vary, depending on the owner of the project, the type of work, size of contract, and the city, county, and state in which the project is located.

Prevailing wage certificates and affidavits are required on most public work where Davis-Bacon regulations or state prevailing wage laws apply. For federal projects, weekly payroll certificates are required for all labor on the project, both contractor and subcontractor. These certified payrolls should be submitted prior to the closeout period, but some may still need to be completed. Each state has different prevailing wage requirements for state-funded work. Some states require notarized affidavits from each contractor and subcontractor stating that wages were paid. All affidavits and certifications must be promptly submitted at the end of the project, as final payment will not be made until release by the monitoring agency.

In public work for states, the states may require release of state agencies prior to final payment. These agencies may include sales tax collection, workmen's compensation, and unemployment insurance funds. Application for these releases must be made immediately at the close of the project to facilitate timely release of the final payment.

Financial Resolution of the Project

All financial issues must be completed, just as the physical attributes of the facility must be. The objective in financial resolution is to resolve all outstanding cost commitments, establish a final project cost, and determine profit or loss. This can be a large task even on a small project. Final payment cannot be issued until financial issues are settled between the contractor and subcontractors and between the contractor and owner.

Subcontractor Payment

At the end of the project, or approaching the end of the project, the contractor should determine the following for each subcontract:

- the amount of work remaining and the value of such work
- the amount of payment, excluding retainage, due
- the amount of back-charges and disputed payment
- the amount of retainage due
- the approximate dates of completion of the remaining work

With this information, the contractor must proceed to make remaining payments to the subcontractor and establish what must be completed prior to final payment. Because the contractor must certify that payment has been made, the remaining amount should be established and resolved. Resolution of disputed issues needs to be negotiated and finalized. Some issues may remain in dispute and should be listed as exceptions to the payment. These disputed items should be pursued, however, to avoid confusing payment and settlement issues.

Resolution With the Owner

The final contract amount needs to be established. To accomplish this, all change orders must be completed and issued, as well as the associated work done. Last-minute change orders, summary change orders, and delayed change orders frequently seem to accumulate at the end of the project. These change orders should be dealt with quickly, expediting the process.

Claims also need to be resolved. The contractor should examine each unresolved issue and determine what can be done to solve the dispute. A procedure for resolution, even if the resolution involves arbitration or litigation, must be established for each disputed item. After procedures have been created, the contractor should purse each issue as quickly as possible. Issues may arise that cannot be immediately resolved, thus third parties may have to resolve them.

For items that need to be completed, or issues that need to be resolved, the contractor can post a **retainage bond** for the amount of retainage or amount being held. The retainage bond is issued by a surety company for the remaining value, allowing the contractor to be paid the retainage. The retainage bond assures the owner that the remainder of contract requirements will be completed. The surety company charges the contractor additional premium for this bond.

The contractor also may need to file a lien on the project for an unpaid balance or disputed items. Each state has finite lien periods after the completion of the work. All liens must be filed within this period or the contractor loses the right to file a lien. This is a serious action and must be done within state regulations.

Cost Control Completion

The final accounting of the project must be done as the final resolutions close. The contractor must analyze the project accounting to determine whether a profit or loss was realized. Beyond the accounting for the project, each activity cost should be compared to the estimate. A variance of 10 percent over or under the estimate should be examined. As the actual costs are used to create historical cost data, large variances need to be examined, and a reason for the variance established. Any unusual condition should be noted, but this should not adversely affect the historical cost data.

Final reports for the project's financial condition should be in a standard format, comparable to other projects. It is important that management review the project and determine its success and failures and reasons for each. Conclusions can relate to personnel, locality, type of work, specific subcontracts used, and a variety of other reasons. Management must understand the successes and failures and apply that knowledge to future projects.

Archiving Records

Records of the project must be stored as a reference. This reference is used for dispute resolution, public agency audits, management reference, and financial (tax) accounting. Most of the records should be stored for at least the specified period of the statute of limitations in the state in which the project is located. Some companies store records for ten years before destroying them.

Most project records today are kept on computer media for the duration of the project. At the project's conclusion, though, all documents must be converted to hard copy. Storing the documents on computer disks is acceptable for short-term use, but with the evolution of computer hardware and software, the computer data may not be accessible in the future. Records should be stored in an environment that will not cause deterioration of the documents.

All project records have some value, but it is impractical to store all documents. Project documents should be examined, anticipating future need. Two major sets of document exist: project documents and financial documents.

Project Documents include:

- drawings, specifications, change orders, as-built drawings
- pre-contract estimates, proposals
- shop drawings, submittals
- contracts, agreements, change orders, change directives (with owner)
- subcontracts, subcontract change orders, purchase orders, delivery records
- correspondence: owner, architect, subcontractors, suppliers, miscellaneous (with issue cross-index)
- photographs, videotapes
- daily reports, diaries, special reports on occurrences
- minutes of meetings
- logs: submittals, telephone, etc.
- safety records: minutes, reports, injury reports, inspections, etc.
- permits and inspection records

Financial Documents include:

- payroll
- cost reports
- vendor invoices and payment records
- subcontractor bonds, releases, and warranties
- billings and receipts from owner

For large projects, documents available for storage are mammoth. Microfilm storage may be desirable in this case to conserve space. All agreements, change orders, and other documents with original signatures should be saved in their original form.

Summary

The closeout phase of the construction project requires diligent and focused efforts by the contractor's personnel to complete contractual requirements, allowing final payment for the project. Careful coordination with subcontractors, the architect, building officials, and the owner is necessary to minimize the duration of the closeout phase.

A sequence of closeout activities, starting with the contractor's punchlist and ending with the issuance of the certificate of substantial completion, is done to complete the physical work at the jobsite. Use of checklists and logs can help organize the complicated process of completing remaining work. Substantial completion occurs at the culmination of the punchlist. Substantial completion starts the warranty period for the affected areas and stops the potential of liquidated damages or further liquidated damages.

In addition to completion of the physical project, several requirements must be met. Inspection and testing is essential prior to occupancy of the facility. The owner's maintenance personnel should be trained in the operation and maintenance of the facility. Operating and Maintenance Manuals must be prepared and give to maintenance personnel. Spare parts, additional material, and keys need to be given to the owner. The record drawings must be completed by the contractor and subcontractors and delivered to the owner. Warranties and guarantees need to be prepared and given to the owner. A number of releases, such as affidavits of payments, lien releases, consent of surety, and miscellaneous agency releases, as required, should be requested, prepared, and delivered to the owner.

Financial resolution must be made for all subcontracts and contracts on the project prior to final payment. Resolution of change orders, claims, back-charges, and any outstanding financial relationships needs to be done to enable release of the final payment. The contractor should determine the project's profitability, as well as analyze its positive and negative aspects. Each project creates data for estimating, controlling, and managing future projects.

When concluding the project, the contractor must store the necessary documents for reference. Organized archives permits efficient audits and document searches.

Efficient closeout procedures allow the contractor to promptly proceed to other work. The contractor should concentrate on this area prior to proceeding to the next undertaking. Closeout ultimately affects whether the project is viewed as successful or unsuccessful.

APPENDIX

THE   AMERICAN   INSTITUTE   OF   ARCHITECTS

AIA Document A201

General Conditions of the Contract for Construction

THIS DOCUMENT HAS IMPORTANT LEGAL CONSEQUENCES; CONSULTATION WITH AN ATTORNEY IS ENCOURAGED WITH RESPECT TO ITS MODIFICATION

1987 EDITION
TABLE OF ARTICLES

1. GENERAL PROVISIONS

2. OWNER

3. CONTRACTOR

4. ADMINISTRATION OF THE CONTRACT

5. SUBCONTRACTORS

6. CONSTRUCTION BY OWNER OR BY SEPARATE CONTRACTORS

7. CHANGES IN THE WORK

8. TIME

9. PAYMENTS AND COMPLETION

10. PROTECTION OF PERSONS AND PROPERTY

11. INSURANCE AND BONDS

12. UNCOVERING AND CORRECTION OF WORK

13. MISCELLANEOUS PROVISIONS

14. TERMINATION OR SUSPENSION OF THE CONTRACT

This document has been approved and endorsed by the Associated General Contractors of America.

INDEX

2 A201-1987

A201-1987 3

AIA DOCUMENT A201 • GENERAL CONDITIONS OF THE CONTRACT FOR CONSTRUCTION • FOURTEENTH EDITION
AIA® • ©1987 THE AMERICAN INSTITUTE OF ARCHITECTS, 1735 NEW YORK AVENUE, N.W., WASHINGTON, D.C. 20006

GENERAL CONDITIONS OF THE CONTRACT FOR CONSTRUCTION

ARTICLE 1
GENERAL PROVISIONS

1.1   BASIC DEFINITIONS

1.1.1   THE CONTRACT DOCUMENTS

The Contract Documents consist of the Agreement between Owner and Contractor (hereinafter the Agreement), Conditions of the Contract (General, Supplementary and other Conditions), Drawings, Specifications, addenda issued prior to execution of the Contract, other documents listed in the Agreement and Modifications issued after execution of the Contract. A Modification is (1) a written amendment to the Contract signed by both parties, (2) a Change Order, (3) a Construction Change Directive or (4) a written order for a minor change in the Work issued by the Architect. Unless specifically enumerated in the Agreement, the Contract Documents do not include other documents such as bidding requirements (advertisement or invitation to bid, Instructions to Bidders, sample forms, the Contractor's bid or portions of addenda relating to bidding requirements).

1.1.2   THE CONTRACT

The Contract Documents form the Contract for Construction. The Contract represents the entire and integrated agreement between the parties hereto and supersedes prior negotiations, representations or agreements, either written or oral. The Contract may be amended or modified only by a Modification. The Contract Documents shall not be construed to create a contractual relationship of any kind (1) between the Architect and Contractor, (2) between the Owner and a Subcontractor or Subsubcontractor or (3) between any persons or entities other than the Owner and Contractor. The Architect shall, however, be entitled to performance and enforcement of obligations under the Contract intended to facilitate performance of the Architect's duties.

1.1.3   THE WORK

The term "Work" means the construction and services required by the Contract Documents, whether completed or partially completed, and includes all other labor, materials, equipment and services provided or to be provided by the Contractor to fulfill the Contractor's obligations. The Work may constitute the whole or a part of the Project.

1.1.4   THE PROJECT

The Project is the total construction of which the Work performed under the Contract Documents may be the whole or a part and which may include construction by the Owner or by separate contractors.

1.1.5   THE DRAWINGS

The Drawings are the graphic and pictorial portions of the Contract Documents, wherever located and whenever issued, showing the design, location and dimensions of the Work, generally including plans, elevations, sections, details, schedules and diagrams.

1.1.6   THE SPECIFICATIONS

The Specifications are that portion of the Contract Documents consisting of the written requirements for materials, equip-ment, construction systems, standards and workmanship for the Work, and performance of related services.

1.1.7   THE PROJECT MANUAL

The Project Manual is the volume usually assembled for the Work which may include the bidding requirements, sample forms, Conditions of the Contract and Specifications.

1.2   EXECUTION, CORRELATION AND INTENT

1.2.1 The Contract Documents shall be signed by the Owner and Contractor as provided in the Agreement. If either the Owner or Contractor or both do not sign all the Contract Documents, the Architect shall identify such unsigned Documents upon request.

1.2.2 Execution of the Contract by the Contractor is a representation that the Contractor has visited the site, become familiar with local conditions under which the Work is to be performed and correlated personal observations with requirements of the Contract Documents.

1.2.3 The intent of the Contract Documents is to include all items necessary for the proper execution and completion of the Work by the Contractor. The Contract Documents are complementary, and what is required by one shall be as binding as if required by all; performance by the Contractor shall be required only to the extent consistent with the Contract Documents and reasonably inferable from them as being necessary to produce the intended results.

1.2.4 Organization of the Specifications into divisions, sections and articles, and arrangement of Drawings shall not control the Contractor in dividing the Work among Subcontractors or in establishing the extent of Work to be performed by any trade.

1.2.5 Unless otherwise stated in the Contract Documents, words which have well-known technical or construction industry meanings are used in the Contract Documents in accordance with such recognized meanings.

1.3   OWNERSHIP AND USE OF ARCHITECT'S DRAWINGS, SPECIFICATIONS AND OTHER DOCUMENTS

1.3.1 The Drawings, Specifications and other documents prepared by the Architect are instruments of the Architect's service through which the Work to be executed by the Contractor is described. The Contractor may retain one contract record set. Neither the Contractor nor any Subcontractor, Subsubcontractor or material or equipment supplier shall own or claim a copyright in the Drawings, Specifications and other documents prepared by the Architect, and unless otherwise indicated the Architect shall be deemed the author of them and will retain all common law, statutory and other reserved rights, in addition to the copyright. All copies of them, except the Contractor's record set, shall be returned or suitably accounted for to the Architect, on request, upon completion of the Work. The Drawings, Specifications and other documents prepared by the Architect, and copies thereof furnished to the Contractor, are for use solely with respect to this Project. They are not to be used by the Contractor or any Subcontractor, Subsubcontractor or material or equipment supplier on other projects or for additions to this Project outside the scope of the

Work without the specific written consent of the Owner and Architect. The Contractor, Subcontractors, Sub-subcontractors and material or equipment suppliers are granted a limited license to use and reproduce applicable portions of the Drawings, Specifications and other documents prepared by the Architect appropriate to and for use in the execution of their Work under the Contract Documents. All copies made under this license shall bear the statutory copyright notice, if any, shown on the Drawings, Specifications and other documents prepared by the Architect. Submittal or distribution to meet official regulatory requirements or for other purposes in connection with this Project is not to be construed as publication in derogation of the Architect's copyright or other reserved rights.

1.4 CAPITALIZATION

1.4.1 Terms capitalized in these General Conditions include those which are (1) specifically defined, (2) the titles of numbered articles and identified references to Paragraphs, Subparagraphs and Clauses in the document or (3) the titles of other documents published by the American Institute of Architects.

1.5 INTERPRETATION

1.5.1 In the interest of brevity the Contract Documents frequently omit modifying words such as "all" and "any" and articles such as "the" and "an," but the fact that a modifier or an article is absent from one statement and appears in another is not intended to affect the interpretation of either statement.

ARTICLE 2
OWNER

2.1 DEFINITION

2.1.1 The Owner is the person or entity identified as such in the Agreement and is referred to throughout the Contract Documents as if singular in number. The term "Owner" means the Owner or the Owner's authorized representative.

2.1.2 The Owner upon reasonable written request shall furnish to the Contractor in writing information which is necessary and relevant for the Contractor to evaluate, give notice of or enforce mechanic's lien rights. Such information shall include a correct statement of the record legal title to the property on which the Project is located, usually referred to as the site, and the Owner's interest therein at the time of execution of the Agreement and, within five days after any change, information of such change in title, recorded or unrecorded.

2.2 INFORMATION AND SERVICES REQUIRED OF THE OWNER

2.2.1 The Owner shall, at the request of the Contractor, prior to execution of the Agreement and promptly from time to time thereafter, furnish to the Contractor reasonable evidence that financial arrangements have been made to fulfill the Owner's obligations under the Contract. *[Note: Unless such reasonable evidence were furnished on request prior to the execution of the Agreement, the prospective contractor would not be required to execute the Agreement or to commence the Work.]*

2.2.2 The Owner shall furnish surveys describing physical characteristics, legal limitations and utility locations for the site of the Project, and a legal description of the site.

2.2.3 Except for permits and fees which are the responsibility of the Contractor under the Contract Documents, the Owner shall secure and pay for necessary approvals, easements, assessments and charges required for construction, use or occupancy of permanent structures or for permanent changes in existing facilities.

2.2.4 Information or services under the Owner's control shall be furnished by the Owner with reasonable promptness to avoid delay in orderly progress of the Work.

2.2.5 Unless otherwise provided in the Contract Documents, the Contractor will be furnished, free of charge, such copies of Drawings and Project Manuals as are reasonably necessary for execution of the Work.

2.2.6 The foregoing are in addition to other duties and responsibilities of the Owner enumerated herein and especially those in respect to Article 6 (Construction by Owner or by Separate Contractors), Article 9 (Payments and Completion) and Article 11 (Insurance and Bonds).

2.3 OWNER'S RIGHT TO STOP THE WORK

2.3.1 If the Contractor fails to correct Work which is not in accordance with the requirements of the Contract Documents as required by Paragraph 12.2 or persistently fails to carry out Work in accordance with the Contract Documents, the Owner, by written order signed personally or by an agent specifically so empowered by the Owner in writing, may order the Contractor to stop the Work, or any portion thereof, until the cause for such order has been eliminated; however, the right of the Owner to stop the Work shall not give rise to a duty on the part of the Owner to exercise this right for the benefit of the Contractor or any other person or entity, except to the extent required by Subparagraph 6.1.3.

2.4 OWNER'S RIGHT TO CARRY OUT THE WORK

2.4.1 If the Contractor defaults or neglects to carry out the Work in accordance with the Contract Documents and fails within a seven-day period after receipt of written notice from the Owner to commence and continue correction of such default or neglect with diligence and promptness, the Owner may after such seven-day period give the Contractor a second written notice to correct such deficiencies within a second seven-day period. If the Contractor within such second seven-day period after receipt of such second notice fails to commence and continue to correct any deficiencies, the Owner may, without prejudice to other remedies the Owner may have, correct such deficiencies. In such case an appropriate Change Order shall be issued deducting from payments then or thereafter due the Contractor the cost of correcting such deficiencies, including compensation for the Architect's additional services and expenses made necessary by such default, neglect or failure. Such action by the Owner and amounts charged to the Contractor are both subject to prior approval of the Architect. If payments then or thereafter due the Contractor are not sufficient to cover such amounts, the Contractor shall pay the difference to the Owner.

ARTICLE 3
CONTRACTOR

3.1 DEFINITION

3.1.1 The Contractor is the person or entity identified as such in the Agreement and is referred to throughout the Contract Documents as if singular in number. The term "Contractor" means the Contractor or the Contractor's authorized representative.

3.2 REVIEW OF CONTRACT DOCUMENTS AND FIELD CONDITIONS BY CONTRACTOR

3.2.1 The Contractor shall carefully study and compare the Contract Documents with each other and with information furnished by the Owner pursuant to Subparagraph 2.2.2 and shall at once report to the Architect errors, inconsistencies or omissions discovered. The Contractor shall not be liable to the Owner or Architect for damage resulting from errors, inconsistencies or omissions in the Contract Documents unless the Contractor recognized such error, inconsistency or omission and knowingly failed to report it to the Architect. If the Contractor performs any construction activity knowing it involves a recognized error, inconsistency or omission in the Contract Documents without such notice to the Architect, the Contractor shall assume appropriate responsibility for such performance and shall bear an appropriate amount of the attributable costs for correction.

3.2.2 The Contractor shall take field measurements and verify field conditions and shall carefully compare such field measurements and conditions and other information known to the Contractor with the Contract Documents before commencing activities. Errors, inconsistencies or omissions discovered shall be reported to the Architect at once.

3.2.3 The Contractor shall perform the Work in accordance with the Contract Documents and submittals approved pursuant to Paragraph 3.12.

3.3 SUPERVISION AND CONSTRUCTION PROCEDURES

3.3.1 The Contractor shall supervise and direct the Work, using the Contractor's best skill and attention. The Contractor shall be solely responsible for and have control over construction means, methods, techniques, sequences and procedures and for coordinating all portions of the Work under the Contract, unless Contract Documents give other specific instructions concerning these matters.

3.3.2 The Contractor shall be responsible to the Owner for acts and omissions of the Contractor's employees, Subcontractors and their agents and employees, and other persons performing portions of the Work under a contract with the Contractor.

3.3.3 The Contractor shall not be relieved of obligations to perform the Work in accordance with the Contract Documents either by activities or duties of the Architect in the Architect's administration of the Contract, or by tests, inspections or approvals required or performed by persons other than the Contractor.

3.3.4 The Contractor shall be responsible for inspection of portions of Work already performed under this Contract to determine that such portions are in proper condition to receive subsequent Work.

3.4 LABOR AND MATERIALS

3.4.1 Unless otherwise provided in the Contract Documents, the Contractor shall provide and pay for labor, materials, equipment, tools, construction equipment and machinery, water, heat, utilities, transportation, and other facilities and services necessary for proper execution and completion of the Work, whether temporary or permanent and whether or not incorporated or to be incorporated in the Work.

3.4.2 The Contractor shall enforce strict discipline and good order among the Contractor's employees and other persons carrying out the Contract. The Contractor shall not permit employment of unfit persons or persons not skilled in tasks assigned to them.

3.5 WARRANTY

3.5.1 The Contractor warrants to the Owner and Architect that materials and equipment furnished under the Contract will be of good quality and new unless otherwise required or permitted by the Contract Documents, that the Work will be free from defects not inherent in the quality required or permitted, and that the Work will conform with the requirements of the Contract Documents. Work not conforming to these requirements, including substitutions not properly approved and authorized, may be considered defective. The Contractor's warranty excludes remedy for damage or defect caused by abuse, modifications not executed by the Contractor, improper or insufficient maintenance, improper operation, or normal wear and tear under normal usage. If required by the Architect, the Contractor shall furnish satisfactory evidence as to the kind and quality of materials and equipment.

3.6 TAXES

3.6.1 The Contractor shall pay sales, consumer, use and similar taxes for the Work or portions thereof provided by the Contractor which are legally enacted when bids are received or negotiations concluded, whether or not yet effective or merely scheduled to go into effect.

3.7 PERMITS, FEES AND NOTICES

3.7.1 Unless otherwise provided in the Contract Documents, the Contractor shall secure and pay for the building permit and other permits and governmental fees, licenses and inspections necessary for proper execution and completion of the Work which are customarily secured after execution of the Contract and which are legally required when bids are received or negotiations concluded.

3.7.2 The Contractor shall comply with and give notices required by laws, ordinances, rules, regulations and lawful orders of public authorities bearing on performance of the Work.

3.7.3 It is not the Contractor's responsibility to ascertain that the Contract Documents are in accordance with applicable laws, statutes, ordinances, building codes, and rules and regulations. However, if the Contractor observes that portions of the Contract Documents are at variance therewith, the Contractor shall promptly notify the Architect and Owner in writing, and necessary changes shall be accomplished by appropriate Modification.

3.7.4 If the Contractor performs Work knowing it to be contrary to laws, statutes, ordinances, building codes, and rules and regulations without such notice to the Architect and Owner, the Contractor shall assume full responsibility for such Work and shall bear the attributable costs.

3.8 ALLOWANCES

3.8.1 The Contractor shall include in the Contract Sum all allowances stated in the Contract Documents. Items covered by allowances shall be supplied for such amounts and by such persons or entities as the Owner may direct, but the Contractor shall not be required to employ persons or entities against which the Contractor makes reasonable objection.

3.8.2 Unless otherwise provided in the Contract Documents:

.1 materials and equipment under an allowance shall be selected promptly by the Owner to avoid delay in the Work;

.2 allowances shall cover the cost to the Contractor of materials and equipment delivered at the site and all required taxes, less applicable trade discounts;

.3 Contractor's costs for unloading and handling at the site, labor, installation costs, overhead, profit and other expenses contemplated for stated allowance amounts shall be included in the Contract Sum and not in the allowances;

.4 whenever costs are more than or less than allowances, the Contract Sum shall be adjusted accordingly by Change Order. The amount of the Change Order shall reflect (1) the difference between actual costs and the allowances under Clause 3.8.2.2 and (2) changes in Contractor's costs under Clause 3.8.2.3.

3.9 SUPERINTENDENT

3.9.1 The Contractor shall employ a competent superintendent and necessary assistants who shall be in attendance at the Project site during performance of the Work. The superintendent shall represent the Contractor, and communications given to the superintendent shall be as binding as if given to the Contractor. Important communications shall be confirmed in writing. Other communications shall be similarly confirmed on written request in each case.

3.10 CONTRACTOR'S CONSTRUCTION SCHEDULES

3.10.1 The Contractor, promptly after being awarded the Contract, shall prepare and submit for the Owner's and Architect's information a Contractor's construction schedule for the Work. The schedule shall not exceed time limits current under the Contract Documents, shall be revised at appropriate intervals as required by the conditions of the Work and Project, shall be related to the entire Project to the extent required by the Contract Documents, and shall provide for expeditious and practicable execution of the Work.

3.10.2 The Contractor shall prepare and keep current, for the Architect's approval, a schedule of submittals which is coordinated with the Contractor's construction schedule and allows the Architect reasonable time to review submittals.

3.10.3 The Contractor shall conform to the most recent schedules.

3.11 DOCUMENTS AND SAMPLES AT THE SITE

3.11.1 The Contractor shall maintain at the site for the Owner one record copy of the Drawings, Specifications, addenda, Change Orders and other Modifications, in good order and marked currently to record changes and selections made during construction, and in addition approved Shop Drawings, Product Data, Samples and similar required submittals. These shall be available to the Architect and shall be delivered to the Architect for submittal to the Owner upon completion of the Work.

3.12 SHOP DRAWINGS, PRODUCT DATA AND SAMPLES

3.12.1 Shop Drawings are drawings, diagrams, schedules and other data specially prepared for the Work by the Contractor or a Subcontractor, Sub-subcontractor, manufacturer, supplier or distributor to illustrate some portion of the Work.

3.12.2 Product Data are illustrations, standard schedules, performance charts, instructions, brochures, diagrams and other information furnished by the Contractor to illustrate materials or equipment for some portion of the Work.

3.12.3 Samples are physical examples which illustrate materials, equipment or workmanship and establish standards by which the Work will be judged.

3.12.4 Shop Drawings, Product Data, Samples and similar submittals are not Contract Documents. The purpose of their submittal is to demonstrate for those portions of the Work for which submittals are required the way the Contractor proposes to conform to the information given and the design concept expressed in the Contract Documents. Review by the Architect is subject to the limitations of Subparagraph 4.2.7.

3.12.5 The Contractor shall review, approve and submit to the Architect Shop Drawings, Product Data, Samples and similar submittals required by the Contract Documents with reasonable promptness and in such sequence as to cause no delay in the Work or in the activities of the Owner or of separate contractors. Submittals made by the Contractor which are not required by the Contract Documents may be returned without action.

3.12.6 The Contractor shall perform no portion of the Work requiring submittal and review of Shop Drawings, Product Data, Samples or similar submittals until the respective submittal has been approved by the Architect. Such Work shall be in accordance with approved submittals.

3.12.7 By approving and submitting Shop Drawings, Product Data, Samples and similar submittals, the Contractor represents that the Contractor has determined and verified materials, field measurements and field construction criteria related thereto, or will do so, and has checked and coordinated the information contained within such submittals with the requirements of the Work and of the Contract Documents.

3.12.8 The Contractor shall not be relieved of responsibility for deviations from requirements of the Contract Documents by the Architect's approval of Shop Drawings, Product Data, Samples or similar submittals unless the Contractor has specifically informed the Architect in writing of such deviation at the time of submittal and the Architect has given written approval to the specific deviation. The Contractor shall not be relieved of responsibility for errors or omissions in Shop Drawings, Product Data, Samples or similar submittals by the Architect's approval thereof.

3.12.9 The Contractor shall direct specific attention, in writing or on resubmitted Shop Drawings, Product Data, Samples or similar submittals, to revisions other than those requested by the Architect on previous submittals.

3.12.10 Informational submittals upon which the Architect is not expected to take responsive action may be so identified in the Contract Documents.

3.12.11 When professional certification of performance criteria of materials, systems or equipment is required by the Contract Documents, the Architect shall be entitled to rely upon the accuracy and completeness of such calculations and certifications.

3.13 USE OF SITE

3.13.1 The Contractor shall confine operations at the site to areas permitted by law, ordinances, permits and the Contract Documents and shall not unreasonably encumber the site with materials or equipment.

3.14 CUTTING AND PATCHING

3.14.1 The Contractor shall be responsible for cutting, fitting or patching required to complete the Work or to make its parts fit together properly.

3.14.2 The Contractor shall not damage or endanger a portion of the Work or fully or partially completed construction of the Owner or separate contractors by cutting, patching or otherwise altering such construction, or by excavation. The Contractor shall not cut or otherwise alter such construction by the

Owner or a separate contractor except with written consent of the Owner and of such separate contractor; such consent shall not be unreasonably withheld. The Contractor shall not unreasonably withhold from the Owner or a separate contractor the Contractor's consent to cutting or otherwise altering the Work.

3.15 CLEANING UP

3.15.1 The Contractor shall keep the premises and surrounding area free from accumulation of waste materials or rubbish caused by operations under the Contract. At completion of the Work the Contractor shall remove from and about the Project waste materials, rubbish, the Contractor's tools, construction equipment, machinery and surplus materials.

3.15.2 If the Contractor fails to clean up as provided in the Contract Documents, the Owner may do so and the cost thereof shall be charged to the Contractor.

3.16 ACCESS TO WORK

3.16.1 The Contractor shall provide the Owner and Architect access to the Work in preparation and progress wherever located.

3.17 ROYALTIES AND PATENTS

3.17.1 The Contractor shall pay all royalties and license fees. The Contractor shall defend suits or claims for infringement of patent rights and shall hold the Owner and Architect harmless from loss on account thereof, but shall not be responsible for such defense or loss when a particular design, process or product of a particular manufacturer or manufacturers is required by the Contract Documents. However, if the Contractor has reason to believe that the required design, process or product is an infringement of a patent, the Contractor shall be responsible for such loss unless such information is promptly furnished to the Architect.

3.18 INDEMNIFICATION

3.18.1 To the fullest extent permitted by law, the Contractor shall indemnify and hold harmless the Owner, Architect, Architect's consultants, and agents and employees of any of them from and against claims, damages, losses and expenses, including but not limited to attorneys' fees, arising out of or resulting from performance of the Work, provided that such claim, damage, loss or expense is attributable to bodily injury, sickness, disease or death, or to injury to or destruction of tangible property (other than the Work itself) including loss of use resulting therefrom, but only to the extent caused in whole or in part by negligent acts or omissions of the Contractor, a Subcontractor, anyone directly or indirectly employed by them or anyone for whose acts they may be liable, regardless of whether or not such claim, damage, loss or expense is caused in part by a party indemnified hereunder. Such obligation shall not be construed to negate, abridge, or reduce other rights or obligations of indemnity which would otherwise exist as to a party or person described in this Paragraph 3.18.

3.18.2 In claims against any person or entity indemnified under this Paragraph 3.18 by an employee of the Contractor, a Subcontractor, anyone directly or indirectly employed by them or anyone for whose acts they may be liable, the indemnification obligation under this Paragraph 3.18 shall not be limited by a limitation on amount or type of damages, compensation or benefits payable by or for the Contractor or a Subcontractor under workers' or workmen's compensation acts, disability benefit acts or other employee benefit acts.

3.18.3 The obligations of the Contractor under this Paragraph 3.18 shall not extend to the liability of the Architect, the Archi-

tect's consultants, and agents and employees of any of them arising out of (1) the preparation or approval of maps, drawings, opinions, reports, surveys, Change Orders, designs or specifications, or (2) the giving of or the failure to give directions or instructions by the Architect, the Architect's consultants, and agents and employees of any of them provided such giving or failure to give is the primary cause of the injury or damage.

ARTICLE 4
ADMINISTRATION OF THE CONTRACT

4.1 ARCHITECT

4.1.1 The Architect is the person lawfully licensed to practice architecture or an entity lawfully practicing architecture identified as such in the Agreement and is referred to throughout the Contract Documents as if singular in number. The term "Architect" means the Architect or the Architect's authorized representative.

4.1.2 Duties, responsibilities and limitations of authority of the Architect as set forth in the Contract Documents shall not be restricted, modified or extended without written consent of the Owner, Contractor and Architect. Consent shall not be unreasonably withheld.

4.1.3 In case of termination of employment of the Architect, the Owner shall appoint an architect against whom the Contractor makes no reasonable objection and whose status under the Contract Documents shall be that of the former architect.

4.1.4 Disputes arising under Subparagraphs 4.1.2 and 4.1.3 shall be subject to arbitration.

4.2 ARCHITECT'S ADMINISTRATION OF THE CONTRACT

4.2.1 The Architect will provide administration of the Contract as described in the Contract Documents, and will be the Owner's representative (1) during construction, (2) until final payment is due and (3) with the Owner's concurrence, from time to time during the correction period described in Paragraph 12.2. The Architect will advise and consult with the Owner. The Architect will have authority to act on behalf of the Owner only to the extent provided in the Contract Documents, unless otherwise modified by written instrument in accordance with other provisions of the Contract.

4.2.2 The Architect will visit the site at intervals appropriate to the stage of construction to become generally familiar with the progress and quality of the completed Work and to determine in general if the Work is being performed in a manner indicating that the Work, when completed, will be in accordance with the Contract Documents. However, the Architect will not be required to make exhaustive or continuous on-site inspections to check quality or quantity of the Work. On the basis of on-site observations as an architect, the Architect will keep the Owner informed of progress of the Work, and will endeavor to guard the Owner against defects and deficiencies in the Work.

4.2.3 The Architect will not have control over or charge of and will not be responsible for construction means, methods, techniques, sequences or procedures, or for safety precautions and programs in connection with the Work, since these are solely the Contractor's responsibility as provided in Paragraph 3.3. The Architect will not be responsible for the Contractor's failure to carry out the Work in accordance with the Contract Documents. The Architect will not have control over or charge of and will not be responsible for acts or omissions of the Con-

AIA DOCUMENT A201 • GENERAL CONDITIONS OF THE CONTRACT FOR CONSTRUCTION • FOURTEENTH EDITION
AIA® • ©1987 THE AMERICAN INSTITUTE OF ARCHITECTS. 1735 NEW YORK AVENUE, N.W., WASHINGTON, D.C. 20006

tractor, Subcontractors, or their agents or employees, or of any other persons performing portions of the Work.

4.2.4 Communications Facilitating Contract Administration. Except as otherwise provided in the Contract Documents or when direct communications have been specially authorized, the Owner and Contractor shall endeavor to communicate through the Architect. Communications by and with the Architect's consultants shall be through the Architect. Communications by and with Subcontractors and material suppliers shall be through the Contractor. Communications by and with separate contractors shall be through the Owner.

4.2.5 Based on the Architect's observations and evaluations of the Contractor's Applications for Payment, the Architect will review and certify the amounts due the Contractor and will issue Certificates for Payment in such amounts.

4.2.6 The Architect will have authority to reject Work which does not conform to the Contract Documents. Whenever the Architect considers it necessary or advisable for implementation of the intent of the Contract Documents, the Architect will have authority to require additional inspection or testing of the Work in accordance with Subparagraphs 13.5.2 and 13.5.3, whether or not such Work is fabricated, installed or completed. However, neither this authority of the Architect nor a decision made in good faith either to exercise or not to exercise such authority shall give rise to a duty or responsibility of the Architect to the Contractor, Subcontractors, material and equipment suppliers, their agents or employees, or other persons performing portions of the Work.

4.2.7 The Architect will review and approve or take other appropriate action upon the Contractor's submittals such as Shop Drawings, Product Data and Samples, but only for the limited purpose of checking for conformance with information given and the design concept expressed in the Contract Documents. The Architect's action will be taken with such reasonable promptness as to cause no delay in the Work or in the activities of the Owner, Contractor or separate contractors, while allowing sufficient time in the Architect's professional judgment to permit adequate review. Review of such submittals is not conducted for the purpose of determining the accuracy and completeness of other details such as dimensions and quantities, or for substantiating instructions for installation or performance of equipment or systems, all of which remain the responsibility of the Contractor as required by the Contract Documents. The Architect's review of the Contractor's submittals shall not relieve the Contractor of the obligations under Paragraphs 3.3, 3.5 and 3.12. The Architect's review shall not constitute approval of safety precautions or, unless otherwise specifically stated by the Architect, of any construction means, methods, techniques, sequences or procedures. The Architect's approval of a specific item shall not indicate approval of an assembly of which the item is a component.

4.2.8 The Architect will prepare Change Orders and Construction Change Directives, and may authorize minor changes in the Work as provided in Paragraph 7.4.

4.2.9 The Architect will conduct inspections to determine the date or dates of Substantial Completion and the date of final completion, will receive and forward to the Owner for the Owner's review and records written warranties and related documents required by the Contract and assembled by the Contractor, and will issue a final Certificate for Payment upon compliance with the requirements of the Contract Documents.

4.2.10 If the Owner and Architect agree, the Architect will provide one or more project representatives to assist in carrying out the Architect's responsibilities at the site. The duties, responsibilities and limitations of authority of such project representatives shall be as set forth in an exhibit to be incorporated in the Contract Documents.

4.2.11 The Architect will interpret and decide matters concerning performance under and requirements of the Contract Documents on written request of either the Owner or Contractor. The Architect's response to such requests will be made with reasonable promptness and within any time limits agreed upon. If no agreement is made concerning the time within which interpretations required of the Architect shall be furnished in compliance with this Paragraph 4.2, then delay shall not be recognized on account of failure by the Architect to furnish such interpretations until 15 days after written request is made for them.

4.2.12 Interpretations and decisions of the Architect will be consistent with the intent of and reasonably inferable from the Contract Documents and will be in writing or in the form of drawings. When making such interpretations and decisions, the Architect will endeavor to secure faithful performance by both Owner and Contractor, will not show partiality to either and will not be liable for results of interpretations or decisions so rendered in good faith.

4.2.13 The Architect's decisions on matters relating to aesthetic effect will be final if consistent with the intent expressed in the Contract Documents.

4.3   CLAIMS AND DISPUTES

4.3.1 Definition. A Claim is a demand or assertion by one of the parties seeking, as a matter of right, adjustment or interpretation of Contract terms, payment of money, extension of time or other relief with respect to the terms of the Contract. The term "Claim" also includes other disputes and matters in question between the Owner and Contractor arising out of or relating to the Contract. Claims must be made by written notice. The responsibility to substantiate Claims shall rest with the party making the Claim.

4.3.2 Decision of Architect. Claims, including those alleging an error or omission by the Architect, shall be referred initially to the Architect for action as provided in Paragraph 4.4. A decision by the Architect, as provided in Subparagraph 4.4.4, shall be required as a condition precedent to arbitration or litigation of a Claim between the Contractor and Owner as to all such matters arising prior to the date final payment is due, regardless of (1) whether such matters relate to execution and progress of the Work or (2) the extent to which the Work has been completed. The decision by the Architect in response to a Claim shall not be a condition precedent to arbitration or litigation in the event (1) the position of Architect is vacant, (2) the Architect has not received evidence or has failed to render a decision within agreed time limits, (3) the Architect has failed to take action required under Subparagraph 4.4.4 within 30 days after the Claim is made, (4) 45 days have passed after the Claim has been referred to the Architect or (5) the Claim relates to a mechanic's lien.

4.3.3 Time Limits on Claims. Claims by either party must be made within 21 days after occurrence of the event giving rise to such Claim or within 21 days after the claimant first recognizes the condition giving rise to the Claim, whichever is later. Claims must be made by written notice. An additional Claim made after the initial Claim has been implemented by Change Order will not be considered unless submitted in a timely manner.

4.3.4 Continuing Contract Performance. Pending final resolution of a Claim including arbitration, unless otherwise agreed in writing the Contractor shall proceed diligently with performance of the Contract and the Owner shall continue to make payments in accordance with the Contract Documents.

4.3.5 Waiver of Claims: Final Payment. The making of final payment shall constitute a waiver of Claims by the Owner except those arising from:

.1 liens, Claims, security interests or encumbrances arising out of the Contract and unsettled;

.2 failure of the Work to comply with the requirements of the Contract Documents; or

.3 terms of special warranties required by the Contract Documents.

4.3.6 Claims for Concealed or Unknown Conditions. If conditions are encountered at the site which are (1) subsurface or otherwise concealed physical conditions which differ materially from those indicated in the Contract Documents or (2) unknown physical conditions of an unusual nature, which differ materially from those ordinarily found to exist and generally recognized as inherent in construction activities of the character provided for in the Contract Documents, then notice by the observing party shall be given to the other party promptly before conditions are disturbed and in no event later than 21 days after first observance of the conditions. The Architect will promptly investigate such conditions and, if they differ materially and cause an increase or decrease in the Contractor's cost of, or time required for, performance of any part of the Work, will recommend an equitable adjustment in the Contract Sum or Contract Time, or both. If the Architect determines that the conditions at the site are not materially different from those indicated in the Contract Documents and that no change in the terms of the Contract is justified, the Architect shall so notify the Owner and Contractor in writing, stating the reasons. Claims by either party in opposition to such determination must be made within 21 days after the Architect has given notice of the decision. If the Owner and Contractor cannot agree on an adjustment in the Contract Sum or Contract Time, the adjustment shall be referred to the Architect for initial determination, subject to further proceedings pursuant to Paragraph 4.4.

4.3.7 Claims for Additional Cost. If the Contractor wishes to make Claim for an increase in the Contract Sum, written notice as provided herein shall be given before proceeding to execute the Work. Prior notice is not required for Claims relating to an emergency endangering life or property arising under Paragraph 10.3. If the Contractor believes additional cost is involved for reasons including but not limited to (1) a written interpretation from the Architect, (2) an order by the Owner to stop the Work where the Contractor was not at fault, (3) a written order for a minor change in the Work issued by the Architect, (4) failure of payment by the Owner, (5) termination of the Contract by the Owner, (6) Owner's suspension or (7) other reasonable grounds, Claim shall be filed in accordance with the procedure established herein.

4.3.8 Claims for Additional Time

4.3.8.1 If the Contractor wishes to make Claim for an increase in the Contract Time, written notice as provided herein shall be given. The Contractor's Claim shall include an estimate of cost and of probable effect of delay on progress of the Work. In the case of a continuing delay only one Claim is necessary.

4.3.8.2 If adverse weather conditions are the basis for a Claim for additional time, such Claim shall be documented by data substantiating that weather conditions were abnormal for the period of time and could not have been reasonably anticipated, and that weather conditions had an adverse effect on the scheduled construction.

4.3.9 Injury or Damage to Person or Property. If either party to the Contract suffers injury or damage to person or property because of an act or omission of the other party, of any of the other party's employees or agents, or of others for whose acts such party is legally liable, written notice of such injury or damage, whether or not insured, shall be given to the other party within a reasonable time not exceeding 21 days after first observance. The notice shall provide sufficient detail to enable the other party to investigate the matter. If a Claim for additional cost or time related to this Claim is to be asserted, it shall be filed as provided in Subparagraphs 4.3.7 or 4.3.8.

4.4   RESOLUTION OF CLAIMS AND DISPUTES

4.4.1 The Architect will review Claims and take one or more of the following preliminary actions within ten days of receipt of a Claim: (1) request additional supporting data from the claimant, (2) submit a schedule to the parties indicating when the Architect expects to take action, (3) reject the Claim in whole or in part stating reasons for rejection, (4) recommend approval of the Claim by the other party or (5) suggest a compromise. The Architect may also, but is not obligated to, notify the surety, if any, of the nature and amount of the Claim.

4.4.2 If a Claim has been resolved, the Architect will prepare or obtain appropriate documentation.

4.4.3 If a Claim has not been resolved, the party making the Claim shall, within ten days after the Architect's preliminary response, take one or more of the following actions: (1) submit additional supporting data requested by the Architect, (2) modify the initial Claim or (3) notify the Architect that the initial Claim stands.

4.4.4 If a Claim has not been resolved after consideration of the foregoing and of further evidence presented by the parties or requested by the Architect, the Architect will notify the parties in writing that the Architect's decision will be made within seven days, which decision shall be final and binding on the parties but subject to arbitration. Upon expiration of such time period, the Architect will render to the parties the Architect's written decision relative to the Claim, including any change in the Contract Sum or Contract Time or both. If there is a surety and there appears to be a possibility of a Contractor's default, the Architect may, but is not obligated to, notify the surety and request the surety's assistance in resolving the controversy.

4.5   ARBITRATION

4.5.1 Controversies and Claims Subject to Arbitration. Any controversy or Claim arising out of or related to the Contract, or the breach thereof, shall be settled by arbitration in accordance with the Construction Industry Arbitration Rules of the American Arbitration Association, and judgment upon the award rendered by the arbitrator or arbitrators may be entered in any court having jurisdiction thereof, except controversies or Claims relating to aesthetic effect and except those waived as provided for in Subparagraph 4.3.5. Such controversies or Claims upon which the Architect has given notice and rendered a decision as provided in Subparagraph 4.4.4 shall be subject to arbitration upon written demand of either party. Arbitration may be commenced when 45 days have passed after a Claim has been referred to the Architect as provided in Paragraph 4.3 and no decision has been rendered.

12   A201-1987   AIA DOCUMENT A201 • GENERAL CONDITIONS OF THE CONTRACT FOR CONSTRUCTION • FOURTEENTH EDITION
AIA® • ©1987 THE AMERICAN INSTITUTE OF ARCHITECTS, 1735 NEW YORK AVENUE, N.W., WASHINGTON, D.C. 20006
WARNING: Unlicensed photocopying violates U.S. copyright laws and is subject to legal prosecution.

4.5.2 Rules and Notices for Arbitration. Claims between the Owner and Contractor not resolved under Paragraph 4.4 shall, if subject to arbitration under Subparagraph 4.5.1, be decided by arbitration in accordance with the Construction Industry Arbitration Rules of the American Arbitration Association currently in effect, unless the parties mutually agree otherwise. Notice of demand for arbitration shall be filed in writing with the other party to the Agreement between the Owner and Contractor and with the American Arbitration Association, and a copy shall be filed with the Architect.

4.5.3 Contract Performance During Arbitration. During arbitration proceedings, the Owner and Contractor shall comply with Subparagraph 4.3.4.

4.5.4 When Arbitration May Be Demanded. Demand for arbitration of any Claim may not be made until the earlier of (1) the date on which the Architect has rendered a final written decision on the Claim, (2) the tenth day after the parties have presented evidence to the Architect or have been given reasonable opportunity to do so, if the Architect has not rendered a final written decision by that date, or (3) any of the five events described in Subparagraph 4.3.2.

4.5.4.1 When a written decision of the Architect states that (1) the decision is final but subject to arbitration and (2) a demand for arbitration of a Claim covered by such decision must be made within 30 days after the date on which the party making the demand receives the final written decision, then failure to demand arbitration within said 30 days' period shall result in the Architect's decision becoming final and binding upon the Owner and Contractor. If the Architect renders a decision after arbitration proceedings have been initiated, such decision may be entered as evidence, but shall not supersede arbitration proceedings unless the decision is acceptable to all parties concerned.

4.5.4.2 A demand for arbitration shall be made within the time limits specified in Subparagraphs 4.5.1 and 4.5.4 and Clause 4.5.4.1 as applicable, and in other cases within a reasonable time after the Claim has arisen, and in no event shall it be made after the date when institution of legal or equitable proceedings based on such Claim would be barred by the applicable statute of limitations as determined pursuant to Paragraph 13.7.

4.5.5 Limitation on Consolidation or Joinder. No arbitration arising out of or relating to the Contract Documents shall include, by consolidation or joinder or in any other manner, the Architect, the Architect's employees or consultants, except by written consent containing specific reference to the Agreement and signed by the Architect, Owner, Contractor and any other person or entity sought to be joined. No arbitration shall include, by consolidation or joinder or in any other manner, parties other than the Owner, Contractor, a separate contractor as described in Article 6 and other persons substantially involved in a common question of fact or law whose presence is required if complete relief is to be accorded in arbitration. No person or entity other than the Owner, Contractor or a separate contractor as described in Article 6 shall be included as an original third party or additional third party to an arbitration whose interest or responsibility is insubstantial. Consent to arbitration involving an additional person or entity shall not constitute consent to arbitration of a dispute not described therein or with a person or entity not named or described therein. The foregoing agreement to arbitrate and other agreements to arbitrate with an additional person or entity duly consented to by parties to the Agreement shall be specifically enforceable under applicable law in any court having jurisdiction thereof.

4.5.6 Claims and Timely Assertion of Claims. A party who files a notice of demand for arbitration must assert in the demand all Claims then known to that party on which arbitration is permitted to be demanded. When a party fails to include a Claim through oversight, inadvertence or excusable neglect, or when a Claim has matured or been acquired subsequently, the arbitrator or arbitrators may permit amendment.

4.5.7 Judgment on Final Award. The award rendered by the arbitrator or arbitrators shall be final, and judgment may be entered upon it in accordance with applicable law in any court having jurisdiction thereof.

ARTICLE 5
SUBCONTRACTORS

5.1 DEFINITIONS

5.1.1 A Subcontractor is a person or entity who has a direct contract with the Contractor to perform a portion of the Work at the site. The term "Subcontractor" is referred to throughout the Contract Documents as if singular in number and means a Subcontractor or an authorized representative of the Subcontractor. The term "Subcontractor" does not include a separate contractor or subcontractors of a separate contractor.

5.1.2 A Sub-subcontractor is a person or entity who has a direct or indirect contract with a Subcontractor to perform a portion of the Work at the site. The term "Sub-subcontractor" is referred to throughout the Contract Documents as if singular in number and means a Sub-subcontractor or an authorized representative of the Sub-subcontractor.

5.2 AWARD OF SUBCONTRACTS AND OTHER CONTRACTS FOR PORTIONS OF THE WORK

5.2.1 Unless otherwise stated in the Contract Documents or the bidding requirements, the Contractor, as soon as practicable after award of the Contract, shall furnish in writing to the Owner through the Architect the names of persons or entities (including those who are to furnish materials or equipment fabricated to a special design) proposed for each principal portion of the Work. The Architect will promptly reply to the Contractor in writing stating whether or not the Owner or the Architect, after due investigation, has reasonable objection to any such proposed person or entity. Failure of the Owner or Architect to reply promptly shall constitute notice of no reasonable objection.

5.2.2 The Contractor shall not contract with a proposed person or entity to whom the Owner or Architect has made reasonable and timely objection. The Contractor shall not be required to contract with anyone to whom the Contractor has made reasonable objection.

5.2.3 If the Owner or Architect has reasonable objection to a person or entity proposed by the Contractor, the Contractor shall propose another to whom the Owner or Architect has no reasonable objection. The Contract Sum shall be increased or decreased by the difference in cost occasioned by such change and an appropriate Change Order shall be issued. However, no increase in the Contract Sum shall be allowed for such change unless the Contractor has acted promptly and responsively in submitting names as required.

5.2.4 The Contractor shall not change a Subcontractor, person or entity previously selected if the Owner or Architect makes reasonable objection to such change.

5.3   SUBCONTRACTUAL RELATIONS

5.3.1 By appropriate agreement, written where legally required for validity, the Contractor shall require each Subcontractor, to the extent of the Work to be performed by the Subcontractor, to be bound to the Contractor by terms of the Contract Documents, and to assume toward the Contractor all the obligations and responsibilities which the Contractor, by these Documents, assumes toward the Owner and Architect. Each subcontract agreement shall preserve and protect the rights of the Owner and Architect under the Contract Documents with respect to the Work to be performed by the Subcontractor so that subcontracting thereof will not prejudice such rights, and shall allow to the Subcontractor, unless specifically provided otherwise in the subcontract agreement, the benefit of all rights, remedies and redress against the Contractor that the Contractor, by the Contract Documents, has against the Owner. Where appropriate, the Contractor shall require each Subcontractor to enter into similar agreements with Sub-sub-contractors. The Contractor shall make available to each proposed Subcontractor, prior to the execution of the subcontract agreement, copies of the Contract Documents to which the Subcontractor will be bound, and, upon written request of the Subcontractor, identify to the Subcontractor terms and conditions of the proposed subcontract agreement which may be at variance with the Contract Documents. Subcontractors shall similarly make copies of applicable portions of such documents available to their respective proposed Sub-subcontractors.

5.4   CONTINGENT ASSIGNMENT OF SUBCONTRACTS

5.4.1 Each subcontract agreement for a portion of the Work is assigned by the Contractor to the Owner provided that:

.1 assignment is effective only after termination of the Contract by the Owner for cause pursuant to Paragraph 14.2 and only for those subcontract agreements which the Owner accepts by notifying the Subcontractor in writing; and

.2 assignment is subject to the prior rights of the surety, if any, obligated under bond relating to the Contract.

5.4.2 If the Work has been suspended for more than 30 days, the Subcontractor's compensation shall be equitably adjusted.

ARTICLE 6

CONSTRUCTION BY OWNER OR BY SEPARATE CONTRACTORS

6.1   OWNER'S RIGHT TO PERFORM CONSTRUCTION AND TO AWARD SEPARATE CONTRACTS

6.1.1 The Owner reserves the right to perform construction or operations related to the Project with the Owner's own forces, and to award separate contracts in connection with other portions of the Project or other construction or operations on the site under Conditions of the Contract identical or substantially similar to these including those portions related to insurance and waiver of subrogation. If the Contractor claims that delay or additional cost is involved because of such action by the Owner, the Contractor shall make such Claim as provided elsewhere in the Contract Documents.

6.1.2 When separate contracts are awarded for different portions of the Project or other construction or operations on the site, the term "Contractor" in the Contract Documents in each case shall mean the Contractor who executes each separate Owner-Contractor Agreement.

6.1.3 The Owner shall provide for coordination of the activities of the Owner's own forces and of each separate contractor with the Work of the Contractor, who shall cooperate with them. The Contractor shall participate with other separate contractors and the Owner in reviewing their construction schedules when directed to do so. The Contractor shall make any revisions to the construction schedule and Contract Sum deemed necessary after a joint review and mutual agreement. The construction schedules shall then constitute the schedules to be used by the Contractor, separate contractors and the Owner until subsequently revised.

6.1.4 Unless otherwise provided in the Contract Documents, when the Owner performs construction or operations related to the Project with the Owner's own forces, the Owner shall be deemed to be subject to the same obligations and to have the same rights which apply to the Contractor under the Conditions of the Contract, including, without excluding others, those stated in Article 3, this Article 6 and Articles 10, 11 and 12.

6.2   MUTUAL RESPONSIBILITY

6.2.1 The Contractor shall afford the Owner and separate contractors reasonable opportunity for introduction and storage of their materials and equipment and performance of their activities and shall connect and coordinate the Contractor's construction and operations with theirs as required by the Contract Documents.

6.2.2 If part of the Contractor's Work depends for proper execution or results upon construction or operations by the Owner or a separate contractor, the Contractor shall, prior to proceeding with that portion of the Work, promptly report to the Architect apparent discrepancies or defects in such other construction that would render it unsuitable for such proper execution and results. Failure of the Contractor so to report shall constitute an acknowledgment that the Owner's or separate contractors' completed or partially completed construction is fit and proper to receive the Contractor's Work, except as to defects not then reasonably discoverable.

6.2.3 Costs caused by delays or by improperly timed activities or defective construction shall be borne by the party responsible therefor.

6.2.4 The Contractor shall promptly remedy damage wrongfully caused by the Contractor to completed or partially completed construction or to property of the Owner or separate contractors as provided in Subparagraph 10.2.5.

6.2.5 Claims and other disputes and matters in question between the Contractor and a separate contractor shall be subject to the provisions of Paragraph 4.3 provided the separate contractor has reciprocal obligations.

6.2.6 The Owner and each separate contractor shall have the same responsibilities for cutting and patching as are described for the Contractor in Paragraph 3.14.

6.3   OWNER'S RIGHT TO CLEAN UP

6.3.1 If a dispute arises among the Contractor, separate contractors and the Owner as to the responsibility under their respective contracts for maintaining the premises and surrounding area free from waste materials and rubbish as described in Paragraph 3.15, the Owner may clean up and allocate the cost among those responsible as the Architect determines to be just.

AIA DOCUMENT A201 • GENERAL CONDITIONS OF THE CONTRACT FOR CONSTRUCTION • FOURTEENTH EDITION
AIA® • ©1987 THE AMERICAN INSTITUTE OF ARCHITECTS, 1735 NEW YORK AVENUE, N.W., WASHINGTON, D.C. 20006

ARTICLE 7

CHANGES IN THE WORK

7.1 CHANGES

7.1.1 Changes in the Work may be accomplished after execution of the Contract, and without invalidating the Contract, by Change Order, Construction Change Directive or order for a minor change in the Work, subject to the limitations stated in this Article 7 and elsewhere in the Contract Documents.

7.1.2 A Change Order shall be based upon agreement among the Owner, Contractor and Architect; a Construction Change Directive requires agreement by the Owner and Architect and may or may not be agreed to by the Contractor; an order for a minor change in the Work may be issued by the Architect alone.

7.1.3 Changes in the Work shall be performed under applicable provisions of the Contract Documents, and the Contractor shall proceed promptly, unless otherwise provided in the Change Order, Construction Change Directive or order for a minor change in the Work.

7.1.4 If unit prices are stated in the Contract Documents or subsequently agreed upon, and if quantities originally contemplated are so changed in a proposed Change Order or Construction Change Directive that application of such unit prices to quantities of Work proposed will cause substantial inequity to the Owner or Contractor, the applicable unit prices shall be equitably adjusted.

7.2 CHANGE ORDERS

7.2.1 A Change Order is a written instrument prepared by the Architect and signed by the Owner, Contractor and Architect, stating their agreement upon all of the following:

.1 a change in the Work;

.2 the amount of the adjustment in the Contract Sum, if any; and

.3 the extent of the adjustment in the Contract Time, if any.

7.2.2 Methods used in determining adjustments to the Contract Sum may include those listed in Subparagraph 7.3.3.

7.3 CONSTRUCTION CHANGE DIRECTIVES

7.3.1 A Construction Change Directive is a written order prepared by the Architect and signed by the Owner and Architect, directing a change in the Work and stating a proposed basis for adjustment, if any, in the Contract Sum or Contract Time, or both. The Owner may by Construction Change Directive, without invalidating the Contract, order changes in the Work within the general scope of the Contract consisting of additions, deletions or other revisions, the Contract Sum and Contract Time being adjusted accordingly.

7.3.2 A Construction Change Directive shall be used in the absence of total agreement on the terms of a Change Order.

7.3.3 If the Construction Change Directive provides for an adjustment to the Contract Sum, the adjustment shall be based on one of the following methods:

.1 mutual acceptance of a lump sum properly itemized and supported by sufficient substantiating data to permit evaluation;

.2 unit prices stated in the Contract Documents or subsequently agreed upon;

.3 cost to be determined in a manner agreed upon by the parties and a mutually acceptable fixed or percentage fee; or

.4 as provided in Subparagraph 7.3.6.

7.3.4 Upon receipt of a Construction Change Directive, the Contractor shall promptly proceed with the change in the Work involved and advise the Architect of the Contractor's agreement or disagreement with the method, if any, provided in the Construction Change Directive for determining the proposed adjustment in the Contract Sum or Contract Time.

7.3.5 A Construction Change Directive signed by the Contractor indicates the agreement of the Contractor therewith, including adjustment in Contract Sum and Contract Time or the method for determining them. Such agreement shall be effective immediately and shall be recorded as a Change Order.

7.3.6 If the Contractor does not respond promptly or disagrees with the method for adjustment in the Contract Sum, the method and the adjustment shall be determined by the Architect on the basis of reasonable expenditures and savings of those performing the Work attributable to the change, including, in case of an increase in the Contract Sum, a reasonable allowance for overhead and profit. In such case, and also under Clause 7.3.3, the Contractor shall keep and present, in such form as the Architect may prescribe, an itemized accounting together with appropriate supporting data. Unless otherwise provided in the Contract Documents, costs for the purposes of this Subparagraph 7.3.6 shall be limited to the following:

.1 costs of labor, including social security, old age and unemployment insurance, fringe benefits required by agreement or custom, and workers' or workmen's compensation insurance;

.2 costs of materials, supplies and equipment, including cost of transportation, whether incorporated or consumed;

.3 rental costs of machinery and equipment, exclusive of hand tools, whether rented from the Contractor or others;

.4 costs of premiums for all bonds and insurance, permit fees, and sales, use or similar taxes related to the Work; and

.5 additional costs of supervision and field office personnel directly attributable to the change.

7.3.7 Pending final determination of cost to the Owner, amounts not in dispute may be included in Applications for Payment. The amount of credit to be allowed by the Contractor to the Owner for a deletion or change which results in a net decrease in the Contract Sum shall be actual net cost as confirmed by the Architect. When both additions and credits covering related Work or substitutions are involved in a change, the allowance for overhead and profit shall be figured on the basis of net increase, if any, with respect to that change.

7.3.8 If the Owner and Contractor do not agree with the adjustment in Contract Time or the method for determining it, the adjustment or the method shall be referred to the Architect for determination.

7.3.9 When the Owner and Contractor agree with the determination made by the Architect concerning the adjustments in the Contract Sum and Contract Time, or otherwise reach agreement upon the adjustments, such agreement shall be effective immediately and shall be recorded by preparation and execution of an appropriate Change Order.

7.4   MINOR CHANGES IN THE WORK

7.4.1 The Architect will have authority to order minor changes in the Work not involving adjustment in the Contract Sum or extension of the Contract Time and not inconsistent with the intent of the Contract Documents. Such changes shall be effected by written order and shall be binding on the Owner and Contractor. The Contractor shall carry out such written orders promptly.

ARTICLE 8

TIME

8.1   DEFINITIONS

8.1.1 Unless otherwise provided, Contract Time is the period of time, including authorized adjustments, allotted in the Contract Documents for Substantial Completion of the Work.

8.1.2 The date of commencement of the Work is the date established in the Agreement. The date shall not be postponed by the failure to act of the Contractor or of persons or entities for whom the Contractor is responsible.

8.1.3 The date of Substantial Completion is the date certified by the Architect in accordance with Paragraph 9.8.

8.1.4 The term "day" as used in the Contract Documents shall mean calendar day unless otherwise specifically defined.

8.2   PROGRESS AND COMPLETION

8.2.1 Time limits stated in the Contract Documents are of the essence of the Contract. By executing the Agreement the Contractor confirms that the Contract Time is a reasonable period for performing the Work.

8.2.2 The Contractor shall not knowingly, except by agreement or instruction of the Owner in writing, prematurely commence operations on the site or elsewhere prior to the effective date of insurance required by Article 11 to be furnished by the Contractor. The date of commencement of the Work shall not be changed by the effective date of such insurance. Unless the date of commencement is established by a notice to proceed given by the Owner, the Contractor shall notify the Owner in writing not less than five days or other agreed period before commencing the Work to permit the timely filing of mortgages, mechanic's liens and other security interests.

8.2.3 The Contractor shall proceed expeditiously with adequate forces and shall achieve Substantial Completion within the Contract Time.

8.3   DELAYS AND EXTENSIONS OF TIME

8.3.1 If the Contractor is delayed at any time in progress of the Work by an act or neglect of the Owner or Architect, or of an employee of either, or of a separate contractor employed by the Owner, or by changes ordered in the Work, or by labor disputes, fire, unusual delay in deliveries, unavoidable casualties or other causes beyond the Contractor's control, or by delay authorized by the Owner pending arbitration, or by other causes which the Architect determines may justify delay, then the Contract Time shall be extended by Change Order for such reasonable time as the Architect may determine.

8.3.2 Claims relating to time shall be made in accordance with applicable provisions of Paragraph 4.3.

8.3.3 This Paragraph 8.3 does not preclude recovery of damages for delay by either party under other provisions of the Contract Documents.

ARTICLE 9

PAYMENTS AND COMPLETION

9.1   CONTRACT SUM

9.1.1 The Contract Sum is stated in the Agreement and, including authorized adjustments, is the total amount payable by the Owner to the Contractor for performance of the Work under the Contract Documents.

9.2   SCHEDULE OF VALUES

9.2.1 Before the first Application for Payment, the Contractor shall submit to the Architect a schedule of values allocated to various portions of the Work, prepared in such form and supported by such data to substantiate its accuracy as the Architect may require. This schedule, unless objected to by the Architect, shall be used as a basis for reviewing the Contractor's Applications for Payment.

9.3   APPLICATIONS FOR PAYMENT

9.3.1 At least ten days before the date established for each progress payment, the Contractor shall submit to the Architect an itemized Application for Payment for operations completed in accordance with the schedule of values. Such application shall be notarized, if required, and supported by such data substantiating the Contractor's right to payment as the Owner or Architect may require, such as copies of requisitions from Subcontractors and material suppliers, and reflecting retainage if provided for elsewhere in the Contract Documents.

9.3.1.1 Such applications may include requests for payment on account of changes in the Work which have been properly authorized by Construction Change Directives but not yet included in Change Orders.

9.3.1.2 Such applications may not include requests for payment of amounts the Contractor does not intend to pay to a Subcontractor or material supplier because of a dispute or other reason.

9.3.2 Unless otherwise provided in the Contract Documents, payments shall be made on account of materials and equipment delivered and suitably stored at the site for subsequent incorporation in the Work. If approved in advance by the Owner, payment may similarly be made for materials and equipment suitably stored off the site at a location agreed upon in writing. Payment for materials and equipment stored on or off the site shall be conditioned upon compliance by the Contractor with procedures satisfactory to the Owner to establish the Owner's title to such materials and equipment or otherwise protect the Owner's interest, and shall include applicable insurance, storage and transportation to the site for such materials and equipment stored off the site.

9.3.3 The Contractor warrants that title to all Work covered by an Application for Payment will pass to the Owner no later than the time of payment. The Contractor further warrants that upon submittal of an Application for Payment all Work for which Certificates for Payment have been previously issued and payments received from the Owner shall, to the best of the Contractor's knowledge, information and belief, be free and clear of liens, claims, security interests or encumbrances in favor of the Contractor, Subcontractors, material suppliers, or other persons or entities making a claim by reason of having provided labor, materials and equipment relating to the Work.

9.4   CERTIFICATES FOR PAYMENT

9.4.1 The Architect will, within seven days after receipt of the Contractor's Application for Payment, either issue to the

Owner a Certificate for Payment, with a copy to the Contractor, for such amount as the Architect determines is properly due, or notify the Contractor and Owner in writing of the Architect's reasons for withholding certification in whole or in part as provided in Subparagraph 9.5.1.

9.4.2 The issuance of a Certificate for Payment will constitute a representation by the Architect to the Owner, based on the Architect's observations at the site and the data comprising the Application for Payment, that the Work has progressed to the point indicated and that, to the best of the Architect's knowledge, information and belief, quality of the Work is in accordance with the Contract Documents. The foregoing representations are subject to an evaluation of the Work for conformance with the Contract Documents upon Substantial Completion, to results of subsequent tests and inspections, to minor deviations from the Contract Documents correctable prior to completion and to specific qualifications expressed by the Architect. The issuance of a Certificate for Payment will further constitute a representation that the Contractor is entitled to payment in the amount certified. However, the issuance of a Certificate for Payment will not be a representation that the Architect has (1) made exhaustive or continuous on-site inspections to check the quality or quantity of the Work, (2) reviewed construction means, techniques, sequences or procedures, (3) reviewed copies of requisitions received from Subcontractors and material suppliers and other data requested by the Owner to substantiate the Contractor's right to payment or (4) made examination to ascertain how or for what purpose the Contractor has used money previously paid on account of the Contract Sum.

9.5   DECISIONS TO WITHHOLD CERTIFICATION

9.5.1 The Architect may decide not to certify payment and may withhold a Certificate for Payment in whole or in part, to the extent reasonably necessary to protect the Owner, if in the Architect's opinion the representations to the Owner required by Subparagraph 9.4.2 cannot be made. If the Architect is unable to certify payment in the amount of the Application, the Architect will notify the Contractor and Owner as provided in Subparagraph 9.4.1. If the Contractor and Architect cannot agree on a revised amount, the Architect will promptly issue a Certificate for Payment for the amount for which the Architect is able to make such representations to the Owner. The Architect may also decide not to certify payment or, because of subsequently discovered evidence or subsequent observations, may nullify the whole or a part of a Certificate for Payment previously issued, to such extent as may be necessary in the Architect's opinion to protect the Owner from loss because of:

   .1 defective Work not remedied;

   .2 third party claims filed or reasonable evidence indicating probable filing of such claims;

   .3 failure of the Contractor to make payments properly to Subcontractors or for labor, materials or equipment;

   .4 reasonable evidence that the Work cannot be completed for the unpaid balance of the Contract Sum;

   .5 damage to the Owner or another contractor;

   .6 reasonable evidence that the Work will not be completed within the Contract Time, and that the unpaid balance would not be adequate to cover actual or liquidated damages for the anticipated delay; or

   .7 persistent failure to carry out the Work in accordance with the Contract Documents.

9.5.2 When the above reasons for withholding certification are removed, certification will be made for amounts previously withheld.

9.6   PROGRESS PAYMENTS

9.6.1 After the Architect has issued a Certificate for Payment, the Owner shall make payment in the manner and within the time provided in the Contract Documents, and shall so notify the Architect.

9.6.2 The Contractor shall promptly pay each Subcontractor, upon receipt of payment from the Owner, out of the amount paid to the Contractor on account of such Subcontractor's portion of the Work, the amount to which said Subcontractor is entitled, reflecting percentages actually retained from payments to the Contractor on account of such Subcontractor's portion of the Work. The Contractor shall, by appropriate agreement with each Subcontractor, require each Subcontractor to make payments to Sub-subcontractors in similar manner.

9.6.3 The Architect will, on request, furnish to a Subcontractor, if practicable, information regarding percentages of completion or amounts applied for by the Contractor and action taken thereon by the Architect and Owner on account of portions of the Work done by such Subcontractor.

9.6.4 Neither the Owner nor Architect shall have an obligation to pay or to see to the payment of money to a Subcontractor except as may otherwise be required by law.

9.6.5 Payment to material suppliers shall be treated in a manner similar to that provided in Subparagraphs 9.6.2, 9.6.3 and 9.6.4.

9.6.6 A Certificate for Payment, a progress payment, or partial or entire use or occupancy of the Project by the Owner shall not constitute acceptance of Work not in accordance with the Contract Documents.

9.7   FAILURE OF PAYMENT

9.7.1 If the Architect does not issue a Certificate for Payment, through no fault of the Contractor, within seven days after receipt of the Contractor's Application for Payment, or if the Owner does not pay the Contractor within seven days after the date established in the Contract Documents the amount certified by the Architect or awarded by arbitration, then the Contractor may, upon seven additional days' written notice to the Owner and Architect, stop the Work until payment of the amount owing has been received. The Contract Time shall be extended appropriately and the Contract Sum shall be increased by the amount of the Contractor's reasonable costs of shut-down, delay and start-up, which shall be accomplished as provided in Article 7.

9.8   SUBSTANTIAL COMPLETION

9.8.1 Substantial Completion is the stage in the progress of the Work when the Work or designated portion thereof is sufficiently complete in accordance with the Contract Documents so the Owner can occupy or utilize the Work for its intended use.

9.8.2 When the Contractor considers that the Work, or a portion thereof which the Owner agrees to accept separately, is substantially complete, the Contractor shall prepare and submit to the Architect a comprehensive list of items to be completed or corrected. The Contractor shall proceed promptly to complete and correct items on the list. Failure to include an item on such list does not alter the responsibility of the Contractor to complete all Work in accordance with the Contract Documents. Upon receipt of the Contractor's list, the Architect will make an inspection to determine whether the Work or designated portion thereof is substantially complete. If the Architect's inspection discloses any item, whether or not included on the Contractor's list, which is not in accordance with the requirements of the Contract Documents, the Contractor shall, before issuance of the Certificate of Substantial Completion, complete or correct such item upon notification by the Architect. The Contractor shall then submit a request for another inspection by the Architect to determine Substantial Completion. When the Work or designated portion thereof is substantially complete, the Architect will prepare a Certificate of Substantial Completion which shall establish the date of Substantial Completion, shall establish responsibilities of the Owner and Contractor for security, maintenance, heat, utilities, damage to the Work and insurance, and shall fix the time within which the Contractor shall finish all items on the list accompanying the Certificate. Warranties required by the Contract Documents shall commence on the date of Substantial Completion of the Work or designated portion thereof unless otherwise provided in the Certificate of Substantial Completion. The Certificate of Substantial Completion shall be submitted to the Owner and Contractor for their written acceptance of responsibilities assigned to them in such Certificate.

9.8.3 Upon Substantial Completion of the Work or designated portion thereof and upon application by the Contractor and certification by the Architect, the Owner shall make payment, reflecting adjustment in retainage, if any, for such Work or portion thereof as provided in the Contract Documents.

9.9   PARTIAL OCCUPANCY OR USE

9.9.1 The Owner may occupy or use any completed or partially completed portion of the Work at any stage when such portion is designated by separate agreement with the Contractor, provided such occupancy or use is consented to by the insurer as required under Subparagraph 11.3.11 and authorized by public authorities having jurisdiction over the Work. Such partial occupancy or use may commence whether or not the portion is substantially complete, provided the Owner and Contractor have accepted in writing the responsibilities assigned to each of them for payments, retainage if any, security, maintenance, heat, utilities, damage to the Work and insurance, and have agreed in writing concerning the period for correction of the Work and commencement of warranties required by the Contract Documents. When the Contractor considers a portion substantially complete, the Contractor shall prepare and submit a list to the Architect as provided under Subparagraph 9.8.2. Consent of the Contractor to partial occupancy or use shall not be unreasonably withheld. The stage of the progress of the Work shall be determined by written agreement between the Owner and Contractor or, if no agreement is reached, by decision of the Architect.

9.9.2 Immediately prior to such partial occupancy or use, the Owner, Contractor and Architect shall jointly inspect the area to be occupied or portion of the Work to be used in order to determine and record the condition of the Work.

9.9.3 Unless otherwise agreed upon, partial occupancy or use of a portion or portions of the Work shall not constitute acceptance of Work not complying with the requirements of the Contract Documents.

9.10   FINAL COMPLETION AND FINAL PAYMENT

9.10.1 Upon receipt of written notice that the Work is ready for final inspection and acceptance and upon receipt of a final Application for Payment, the Architect will promptly make

such inspection and, when the Architect finds the Work acceptable under the Contract Documents and the Contract fully performed, the Architect will promptly issue a final Certificate for Payment stating that to the best of the Architect's knowledge, information and belief, and on the basis of the Architect's observations and inspections, the Work has been completed in accordance with terms and conditions of the Contract Documents and that the entire balance found to be due the Contractor and noted in said final Certificate is due and payable. The Architect's final Certificate for Payment will constitute a further representation that conditions listed in Subparagraph 9.10.2 as precedent to the Contractor's being entitled to final payment have been fulfilled.

9.10.2 Neither final payment nor any remaining retained percentage shall become due until the Contractor submits to the Architect (1) an affidavit that payrolls, bills for materials and equipment, and other indebtedness connected with the Work for which the Owner or the Owner's property might be responsible or encumbered (less amounts withheld by Owner) have been paid or otherwise satisfied, (2) a certificate evidencing that insurance required by the Contract Documents to remain in force after final payment is currently in effect and will not be cancelled or allowed to expire until at least 30 days' prior written notice has been given to the Owner, (3) a written statement that the Contractor knows of no substantial reason that the insurance will not be renewable to cover the period required by the Contract Documents, (4) consent of surety, if any, to final payment and (5), if required by the Owner, other data establishing payment or satisfaction of obligations, such as receipts, releases and waivers of liens, claims, security interests or encumbrances arising out of the Contract, to the extent and in such form as may be designated by the Owner. If a Subcontractor refuses to furnish a release or waiver required by the Owner, the Contractor may furnish a bond satisfactory to the Owner to indemnify the Owner against such lien. If such lien remains unsatisfied after payments are made, the Contractor shall refund to the Owner all money that the Owner may be compelled to pay in discharging such lien, including all costs and reasonable attorneys' fees.

9.10.3 If, after Substantial Completion of the Work, final completion thereof is materially delayed through no fault of the Contractor or by issuance of Change Orders affecting final completion, and the Architect so confirms, the Owner shall, upon application by the Contractor and certification by the Architect, and without terminating the Contract, make payment of the balance due for that portion of the Work fully completed and accepted. If the remaining balance for Work not fully completed or corrected is less than retainage stipulated in the Contract Documents, and if bonds have been furnished, the written consent of surety to payment of the balance due for that portion of the Work fully completed and accepted shall be submitted by the Contractor to the Architect prior to certification of such payment. Such payment shall be made under terms and conditions governing final payment, except that it shall not constitute a waiver of claims. The making of final payment shall constitute a waiver of claims by the Owner as provided in Subparagraph 4.3.5.

9.10.4 Acceptance of final payment by the Contractor, a Subcontractor or material supplier shall constitute a waiver of claims by that payee except those previously made in writing and identified by that payee as unsettled at the time of final Application for Payment. Such waivers shall be in addition to the waiver described in Subparagraph 4.3.5.

AIA DOCUMENT A201 • GENERAL CONDITIONS OF THE CONTRACT FOR CONSTRUCTION • FOURTEENTH EDITION
AIA® • ©1987 THE AMERICAN INSTITUTE OF ARCHITECTS, 1735 NEW YORK AVENUE, N.W., WASHINGTON, D.C. 20006

ARTICLE 10

PROTECTION OF PERSONS AND PROPERTY

10.1 SAFETY PRECAUTIONS AND PROGRAMS

10.1.1 The Contractor shall be responsible for initiating, maintaining and supervising all safety precautions and programs in connection with the performance of the Contract.

10.1.2 In the event the Contractor encounters on the site material reasonably believed to be asbestos or polychlorinated biphenyl (PCB) which has not been rendered harmless, the Contractor shall immediately stop Work in the area affected and report the condition to the Owner and Architect in writing. The Work in the affected area shall not thereafter be resumed except by written agreement of the Owner and Contractor if in fact the material is asbestos or polychlorinated biphenyl (PCB) and has not been rendered harmless. The Work in the affected area shall be resumed in the absence of asbestos or polychlorinated biphenyl (PCB), or when it has been rendered harmless, by written agreement of the Owner and Contractor, or in accordance with final determination by the Architect on which arbitration has not been demanded, or by arbitration under Article 4.

10.1.3 The Contractor shall not be required pursuant to Article 7 to perform without consent any Work relating to asbestos or polychlorinated biphenyl (PCB).

10.1.4 To the fullest extent permitted by law, the Owner shall indemnify and hold harmless the Contractor, Architect, Architect's consultants and agents and employees of any of them from and against claims, damages, losses and expenses, including but not limited to attorneys' fees, arising out of or resulting from performance of the Work in the affected area if in fact the material is asbestos or polychlorinated biphenyl (PCB) and has not been rendered harmless, provided that such claim, damage, loss or expense is attributable to bodily injury, sickness, disease or death, or to injury to or destruction of tangible property (other than the Work itself) including loss of use resulting therefrom, but only to the extent caused in whole or in part by negligent acts or omissions of the Owner, anyone directly or indirectly employed by the Owner or anyone for whose acts the Owner may be liable, regardless of whether or not such claim, damage, loss or expense is caused in part by a party indemnified hereunder. Such obligation shall not be construed to negate, abridge, or reduce other rights or obligations of indemnity which would otherwise exist as to a party or person described in this Subparagraph 10.1.4.

10.2 SAFETY OF PERSONS AND PROPERTY

10.2.1 The Contractor shall take reasonable precautions for safety of, and shall provide reasonable protection to prevent damage, injury or loss to:

> .1 employees on the Work and other persons who may be affected thereby;
>
> .2 the Work and materials and equipment to be incorporated therein, whether in storage on or off the site, under care, custody or control of the Contractor or the Contractor's Subcontractors or Sub-subcontractors; and
>
> .3 other property at the site or adjacent thereto, such as trees, shrubs, lawns, walks, pavements, roadways, structures and utilities not designated for removal, relocation or replacement in the course of construction.

10.2.2 The Contractor shall give notices and comply with applicable laws, ordinances, rules, regulations and lawful orders of public authorities bearing on safety of persons or property or their protection from damage, injury or loss.

10.2.3 The Contractor shall erect and maintain, as required by existing conditions and performance of the Contract, reasonable safeguards for safety and protection, including posting danger signs and other warnings against hazards, promulgating safety regulations and notifying owners and users of adjacent sites and utilities.

10.2.4 When use or storage of explosives or other hazardous materials or equipment or unusual methods are necessary for execution of the Work, the Contractor shall exercise utmost care and carry on such activities under supervision of properly qualified personnel.

10.2.5 The Contractor shall promptly remedy damage and loss (other than damage or loss insured under property insurance required by the Contract Documents) to property referred to in Clauses 10.2.1.2 and 10.2.1.3 caused in whole or in part by the Contractor, a Subcontractor, a Sub-subcontractor, or anyone directly or indirectly employed by any of them, or by anyone for whose acts they may be liable and for which the Contractor is responsible under Clauses 10.2.1.2 and 10.2.1.3, except damage or loss attributable to acts or omissions of the Owner or Architect or anyone directly or indirectly employed by either of them, or by anyone for whose acts either of them may be liable, and not attributable to the fault or negligence of the Contractor. The foregoing obligations of the Contractor are in addition to the Contractor's obligations under Paragraph 3.18.

10.2.6 The Contractor shall designate a responsible member of the Contractor's organization at the site whose duty shall be the prevention of accidents. This person shall be the Contractor's superintendent unless otherwise designated by the Contractor in writing to the Owner and Architect.

10.2.7 The Contractor shall not load or permit any part of the construction or site to be loaded so as to endanger its safety.

10.3 EMERGENCIES

10.3.1 In an emergency affecting safety of persons or property, the Contractor shall act, at the Contractor's discretion, to prevent threatened damage, injury or loss. Additional compensation or extension of time claimed by the Contractor on account of an emergency shall be determined as provided in Paragraph 4.3 and Article 7.

ARTICLE 11

INSURANCE AND BONDS

11.1 CONTRACTOR'S LIABILITY INSURANCE

11.1.1 The Contractor shall purchase from and maintain in a company or companies lawfully authorized to do business in the jurisdiction in which the Project is located such insurance as will protect the Contractor from claims set forth below which may arise out of or result from the Contractor's operations under the Contract and for which the Contractor may be legally liable, whether such operations be by the Contractor or by a Subcontractor or by anyone directly or indirectly employed by any of them, or by anyone for whose acts any of them may be liable:

> .1 claims under workers' or workmen's compensation, disability benefit and other similar employee benefit acts which are applicable to the Work to be performed;

.2 claims for damages because of bodily injury, occupational sickness or disease, or death of the Contractor's employees;

.3 claims for damages because of bodily injury, sickness or disease, or death of any person other than the Contractor's employees;

.4 claims for damages insured by usual personal injury liability coverage which are sustained (1) by a person as a result of an offense directly or indirectly related to employment of such person by the Contractor, or (2) by another person;

.5 claims for damages, other than to the Work itself, because of injury to or destruction of tangible property, including loss of use resulting therefrom;

.6 claims for damages because of bodily injury, death of a person or property damage arising out of ownership, maintenance or use of a motor vehicle; and

.7 claims involving contractual liability insurance applicable to the Contractor's obligations under Paragraph 3.18.

11.1.2 The insurance required by Subparagraph 11.1.1 shall be written for not less than limits of liability specified in the Contract Documents or required by law, whichever coverage is greater. Coverages, whether written on an occurrence or claims-made basis, shall be maintained without interruption from date of commencement of the Work until date of final payment and termination of any coverage required to be maintained after final payment.

11.1.3 Certificates of Insurance acceptable to the Owner shall be filed with the Owner prior to commencement of the Work. These Certificates and the insurance policies required by this Paragraph 11.1 shall contain a provision that coverages afforded under the policies will not be cancelled or allowed to expire until at least 30 days' prior written notice has been given to the Owner. If any of the foregoing insurance coverages are required to remain in force after final payment and are reasonably available, an additional certificate evidencing continuation of such coverage shall be submitted with the final Application for Payment as required by Subparagraph 9.10.2. Information concerning reduction of coverage shall be furnished by the Contractor with reasonable promptness in accordance with the Contractor's information and belief.

11.2   OWNER'S LIABILITY INSURANCE

11.2.1 The Owner shall be responsible for purchasing and maintaining the Owner's usual liability insurance. Optionally, the Owner may purchase and maintain other insurance for self-protection against claims which may arise from operations under the Contract. The Contractor shall not be responsible for purchasing and maintaining this optional Owner's liability insurance unless specifically required by the Contract Documents.

11.3   PROPERTY INSURANCE

11.3.1 Unless otherwise provided, the Owner shall purchase and maintain, in a company or companies lawfully authorized to do business in the jurisdiction in which the Project is located, property insurance in the amount of the initial Contract Sum as well as subsequent modifications thereto for the entire Work at the site on a replacement cost basis without voluntary deductibles. Such property insurance shall be maintained, unless otherwise provided in the Contract Documents or otherwise agreed in writing by all persons and entities who are beneficiaries of such insurance, until final payment has been made as provided in Paragraph 9.10 or until no person or entity other than the Owner has an insurable interest in the property required by this Paragraph 11.3 to be covered, whichever is earlier. This insurance shall include interests of the Owner, the Contractor, Subcontractors and Sub-subcontractors in the Work.

11.3.1.1 Property insurance shall be on an all-risk policy form and shall insure against the perils of fire and extended coverage and physical loss or damage including, without duplication of coverage, theft, vandalism, malicious mischief, collapse, falsework, temporary buildings and debris removal including demolition occasioned by enforcement of any applicable legal requirements, and shall cover reasonable compensation for Architect's services and expenses required as a result of such insured loss. Coverage for other perils shall not be required unless otherwise provided in the Contract Documents.

11.3.1.2 If the Owner does not intend to purchase such property insurance required by the Contract and with all of the coverages in the amount described above, the Owner shall so inform the Contractor in writing prior to commencement of the Work. The Contractor may then effect insurance which will protect the interests of the Contractor, Subcontractors and Sub-subcontractors in the Work, and by appropriate Change Order the cost thereof shall be charged to the Owner. If the Contractor is damaged by the failure or neglect of the Owner to purchase or maintain insurance as described above, without so notifying the Contractor, then the Owner shall bear all reasonable costs properly attributable thereto.

11.3.1.3 If the property insurance requires minimum deductibles and such deductibles are identified in the Contract Documents, the Contractor shall pay costs not covered because of such deductibles. If the Owner or insurer increases the required minimum deductibles above the amounts so identified or if the Owner elects to purchase this insurance with voluntary deductible amounts, the Owner shall be responsible for payment of the additional costs not covered because of such increased or voluntary deductibles. If deductibles are not identified in the Contract Documents, the Owner shall pay costs not covered because of deductibles.

11.3.1.4 Unless otherwise provided in the Contract Documents, this property insurance shall cover portions of the Work stored off the site after written approval of the Owner at the value established in the approval, and also portions of the Work in transit.

11.3.2 Boiler and Machinery Insurance. The Owner shall purchase and maintain boiler and machinery insurance required by the Contract Documents or by law, which shall specifically cover such insured objects during installation and until final acceptance by the Owner; this insurance shall include interests of the Owner, Contractor, Subcontractors and Sub-subcontractors in the Work, and the Owner and Contractor shall be named insureds.

11.3.3 Loss of Use Insurance. The Owner, at the Owner's option, may purchase and maintain such insurance as will insure the Owner against loss of use of the Owner's property due to fire or other hazards, however caused. The Owner waives all rights of action against the Contractor for loss of use of the Owner's property, including consequential losses due to fire or other hazards however caused.

11.3.4 If the Contractor requests in writing that insurance for risks other than those described herein or for other special hazards be included in the property insurance policy, the Owner shall, if possible, include such insurance, and the cost thereof shall be charged to the Contractor by appropriate Change Order.

AIA DOCUMENT A201 • GENERAL CONDITIONS OF THE CONTRACT FOR CONSTRUCTION • FOURTEENTH EDITION
AIA® • ©1987 THE AMERICAN INSTITUTE OF ARCHITECTS, 1735 NEW YORK AVENUE, N.W., WASHINGTON, D.C. 20006

11.3.5 If during the Project construction period the Owner insures properties, real or personal or both, adjoining or adjacent to the site by property insurance under policies separate from those insuring the Project, or if after final payment property insurance is to be provided on the completed Project through a policy or policies other than those insuring the Project during the construction period, the Owner shall waive all rights in accordance with the terms of Subparagraph 11.3.7 for damages caused by fire or other perils covered by this separate property insurance. All separate policies shall provide this waiver of subrogation by endorsement or otherwise.

11.3.6 Before an exposure to loss may occur, the Owner shall file with the Contractor a copy of each policy that includes insurance coverages required by this Paragraph 11.3. Each policy shall contain all generally applicable conditions, definitions, exclusions and endorsements related to this Project. Each policy shall contain a provision that the policy will not be cancelled or allowed to expire until at least 30 days' prior written notice has been given to the Contractor.

11.3.7 Waivers of Subrogation. The Owner and Contractor waive all rights against (1) each other and any of their subcontractors, sub-subcontractors, agents and employees, each of the other, and (2) the Architect, Architect's consultants, separate contractors described in Article 6, if any, and any of their sub-contractors, sub-subcontractors, agents and employees, for damages caused by fire or other perils to the extent covered by property insurance obtained pursuant to this Paragraph 11.3 or other property insurance applicable to the Work, except such rights as they have to proceeds of such insurance held by the Owner as fiduciary. The Owner or Contractor, as appropriate, shall require of the Architect, Architect's consultants, separate contractors described in Article 6, if any, and the subcontractors, sub-subcontractors, agents and employees of any of them, by appropriate agreements, written where legally required for validity, similar waivers each in favor of other parties enumerated herein. The policies shall provide such waivers of subrogation by endorsement or otherwise. A waiver of subrogation shall be effective as to a person or entity even though that person or entity would otherwise have a duty of indemnification, contractual or otherwise, did not pay the insurance premium directly or indirectly, and whether or not the person or entity had an insurable interest in the property damaged.

11.3.8 A loss insured under Owner's property insurance shall be adjusted by the Owner as fiduciary and made payable to the Owner as fiduciary for the insureds, as their interests may appear, subject to requirements of any applicable mortgagee clause and of Subparagraph 11.3.10. The Contractor shall pay Subcontractors their just shares of insurance proceeds received by the Contractor, and by appropriate agreements, written where legally required for validity, shall require Subcontractors to make payments to their Sub-subcontractors in similar manner.

11.3.9 If required in writing by a party in interest, the Owner as fiduciary shall, upon occurrence of an insured loss, give bond for proper performance of the Owner's duties. The cost of required bonds shall be charged against proceeds received as fiduciary. The Owner shall deposit in a separate account proceeds so received, which the Owner shall distribute in accordance with such agreement as the parties in interest may reach, or in accordance with an arbitration award in which case the procedure shall be as provided in Paragraph 4.5. If after such loss no other special agreement is made, replacement of damaged property shall be covered by appropriate Change Order.

11.3.10 The Owner as fiduciary shall have power to adjust and settle a loss with insurers unless one of the parties in interest shall object in writing within five days after occurrence of loss to the Owner's exercise of this power; if such objection be made, arbitrators shall be chosen as provided in Paragraph 4.5. The Owner as fiduciary shall, in that case, make settlement with insurers in accordance with directions of such arbitrators. If distribution of insurance proceeds by arbitration is required, the arbitrators will direct such distribution.

11.3.11 Partial occupancy or use in accordance with Paragraph 9.9 shall not commence until the insurance company or companies providing property insurance have consented to such partial occupancy or use by endorsement or otherwise. The Owner and the Contractor shall take reasonable steps to obtain consent of the insurance company or companies and shall, without mutual written consent, take no action with respect to partial occupancy or use that would cause cancellation, lapse or reduction of insurance.

11.4   PERFORMANCE BOND AND PAYMENT BOND

11.4.1 The Owner shall have the right to require the Contractor to furnish bonds covering faithful performance of the Contract and payment of obligations arising thereunder as stipulated in bidding requirements or specifically required in the Contract Documents on the date of execution of the Contract.

11.4.2 Upon the request of any person or entity appearing to be a potential beneficiary of bonds covering payment of obligations arising under the Contract, the Contractor shall promptly furnish a copy of the bonds or shall permit a copy to be made.

ARTICLE 12

UNCOVERING AND CORRECTION OF WORK

12.1   UNCOVERING OF WORK

12.1.1 If a portion of the Work is covered contrary to the Architect's request or to requirements specifically expressed in the Contract Documents, it must, if required in writing by the Architect, be uncovered for the Architect's observation and be replaced at the Contractor's expense without change in the Contract Time.

12.1.2 If a portion of the Work has been covered which the Architect has not specifically requested to observe prior to its being covered, the Architect may request to see such Work and it shall be uncovered by the Contractor. If such Work is in accordance with the Contract Documents, costs of uncovering and replacement shall, by appropriate Change Order, be charged to the Owner. If such Work is not in accordance with the Contract Documents, the Contractor shall pay such costs unless the condition was caused by the Owner or a separate contractor in which event the Owner shall be responsible for payment of such costs.

12.2   CORRECTION OF WORK

12.2.1 The Contractor shall promptly correct Work rejected by the Architect or failing to conform to the requirements of the Contract Documents, whether observed before or after Substantial Completion and whether or not fabricated, installed or completed. The Contractor shall bear costs of correcting such rejected Work, including additional testing and inspections and compensation for the Architect's services and expenses made necessary thereby.

12.2.2 If, within one year after the date of Substantial Completion of the Work or designated portion thereof, or after the date for commencement of warranties established under Sub-paragraph 9.9.1, or by terms of an applicable special warranty required by the Contract Documents, any of the Work is found to be not in accordance with the requirements of the Contract Documents, the Contractor shall correct it promptly after receipt of written notice from the Owner to do so unless the Owner has previously given the Contractor a written acceptance of such condition. This period of one year shall be extended with respect to portions of Work first performed after Substantial Completion by the period of time between Substantial Completion and the actual performance of the Work. This obligation under this Subparagraph 12.2.2 shall survive acceptance of the Work under the Contract and termination of the Contract. The Owner shall give such notice promptly after discovery of the condition.

12.2.3 The Contractor shall remove from the site portions of the Work which are not in accordance with the requirements of the Contract Documents and are neither corrected by the Contractor nor accepted by the Owner.

12.2.4 If the Contractor fails to correct nonconforming Work within a reasonable time, the Owner may correct it in accordance with Paragraph 2.4. If the Contractor does not proceed with correction of such nonconforming Work within a reasonable time fixed by written notice from the Architect, the Owner may remove it and store the salvable materials or equipment at the Contractor's expense. If the Contractor does not pay costs of such removal and storage within ten days after written notice, the Owner may upon ten additional days' written notice sell such materials and equipment at auction or at private sale and shall account for the proceeds thereof, after deducting costs and damages that should have been borne by the Contractor, including compensation for the Architect's services and expenses made necessary thereby. If such proceeds of sale do not cover costs which the Contractor should have borne, the Contract Sum shall be reduced by the deficiency. If payments then or thereafter due the Contractor are not sufficient to cover such amount, the Contractor shall pay the difference to the Owner.

12.2.5 The Contractor shall bear the cost of correcting destroyed or damaged construction, whether completed or partially completed, of the Owner or separate contractors caused by the Contractor's correction or removal of Work which is not in accordance with the requirements of the Contract Documents.

12.2.6 Nothing contained in this Paragraph 12.2 shall be construed to establish a period of limitation with respect to other obligations which the Contractor might have under the Contract Documents. Establishment of the time period of one year as described in Subparagraph 12.2.2 relates only to the specific obligation of the Contractor to correct the Work, and has no relationship to the time within which the obligation to comply with the Contract Documents may be sought to be enforced, nor to the time within which proceedings may be commenced to establish the Contractor's liability with respect to the Contractor's obligations other than specifically to correct the Work.

12.3   ACCEPTANCE OF NONCONFORMING WORK

12.3.1 If the Owner prefers to accept Work which is not in accordance with the requirements of the Contract Documents, the Owner may do so instead of requiring its removal and correction, in which case the Contract Sum will be reduced as appropriate and equitable. Such adjustment shall be effected whether or not final payment has been made.

ARTICLE 13

MISCELLANEOUS PROVISIONS

13.1   GOVERNING LAW

13.1.1 The Contract shall be governed by the law of the place where the Project is located.

13.2   SUCCESSORS AND ASSIGNS

13.2.1 The Owner and Contractor respectively bind themselves, their partners, successors, assigns and legal representatives to the other party hereto and to partners, successors, assigns and legal representatives of such other party in respect to covenants, agreements and obligations contained in the Contract Documents. Neither party to the Contract shall assign the Contract as a whole without written consent of the other. If either party attempts to make such an assignment without such consent, that party shall nevertheless remain legally responsible for all obligations under the Contract.

13.3   WRITTEN NOTICE

13.3.1 Written notice shall be deemed to have been duly served if delivered in person to the individual or a member of the firm or entity or to an officer of the corporation for which it was intended, or if delivered at or sent by registered or certified mail to the last business address known to the party giving notice.

13.4   RIGHTS AND REMEDIES

13.4.1 Duties and obligations imposed by the Contract Documents and rights and remedies available thereunder shall be in addition to and not a limitation of duties, obligations, rights and remedies otherwise imposed or available by law.

13.4.2 No action or failure to act by the Owner, Architect or Contractor shall constitute a waiver of a right or duty afforded them under the Contract, nor shall such action or failure to act constitute approval of or acquiescence in a breach thereunder, except as may be specifically agreed in writing.

13.5   TESTS AND INSPECTIONS

13.5.1 Tests, inspections and approvals of portions of the Work required by the Contract Documents or by laws, ordinances, rules, regulations or orders of public authorities having jurisdiction shall be made at an appropriate time. Unless otherwise provided, the Contractor shall make arrangements for such tests, inspections and approvals with an independent testing laboratory or entity acceptable to the Owner, or with the appropriate public authority, and shall bear all related costs of tests, inspections and approvals. The Contractor shall give the Architect timely notice of when and where tests and inspections are to be made so the Architect may observe such procedures. The Owner shall bear costs of tests, inspections or approvals which do not become requirements until after bids are received or negotiations concluded.

13.5.2 If the Architect, Owner or public authorities having jurisdiction determine that portions of the Work require additional testing, inspection or approval not included under Subparagraph 13.5.1, the Architect will, upon written authorization from the Owner, instruct the Contractor to make arrangements for such additional testing, inspection or approval by an entity acceptable to the Owner, and the Contractor shall give timely notice to the Architect of when and where tests and inspections are to be made so the Architect may observe such procedures.

AIA DOCUMENT A201 • GENERAL CONDITIONS OF THE CONTRACT FOR CONSTRUCTION • FOURTEENTH EDITION
AIA® • ©1987 THE AMERICAN INSTITUTE OF ARCHITECTS, 1735 NEW YORK AVENUE, N.W., WASHINGTON, D.C. 20006

The Owner shall bear such costs except as provided in Sub-paragraph 13.5.3.

13.5.3 If such procedures for testing, inspection or approval under Subparagraphs 13.5.1 and 13.5.2 reveal failure of the portions of the Work to comply with requirements established by the Contract Documents, the Contractor shall bear all costs made necessary by such failure including those of repeated procedures and compensation for the Architect's services and expenses.

13.5.4 Required certificates of testing, inspection or approval shall, unless otherwise required by the Contract Documents, be secured by the Contractor and promptly delivered to the Architect.

13.5.5 If the Architect is to observe tests, inspections or approvals required by the Contract Documents, the Architect will do so promptly and, where practicable, at the normal place of testing.

13.5.6 Tests or inspections conducted pursuant to the Contract Documents shall be made promptly to avoid unreasonable delay in the Work.

13.6   INTEREST

13.6.1 Payments due and unpaid under the Contract Documents shall bear interest from the date payment is due at such rate as the parties may agree upon in writing or, in the absence thereof, at the legal rate prevailing from time to time at the place where the Project is located.

13.7   COMMENCEMENT OF STATUTORY LIMITATION PERIOD

13.7.1 As between the Owner and Contractor:

.1 **Before Substantial Completion.** As to acts or failures to act occurring prior to the relevant date of Substantial Completion, any applicable statute of limitations shall commence to run and any alleged cause of action shall be deemed to have accrued in any and all events not later than such date of Substantial Completion;

.2 **Between Substantial Completion and Final Certificate for Payment.** As to acts or failures to act occurring subsequent to the relevant date of Substantial Completion and prior to issuance of the final Certificate for Payment, any applicable statute of limitations shall commence to run and any alleged cause of action shall be deemed to have accrued in any and all events not later than the date of issuance of the final Certificate for Payment; and

.3 **After Final Certificate for Payment.** As to acts or failures to act occurring after the relevant date of issuance of the final Certificate for Payment, any applicable statute of limitations shall commence to run and any alleged cause of action shall be deemed to have accrued in any and all events not later than the date of any act or failure to act by the Contractor pursuant to any warranty provided under Paragraph 3.5, the date of any correction of the Work or failure to correct the Work by the Contractor under Paragraph 12.2, or the date of actual commission of any other act or failure to perform any duty or obligation by the Contractor or Owner, whichever occurs last.

ARTICLE 14

TERMINATION OR SUSPENSION OF THE CONTRACT

14.1   TERMINATION BY THE CONTRACTOR

14.1.1 The Contractor may terminate the Contract if the Work is stopped for a period of 30 days through no act or fault of the Contractor or a Subcontractor, Sub-subcontractor or their agents or employees or any other persons performing portions of the Work under contract with the Contractor, for any of the following reasons:

.1 issuance of an order of a court or other public authority having jurisdiction;

.2 an act of government, such as a declaration of national emergency, making material unavailable;

.3 because the Architect has not issued a Certificate for Payment and has not notified the Contractor of the reason for withholding certification as provided in Subparagraph 9.4.1, or because the Owner has not made payment on a Certificate for Payment within the time stated in the Contract Documents;

.4 if repeated suspensions, delays or interruptions by the Owner as described in Paragraph 14.3 constitute in the aggregate more than 100 percent of the total number of days scheduled for completion, or 120 days in any 365-day period, whichever is less; or

.5 the Owner has failed to furnish to the Contractor promptly, upon the Contractor's request, reasonable evidence as required by Subparagraph 2.2.1.

14.1.2 If one of the above reasons exists, the Contractor may, upon seven additional days' written notice to the Owner and Architect, terminate the Contract and recover from the Owner payment for Work executed and for proven loss with respect to materials, equipment, tools, and construction equipment and machinery, including reasonable overhead, profit and damages.

14.1.3 If the Work is stopped for a period of 60 days through no act or fault of the Contractor or a Subcontractor or their agents or employees or any other persons performing portions of the Work under contract with the Contractor because the Owner has persistently failed to fulfill the Owner's obligations under the Contract Documents with respect to matters important to the progress of the Work, the Contractor may, upon seven additional days' written notice to the Owner and the Architect, terminate the Contract and recover from the Owner as provided in Subparagraph 14.1.2.

14.2   TERMINATION BY THE OWNER FOR CAUSE

14.2.1 The Owner may terminate the Contract if the Contractor:

.1 persistently or repeatedly refuses or fails to supply enough properly skilled workers or proper materials;

.2 fails to make payment to Subcontractors for materials or labor in accordance with the respective agreements between the Contractor and the Subcontractors;

.3 persistently disregards laws, ordinances, or rules, regulations or orders of a public authority having jurisdiction; or

.4 otherwise is guilty of substantial breach of a provision of the Contract Documents.

14.2.2 When any of the above reasons exist, the Owner, upon certification by the Architect that sufficient cause exists to justify such action, may without prejudice to any other rights or remedies of the Owner and after giving the Contractor and the Contractor's surety, if any, seven days' written notice, terminate employment of the Contractor and may, subject to any prior rights of the surety:

.1 take possession of the site and of all materials, equipment, tools, and construction equipment and machinery thereon owned by the Contractor;

.2 accept assignment of subcontracts pursuant to Paragraph 5.4; and

.3 finish the Work by whatever reasonable method the Owner may deem expedient.

14.2.3 When the Owner terminates the Contract for one of the reasons stated in Subparagraph 14.2.1, the Contractor shall not be entitled to receive further payment until the Work is finished.

14.2.4 If the unpaid balance of the Contract Sum exceeds costs of finishing the Work, including compensation for the Architect's services and expenses made necessary thereby, such excess shall be paid to the Contractor. If such costs exceed the unpaid balance, the Contractor shall pay the difference to the Owner. The amount to be paid to the Contractor or Owner, as the case may be, shall be certified by the Architect, upon application, and this obligation for payment shall survive termination of the Contract.

**14.3   SUSPENSION BY THE OWNER
FOR CONVENIENCE**

14.3.1 The Owner may, without cause, order the Contractor in writing to suspend, delay or interrupt the Work in whole or in part for such period of time as the Owner may determine.

14.3.2 An adjustment shall be made for increases in the cost of performance of the Contract, including profit on the increased cost of performance, caused by suspension, delay or interruption. No adjustment shall be made to the extent:

.1 that performance is, was or would have been so suspended, delayed or interrupted by another cause for which the Contractor is responsible; or

.2 that an equitable adjustment is made or denied under another provision of this Contract.

14.3.3 Adjustments made in the cost of performance may have a mutually agreed fixed or percentage fee.

AIA DOCUMENT A201 • GENERAL CONDITIONS OF THE CONTRACT FOR CONSTRUCTION • FOURTEENTH EDITION
AIA® • ©1987 THE AMERICAN INSTITUTE OF ARCHITECTS, 1735 NEW YORK AVENUE, N.W., WASHINGTON, D.C. 20006

3/87

INDEX